Communications in Computer and Information Science 740

Commenced Publication in 2007
Founding and Former Series Editors:
Alfredo Cuzzocrea, Orhun Kara, Dominik Ślęzak, and Xiaokang Yang

More information about this series at http://www.springer.com/series/7899

Markus Helfert · Donald Ferguson
Victor Méndez Muñoz · Jorge Cardoso (Eds.)

Cloud Computing and Services Science

6th International Conference, CLOSER 2016
Rome, Italy, April 23–25, 2016
Revised Selected Papers

 Springer

Editors
Markus Helfert
Dublin City University
Dublin
Ireland

Donald Ferguson
Columbia University
New York, NY
USA

Victor Méndez Muñoz
Escola d'Enginyeria
Bellaterra, Barcelona
Spain

Jorge Cardoso
Departamento de Engenharia Informática
Universidade de Coimbra
Coimbra
Portugal

ISSN 1865-0929 ISSN 1865-0937 (electronic)
Communications in Computer and Information Science
ISBN 978-3-319-62593-5 ISBN 978-3-319-62594-2 (eBook)
DOI 10.1007/978-3-319-62594-2

Library of Congress Control Number: 2017945726

Printed on acid-free paper

This Springer imprint is published by Springer Nature
The registered company is Springer International Publishing AG
The registered company address is: Gewerbestrasse 11, 6330 Cham, Switzerland

Preface

The present book includes extended and revised versions of a set of selected papers from the 6th International Conference on Cloud Computing and Services Science (CLOSER 2016), held in Rome, Italy, during April 23–25, 2016.

CLOSER 2016 received 123 paper submissions from 35 countries, of which 15% were included in this book. The papers were selected by the event chairs and their selection was based on a number of criteria including the classifications and comments provided by the Program Committee members, the session chairs' assessment, and also the program chairs' global view of all papers included in the technical program. The authors of selected papers were then invited to submit a revised and extended version of their papers having at least 30% innovative material.

CLOSER 2016 focused on the emerging area of cloud computing, inspired by recent advances related to infrastructures, operations, and service availability through global networks. Furthermore, the conference considers the link to service science fundamental, acknowledging the service-orientation of most IT-driven collaborations. The conference is nevertheless not about the union of these two (already broad) fields, but about understanding how service science can provide theories, methods, and techniques with which to design, analyze, manage, market, and study the various aspects surrounding cloud computing.

With the ongoing digitization in all aspects of society, the importance and contributions of service science is expected to grow. Many researchers have shown that there are specific challenges in applying service orientation to sectors such as energy, security, finance, payments, and smart cities. In addition, the continuation in the consumer focus has intensified service orientation, with analysis of customer experiences and service behavior as a promising new route. In addition, novel IoT and big data-based technologies have resulted in many service-oriented business models, whereby security and privacy protection remains a constant concern.

The papers selected to be included in this book contribute to the understanding of relevant trends of current research on cloud computing and service science, including:

- Cloud interoperability and migration
- Cloud-native applications, microservices, and containers
- Auditing and SLA management for cloud applications
- Performance, cost, and green consumption management
- Detecting anomalies in cloud platforms
- Deployment and adaptation of cloud data centers
- Public HPC and private cloud computing

We would like to thank all the authors for their contributions and also the reviewers who helped ensure the quality of this publication.

February 2017

Markus Helfert
Donald Ferguson
Víctor Méndez Muñoz
Jorge Cardoso

Organization

Conference Chair

Markus Helfert   Dublin City University, Ireland

Program Chairs

Jorge Cardoso   University of Coimbra, Portugal and Huawei European
     Research Center, Germany
Donald Ferguson   Columbia University, USA
Víctor Méndez Muñoz   Universitat Autònoma de Barcelona, UAB, Spain

Program Committee

Jörn Altmann   Seoul National University, Korea, Republic of
Vasilios Andrikopoulos   University of Groningen, The Netherlands
Claudio Ardagna   Università degli Studi di Milano, Italy
Marcos Dias de Assunção   Institut National de Recherche en Informatique
     et Automatique (Inria), France
Amelia Badica   University of Craiova, Romania
Boualem Benatallah   University of New South Wales, Australia
Karin Bernsmed   SINTEF ICT, Norway
Luiz F. Bittencourt   IC/UNICAMP, Brazil
Ivona Brandić   Vienna UT, Austria
Andrey Brito   Universidade Federal de Campina Grande, Brazil
Ralf Bruns   Hannover University of Applied Sciences and Arts,
     Germany
Anna Brunstrom   Karlstad University, Sweden
Rebecca Bulander   Pforzheim University of Applied Science, Germany
Tomas Bures   Charles University in Prague, Czech Republic
Massimo Cafaro   University of Salento, Italy
Manuel Isidoro   University of Granada, Spain
  Capel-Tuñón
Miriam Capretz   University of Western Ontario, Canada
Jorge Cardoso   University of Coimbra, Portugal and Huawei European
     Research Center, Germany
Eddy Caron   École Normale Supérieure de Lyon, France
John Cartlidge   University of Nottingham in Ningbo China, UK
Augusto Ciuffoletti   Università di Pisa, Italy
Daniela Barreiro Claro   Universidade Federal da Bahia (UFBA), Brazil

Thierry Coupaye	Orange, France
António Miguel Rosado da Cruz	Instituto Politécnico de Viana do Castelo, Portugal
Eduardo Huedo Cuesta	Universidad Complutense de Madrid, Spain
Eliezer Dekel	IBM Research Haifa, Israel
Yuri Demchenko	University of Amsterdam, The Netherlands
Frédéric Desprez	Antenne Inria Giant, Minatec Campus, France
Robert van Engelen	Florida State University, USA
Donald Ferguson	Columbia University, USA
Stefano Ferretti	University of Bologna, Italy
Mike Fisher	BT, UK
Geoffrey Charles Fox	Indiana University, USA
Fabrizio Gagliardi	Barcelona Supercomputing Centre, Spain
David Genest	Université d'Angers, France
Chirine Ghedira	IAE, University Jean Moulin Lyon 3, France
Lee Gillam	University of Surrey, UK
Katja Gilly	Miguel Hernandez University, Spain
Patrizia Grifoni	CNR, Italy
Stephan Groß	T-Systems Multimedia Solutions, Germany
Souleiman Hasan	National University of Ireland, Ireland
Benjamin Heckmann	University of Applied Sciences Darmstadt, Germany
Louis O. Hertzberger	University of Amsterdam, The Netherlands
Dong Huang	Chinese Academy of Sciences, China
Mohamed Hussien	Suez Canal University, Egypt
Sorin M. Iacob	Thales Nederland B.V., The Netherlands
Ilian Ilkov	IBM Nederland B.V., The Netherlands
Anca Daniela Ionita	University Politehnica of Bucharest, Romania
Fuyuki Ishikawa	National Institute of Informatics, Japan
Hiroshi Ishikawa	Tokyo Metropolitan University, Japan
Ivan Ivanov	SUNY Empire State College, USA
Martin Gilje Jaatun	SINTEF ICT, Norway
Keith Jeffery	Independent Consultant (previously Science and Technology Facilities Council), UK
Meiko Jensen	Kiel University of Applied Sciences, Germany
Yiming Ji	University of South Carolina Beaufort, USA
Ming Jiang	University of Sunderland, UK
Xiaolong Jin	Chinese Academy of Sciences, China
Carlos Juiz	Universitat de les Illes Balears, Spain
David R. Kaeli	Northeastern University, USA
Gabor Kecskemeti	Liverpool John Moores University, UK
Attila Kertesz	University of Szeged, Hungary
Carsten Kleiner	University of Applied Sciences and Arts Hannover, Germany
Geir M. Køien	University of Agder, Norway
George Kousiouris	National Technical University of Athens, Greece
Wilfried Lemahieu	KU Leuven, Belgium

Fei Li	Siemens AG, Austria, Austria
Donghui Lin	Kyoto University, Japan
Shijun Liu	School of Computer Science and Technology, Shandong University, China
Francesco Longo	Università degli Studi di Messina, Italy
Antonio García Loureiro	University of Santiago de Compostela, Spain
Joseph P. Loyall	BBN Technologies, USA
Simone Ludwig	North Dakota State University, USA
Glenn Luecke	Iowa State University, USA
Hanan Lutfiyya	University of Western Ontario, Canada
Theo Lynn	Dublin City University, Ireland
Shikharesh Majumdar	Carleton University, Canada
Elisa Marengo	Free University of Bozen-Bolzano, Italy
Ioannis Mavridis	University of Macedonia, Greece
Richard McClatchey	UWE, Bristol, UK
Nouredine Melab	CRIStAL, Université Lille 1, Inria Lille, France
Jose Ramon Gonzalez de Mendivil	Universidad Publica de Navarra, Spain
Mohamed Mohamed	IBM Research, Almaden, USA
Marco Casassa Mont	Hewlett-Packard Laboratories, UK
Rubén Santiago Montero	Universidad Complutense de Madrid, Spain
Kamran Munir	University of the West of England, Bristol, UK
Hidemoto Nakada	National Institute of Advanced Industrial Science and Technology (AIST), Japan
Philippe Navaux	UFRGS, Federal University of Rio Grande Do Sul, Brazil
Jean-Marc Nicod	Institut FEMTO-ST, France
Bogdan Nicolae	IBM Research, Ireland
Mara Nikolaidou	Harokopio University of Athens, Greece
Sebastian Obermeier	ABB Corporate Research, Switzerland
Claus Pahl	Free University of Bozen-Bolzano, Italy
Michael A. Palis	Rutgers University, USA
Nikos Parlavantzas	IRISA, France
David Paul	The University of New England, Australia
Siani Pearson	HP Labs, Bristol, UK
Tomás Fernández Pena	Universidad Santiago de Compostela, Spain
Agostino Poggi	University of Parma, Italy
Francesco Quaglia	Sapienza Università di Roma, Italy
Rajendra Raj	Rochester Institute of Technology, USA
Arcot Rajasekar	University of North Carolina at Chapel Hill, USA
Arkalgud Ramaprasad	University of Illinois at Chicago, USA
Manuel Ramos-Cabrer	University of Vigo, Spain
Christoph Reich	Hochschule Furtwangen University, Germany
Kun Ren	Yale University, USA
Norbert Ritter	University of Hamburg, Germany
Rizos Sakellariou	University of Manchester, UK
Elena Sanchez-Nielsen	Universidad De La Laguna, Spain

Patrizia Scandurra   University of Bergamo, Italy
Erich Schikuta   Universität Wien, Austria
Uwe Schwiegelshohn   TU Dortmund University, Germany
Wael Sellami   Higher Institute of Computer Sciences of Mahdia, Tunisia
Giovanni Semeraro   University of Bari Aldo Moro, Italy
Carlos Serrao   ISCTE-IUL, Portugal
Keiichi Shima   IIJ Innovation Institute, Japan
Marten van Sinderen   University of Twente, The Netherlands
Frank Siqueira   Universidade Federal de Santa Catarina, Brazil
Josef Spillner   Zurich University of Applied Sciences, Switzerland
Ralf Steinmetz   Technische Universität Darmstadt, Germany
Burkhard Stiller   University of Zürich, Switzerland
Yasuyuki Tahara   The University of Electro-Communications, Japan
Samir Tata   TELECOM SudParis, France
Cedric Tedeschi   IRISA - University of Rennes 1, France
Gilbert Tekli   Nobatek, France, Lebanon
Joe Tekli   Lebanese American University, Lebanon
Guy Tel-Zur   Ben-Gurion University of the Negev, Israel
Orazio Tomarchio   University of Catania, Italy
Francesco Tusa   University College London, UK
Astrid Undheim   Telenor ASA, Norway
Luis M. Vaquero   HP Labs, UK
Sabrina de Capitani   Università degli Studi di Milano, Italy
  di Vimercati
Bruno Volckaert   Ghent University - imec, Belgium
Mladen A. Vouk   N.C. State University, USA
Hiroshi Wada   NICTA, Australia
Jan-Jan Wu   Academia Sinica, Taiwan
Bo Yang   University of Electronic Science and Technology
                              of China, China
Ustun Yildiz   University of California, San Diego, USA
Michael Zapf   Georg Simon Ohm University of Applied Sciences,
                              Germany
Wolfgang Ziegler   Fraunhofer Institute SCAI, Germany
Farhana Zulkernine   Queen's University, Canada

Additional Reviewers

Marcio Roberto Miranda   UNICAMP, Brazil
  Assis
Belen Bermejo   Universitat de les Illes Balears, Spain
Babacar Mane   UFBA, Brazil
Hendrik Moens   IBCN iMinds, Belgium
Eduardo Roloff   UFRGS, Brazil
Nikolaos Tsigganos   University of Macedonia, Greece

Invited Speakers

Pierangela Samarati	Università degli Studi di Milano, Italy
Frank Leymann	University of Stuttgart, Germany
Peter Sloot	Complexity Institute Singapore, Singapore;
	ITMO St. Petersburg, Russion Federation;
	University of Amsterdam, The Netherlands
Verena Kantere	University of Geneva, Switzerland
Mohammed Atiquzzaman	University of Oklahoma, USA

Contents

Invited Papers

Supporting Users in Data Outsourcing and Protection in the Cloud

S. De Capitani di Vimercati[✉], S. Foresti, G. Livraga, and P. Samarati

Dipartimento di Informatica, Università degli Studi di Milano, 26013 Crema, Italy
{sabrina.decapitani,sara.foresti,giovanni.livraga,
pierangela.samarati}@unimi.it

Abstract. Moving data and applications to the cloud allows users and companies to enjoy considerable benefits. However, these benefits are also accompanied by a number of security issues that should be addressed. Among these, the need to ensure that possible requirements on security, costs, and quality of services are satisfied by the cloud providers, and the need to adopt techniques ensuring the proper protection of their data and applications. In this paper, we present different strategies and solutions that can be applied to address these issues.

1 Introduction

The 'Cloud' has emerged as a successful paradigm that enables users and companies to outsource data and applications to cloud providers, enjoying the availability of virtually unlimited storage and computation resources at competitive prices. An ever-growing number of cloud providers offer a variety of service options in terms of pricing, performance, and features. The advantages in adopting cloud services, however, come also with new security and privacy problems [24,54]. A first important problem consists in selecting, among the wide variety of cloud providers available on the market, the cloud provider most suitable for the needs of applications and data to be outsourced. This requires to properly model the requirements and/or preferences and to match such requirements with the characteristics and the service options offered by the cloud providers. Clearly, these requirements may differ for different users and/or for different data and applications that are moved to the cloud. Users therefore need flexible and expressive techniques supporting both the definition of their requirements and preferences, and the matching of these requirements with the characteristics of the different cloud providers (e.g., [6,28]).

Another problem (which often may lead users and companies to refrain from adopting the cloud for managing their data and applications) is related to ensuring proper protection of the outsourced data. As a matter of fact, when data and applications are moved to the cloud, they are no more under the direct control of their owner and must be properly protected from unauthorized parties or the cloud provider itself. Cloud providers may be *honest-but-curious* (i.e., trusted for managing data but not to access their content) or may even have a *lazy* or *malicious* behavior. Depending on the trust assumption that the user has on

M. Helfert et al. (Eds.): CLOSER 2016, CCIS 740, pp. 3–15, 2017.
DOI: 10.1007/978-3-319-62594-2_1

the provider running the selected cloud service, different problems might need to be addresses, including data confidentiality, integrity, and availability protection (e.g., [24]), the enforcement of access control (e.g., [16,19,34,59]), and the management of fault tolerance (e.g., [38–40]).

The remainder of this paper is organized as follows. Section 2 describes some approaches enabling users to express their requirements and preferences as well as possible strategies for selecting a cloud provider (or set thereof) that satisfies such requirements and preferences. Section 3 discusses the main data security issues that arise when using the cloud, and possible solutions to them. Finally, Sect. 4 concludes the paper.

2   Supporting Users in Cloud Provider Selection

Due to the growing number of cloud providers offering services that differ in their costs, security mechanisms, and Quality of Service (QoS), the research and industry communities have dedicated attention to the problem of improving the user experiences and the use of available cloud services. Industry security standards such as the Cloud Security Alliance Cloud Controls Matrix [13] have been proposed to allow cloud vendors and users to assess the overall security risks of cloud providers. Also several techniques have been recently proposed to assist data owners in selecting the cloud provider that better satisfies their needs. In this section, we provide an overview of such techniques.

User-side QoS. A natural strategy to compare a set of candidate cloud providers is based on measuring their Quality of Service (QoS) and comparing the results. Cloud providers typically publish indicators about the performance of their services in their Service Level Agreements (SLAs). Some of these indicators are used as parameters to measure the QoS of the provider (e.g., the response time). However, the value declared by the provider in its SLA (*provider-side QoS*) can differ from the value observed by a user (*user-side QoS*). Also, different users can experience different user-side QoS values for the same provider. For instance, the response time observed by a user can differ from the one experienced by another user operating in a different geographical area because they operate on networks with different latencies. Therefore, if a user selects a cloud provider on the basis of the provider-side QoS, she may end up with a choice not optimal for her. To overcome this problem, some techniques introduced the idea of selecting the cloud provider(s) based on the user-side QoS (e.g., [61]). A precise evaluation of user-side QoS values can however be a difficult task. In fact, in many situations, it requires to measure the parameters of interest (e.g., the response time) by actually invoking/using the services offered by the provider. This practice may cause communication overheads and possible economic charges that might not be always acceptable. To solve this issue, QoS parameters could be predicted by defining an automated QoS prediction framework that considers past service usage experiences of other "similar users". Measured or estimated QoS parameters are finally used to rank all the (functionally equivalent) providers among which the user can choose (e.g., [61]).

User Requirements. A user may have different requirements (i.e., conditions) that cloud providers should satisfy to meet the user's needs (e.g., at least four backup copies of the outsourced data should be maintained), or preferences on the values of the characteristics of the cloud service (e.g., a user may specify a value for the required availability and for the response time). User requirements might depend, for example, on the specific application scenario (e.g., data that need timely retrieval have to be stored at a provider with negligible downtime and fast response time) as well as by laws and regulations (e.g., sensitive data have to be stored at a provider applying appropriate security measures).

A line of research has investigated the definition of approaches to select the cloud provider (or set thereof) that better satisfies all the users requirements (i.e., optimizing the values of the attributes of interest for the user). The proposed solutions are typically based on the presence of a trusted middleware/interface playing the role of a *broker* [33] in the system architecture. The broker takes both the requirements of the user and the characteristics of the candidate cloud providers (possibly expressed in a machine-readable format [52]) as input, and tries to find the best match between the user requirements and the characteristics of the cloud providers. This matching problem can become more complex when a user defines multiple, may be even contrasting, requirements. In this case, it is necessary to quantify the satisfaction of each requirement by each provider, and to properly combine the measures associated with each provider. Multi-Criteria Decision Making (MCDM) techniques have been proposed as one of the basic approaches to address such a problem (e.g., [51]). Among the solutions relying on MCDM techniques, SMICloud [30,31] adopts a hierarchical decision-making technique, called Analytic Hierarchy Process, to compare and rank cloud providers on the basis of the satisfaction of user requirements. SMICloud models user requirements as key performance indicators (KPIs) that include, for example, the response time of a service, its sustainability (e.g., environmental impact), and its economic costs. MCDM, possibly coupled with machine learning, has also been proposed as a method for selecting the *instance type* (i.e., the configuration of compute, memory, and storage capabilities) that has the best trade-off between economic costs and performance and that satisfies the user requirements (e.g., [47,55]). For each of the resources to be employed (e.g., memory, CPU), these proposals select the provider (or set thereof) to be used for its provisioning as well as the amount of the resource to be obtained from each cloud provider so that the user requirements are satisfied.

Besides expressing requirements over QoS indicators and KPIs, a user might wish to formulate her requirements as generic conditions over a non-predefined set of attributes characterizing the service delivery and the cloud provider. For instance, to obey security regulations, a user may require that sensitive data be stored only in servers located in a given country, even if the physical location of a server is not an attribute explicitly represented in the provider SLAs. The satisfaction of user-based requirements might depend on the (joint) satisfaction of multiple conditions expressed over the attributes declared by cloud providers. Also, certain characteristics of a cloud provider may *depend* on other characteristics. For instance, the response time of a system may depend on the incoming request rate (i.e., the number of incoming requests per second), meaning that it can be ensured only if an upper bound is enforced on the number of requests per time unit. When checking whether cloud providers satisfy the user requirements it is also important to consider such dependencies, accounting for the fact that different services might entail different dependencies (e.g., two services with different hardware/software configurations might accept different request rates to guarantee the same response time). Recent approaches have designed solutions for establishing an SLA between a user and a cloud provider based on generic user requirements and on the automatic evaluation of dependencies existing for the provider (e.g., [26]). For instance, a dependency can state that a response time of 5 ms can be guaranteed only if the request rate is lower than 1 per second. The solution in [26] takes as input a set of generic user requirements and a set of dependencies for a provider. By adopting off-the-shelves Constraint Satisfaction Problem (CSP) solvers, this technique determines a *valid* SLA (if it exists), denoted vSLA, that satisfies the conditions expressed by the user as well as further conditions possibly triggered by dependencies. With reference to the example above, if the requirement of the user includes a value for the response time, then the generated vSLA will also include a condition on the request rate that the candidate provider should also satisfy. Given a set of requirements and a set of dependencies, different valid SLAs might exist. The approach in [25] extends the work in [26] by allowing users to specify preferences over conditions that can be considered for selecting, among the valid SLAs, the one that the user prefers. Preferences are expressed over the values that can be assumed by the attributes involved in requirements and dependencies (e.g., the one between the response time and the request rate). Building on the CSP-based approach in [26], these preferences are used to automatically evaluate vSLAs, ranking higher those that better satisfy the preferences of the user.

Multi-application Requirements. Many of the approaches for selecting cloud provider(s) operate under the implicit assumption that a user is moving to the cloud a single application at a time (or a set of applications with the same requirements). Hence, the user requirements reflect the application needs. However, if a user wishes to outsource multiple applications to a single cloud provider, the selection process may be complicated by the fact that different applications might have different, even conflicting, requirements. Conflicting requirements then need to be reconciled to find a cloud provider that better suits the needs of all the applications. For instance, for an application operating with sensitive/personal information, a user will likely have a strong requirement on the security measures applied by the cloud provider (e.g., encryption algorithms, access control), while for an application performing data-intensive computations on non-sensitive information the same user will be more likely interested in performance (e.g., processing speed and network latency). An intuitive approach for considering the requirements of multiple applications in the selection of a cloud provider consists in first identifying the provider that would be preferred by each application singularly taken, and then in selecting the provider chosen by the majority of the applications. While such an approach would certainly choose a single provider, it may still leave the requirements of some applications completely unsatisfied. To overcome this problem, alternative approaches based on MCDM techniques (e.g., [6]) aim at selecting cloud provider(s) in such a way that the chosen provider(s) balances the satisfaction of the application requirements, thus ensuring not to leave any application unsatisfied. Given the requirements characterizing the needs of each application and a set of cloud plans (among those available from cloud providers), the solution in [6] first adopts a MCDM technique to produce, for each application, a ranking of the cloud plans. Then, a consensus-based process is adopted to select the plan that is considered more acceptable by all applications. In particular, the consensus-voting algorithm takes as input the rank of each application and chooses the plan that better balances the preferences of all applications. Note that this approach considers all applications, and rankings produced by them, to be equally important when computing a solution to reach consensus. This implies that the proposed solution only considers the position of the plans in the ranking and does not take into consideration their relative distance. The cloud provider chosen by applying this technique may not be the first choice of any of the considered applications, but the technique guarantees that no application is left completely unsatisfied. An interesting alternative to be investigated can consider the application requirements by different applications as a single set of requirements to be globally optimized, considering not only the rankings but the distance among plans, possibly evaluating also preferences among applications.

Fuzzy User Requirements. To express precise requirements on the characteristics that should be satisfied by cloud providers, a user needs to have a technical understanding of both the service characteristics and the needs of the applications. However, the identification of 'good' value(s) for a given attribute might be challenging for some users, due to either a lack of technical skills, or the difficulty in identifying the precise needs for their applications. Also, the choice of a specific value or of a precise threshold for an attribute may be an overkill in many situations, imposing a too strong constraint. As an example, a requirement imposing a minimum service availability of 99.995% would exclude a service with a guaranteed availability of 99.990%, which – while less desirable – may still be considered acceptable by the user. To provide users with higher flexibility in the definition of their requirements, some works have considered fuzzy logic for the specification and consideration of requirements (e.g., [15,28]). Fuzzy logic can help users in defining their requirements in a flexible way whenever these requirements cannot be expressed through crisp values over attributes or are not easily definable. For instance, it is simpler for a non-skilled user to generically require 'high availability' rather than specifying precisely which values are good and which are not. In this example, 'high availability' maps to a set of values for the availability attribute, which fit the definition of 'high availability' according to the fuzzy membership function to be applied. In fact, the user and the cloud provider must agree on the meaning of the fuzzy values that a user can use in the definition of her requirements. Fuzzy logic can be used also to address other issues in cloud scenarios, including the evaluation of cloud service performances (e.g., [49]) and the allocation by the provider of its resources to users applications (e.g., [4,15,29,50]). In this context, the allocation of resources to applications needs to take into account several aspects (e.g., the performance of applications, users costs, energy consumption, and security). Hence, finding an allocation that optimizes all these aspects is a difficult task that is usually addressed through MCDM techniques. Also, fuzzy logic can be useful for supporting flexible reasoning in resource allocation, especially in dynamic scenarios where applications are frequently activated/deactivated.

3   Protecting Data in the Cloud

Moving data and applications to the cloud implies a loss of control of the data owner over them and consequent concerns about their security. Guarantee data and applications security requires to address several problems such as the protection of the confidentiality and integrity of data and computations as well as data availability (e.g., [16,19,34,38–40,59]), the enforcement of access control (e.g., [16,19,34,59]), and query privacy (e.g., [22,23]). In this section, we focus on the solutions addressing the confidentiality, integrity, and availability of data.

3.1   Data Confidentiality

The protection of the confidentiality of (sensitive) data is the first problem that has to be considered when storing data at an external cloud provider. Sensitive data must be protected from untrusted/unauthorized parties, including the storing cloud provider, which can be considered *honest-but-curious* (i.e., trusted to properly operate over the data but not to see their content). Current solutions addressing this problem are based on the (possibly combined) adoption of *encryption* and *fragmentation*.

Encryption. Wrapping data with a layer of encryption before outsourcing represents a natural and effective solution to protect the confidentiality of outsourced data [53]. Indeed, only the data owner and authorized users, knowing the encryption key, can access the plaintext data. Encrypting data before outsourcing them guarantees that neither the cloud provider nor external third parties (possibly gaining access to the provider storage devices) can access the data content in the clear. However, encryption makes query execution difficult because data cannot be decrypted at the provider side. To address this problem, different solutions have been proposed, including encrypted database systems supporting SQL queries over encrypted data (e.g., [5,48]) and indexes for query execution (e.g., [10,37]). CryptDB [48] is an example of encrypted database system that supports queries on encrypted data. The idea is to encrypt the values of each attribute of a relational table with different layers of encryption, computed using different kinds of encryption (i.e., random, deterministic, order-preserving, homomorphic, join, order-preserving join, and word search), which depend on the queries to be executed. For instance, the values of an attribute can be first encrypted using an order-preserving encryption schema, then a deterministic encryption schema, and then a random encryption. Proceeding from the outermost layer to the innermost layer, the adopted encryption scheme provides weaker security guarantees but supports more computations over the encrypted data. As an example, if the cloud provider has to execute a GROUP BY on such attribute, the random encryption layer is removed since such encryption does not allow to determine which values of the attribute are equal to each other. Instead, deterministic encryption supports grouping operations, and therefore it is not necessary to remove the deterministic encryption layer.

An index is a metadata associated with the encrypted data that can be used by the cloud provider to select the data to be returned in response to a query. Indexing techniques differ in the kind of queries supported (e.g., [2, 14,37,57]). In general, indexes can be classified in three main categories: *(1) direct indexes* map each plaintext value to a different index value and vice versa (e.g., [14]); *(2) bucket-based indexes* map each plaintext value to one index value but different plaintext values are mapped to the same index value, generating collisions (e.g., [37]); *(3) flattened indexes* map each plaintext value to different index values, each characterized by the same number of occurrences (flattening), and each index value represents one plaintext value only (e.g., [57]).

Besides indexing techniques classified as discussed, many other approaches have been proposed. For instance, indexes based on order preserving encryption support range conditions as well as grouping and ordering clauses (e.g., [2,57]), B+-tree indexes [14] support range queries, and indexes based on homomorphic encryption techniques (e.g., [32,36]) support the execution of aggregate functions.

While promising, encrypted database systems and indexes present open problems, such as the possible information leakage and the still limited support for query execution (e.g., [9,45]).

Fragmentation. Approaches based on encryption for protecting data confidentiality work under the assumption that all data need protection. However, in many scenarios, what is sensitive is the association among data values, rather than values singularly taken. For instance, with reference to a relational table, while the name of patients or the possible values of illness can be considered public, the association among them (i.e., the fact that a patient has a given illness) is clearly sensitive. Confidentiality can be guaranteed in this case by breaking the association storing the involved attributes in separate (unlinkable) *data fragments* (e.g., [1,10–12,17]). The application of fragmentation-based techniques requires first the identification of the sets of attributes whose joint visibility (i.e., association) is considered sensitive. A sensitive association can be modeled as a *confidentiality constraint* corresponding to sets of attributes that should not be publicly visible in the same fragment. For instance, with respect to the previous example, confidentiality constraint ⟨*Name,Illness*⟩ states that the values of attribute *Name* cannot be visible together with the values of attribute *Illness*. Fragmentation splits attributes in different data fragments in such a way that no fragment covers completely any of the confidentiality constraint.

Fragments need to be unlinkable for non-authorized users (including cloud providers) to avoid the reconstruction of the sensitive associations. Different approaches have been proposed to fragment data, as summarized in the following.

– *Two can keep a secret* [1]. The original table is split in two fragments to be stored at two non-communicating cloud providers. Sensitive attributes (i.e., singleton confidentiality constraints) are protected by encoding (e.g., encrypting) them. Sensitive associations are protected by splitting the involved attributes in the two fragments. If an attribute cannot be placed in any of the two fragments without violating a confidentiality constraint, it is encoded. Encoded attributes are stored in both fragments. Only authorized users can access both fragments as well as the encoded attributes, and reconstruct the original relation by joining the two fragments through a common key attribute stored in both fragments.
– *Multiple fragments* [10,12]. The original table can be split into an arbitrary number of disjoint fragments (i.e., fragments that do not have any common attribute). The idea is that sensitive attributes are stored in encrypted form while sensitive associations can always be protected by splitting the involved attributes in different fragments. Each fragment stores a set of attributes in plaintext and all the other attributes of the original table in encrypted form. Authorized users know the encryption key and can reconstruct the original table by accessing a single fragment (any would work) at the cloud provider. The use of multiple fragments guarantees that all the attributes in the original relation that are not considered sensitive by themselves can be represented in plaintext in some fragments.

  Unlinkability among fragments is ensured by the absence of common attributes in fragments. Such a protection may however be put at risk by *data dependencies* among attribute values since the value of some attributes may disclose information about the value of others attributes. For instance, knowing the treatment with which an individual is treated can reduce the uncertainty over her disease. If these attributes are stored in different fragments, they could be exploited for (loosely) joining fragments, thus possibly violating confidentiality constraints. Data dependencies have then to be considered in the fragmentation design [17].
– *Keep a few* [11]. The original table is split into two fragments, one of which is stored at the data owner side. Sensitive attributes are stored at the data owner side while sensitive associations are protected by storing, for each association, at least one attribute at the data owner side. This approach permits to completely depart from encryption. An identifier is maintained in both fragments to allow the data owner to correctly reconstruct the original table. Since one fragment is kept by the data owner, an access request may require her involvement.

3.2   Data Integrity and Availability

In addition to data confidentiality, data integrity and availability are also critical aspects. Data integrity means that the data owner needs guarantees on the fact that cloud providers (and non-authorized users) do not improperly modify the data without being detected. Verifying data integrity consists not only in verifying whether the stored data have not been tampered with or removed (integrity of stored data) but also in verifying the integrity of query results. The integrity of a query result means that the result is *correct* (i.e., the result is computed on the original data and is correct), *complete* (i.e., the computation has been performed on the whole data collection and the result includes all data satisfying the computation), and *fresh* (i.e., the result is computed on the most recent version of the data). Existing solutions addressing these issues can provide *deterministic* or *probabilistic* guarantees.

– *Deterministic techniques* provide guarantees of data integrity with full confidence. Techniques for the integrity of stored data can be based on hashing and digital signatures as building blocks (e.g., [8,35,44]). These solutions require data owners to access their data in the cloud to check their integrity (which may imply high communication overhead). Deterministic guarantees on the integrity of computation results can be achieved by building *authenticated data structures* on the data (e.g., Merkle hash trees [43,46] and skip lists [3,27]). Every computation result is then complemented with a *verification object* VO, extracted by the cloud provider from the authenticated data structure. By checking the VO, the requesting user can easily and efficiently verify if the computation result is correct, complete, and fresh. While the adoption of authenticated data structures has the advantage of providing full confidence on the integrity of computation results, they are defined on a specific attribute and hence provide guarantees only for computations over it.
– *Probabilistic techniques* offer only probabilistic integrity guarantees, but are more flexible than deterministic approaches. Traditional solutions providing probabilistic integrity guarantees of stored data are Proof of Retrievability (POR) and/or Provable Data Possession (PDP) schemes [7,41]. These solutions either include sentinels in the encrypted outsourced data (POR) or pre-compute tokens over encrypted or plaintext data (PDP) to provide the owner with a probabilistic guarantee that the data have not been modified by non-authorized parties.

Probabilistic guarantees on the integrity of computation results can be obtained by inserting fake tuples as sentinels/markers in the original dataset before outsourcing (e.g., [42,60]) or by duplicating (twinning) a portion of the original dataset (e.g., [20,58]). If sentinels/markers and duplicates/twins are not recognizable as such by the cloud provider, their absence in the computation result signals to the requesting user its incompleteness. Clearly, the higher the number of marker/twin tuples the higher the probabilistic guarantees obtained. It is interesting to note that these probabilistic strategies can be jointly used as their protection guarantees nicely complement each other [20,21] and can also be extended to work in a MapReduce scenario [18].

Data availability in the cloud can be interpreted as the ability of verifying whether the cloud provider satisfies users' requirements. Typically, the expected behaviors of a cloud provider can be formalized using a Service Level Agreement (SLA) stipulated between a user and the cloud provider itself. An SLA can include confidentiality, integrity, and availability guarantees that the provider undertook to provide. Some proposals have then investigated the problem of how users can verify whether a cloud provider satisfies the security guarantees declared in an SLA (e.g., [56]). Also the PDP and POR techniques previously discussed for data integrity can also be used for verifying whether the cloud provider stores the data as declared.

4   Conclusions

In this paper, we have discussed the problems of enabling users to select cloud providers that best match their needs, and of empowering users with solutions to protect data outsourced to cloud providers. In particular, we have described techniques that allow users to express their security requirements, possibly defined for multiple applications and also using fuzzy logic, and to ensure confidentiality, integrity, and availability of outsourced data.

Acknowledgments. This work was supported in part by the EC within the FP7 under grant agreement 312797 (ABC4EU), and within the H2020 under grant agreement 644579 (ESCUDO-CLOUD).

References

1. Aggarwal, G., Bawa, M., Ganesan, P., Garcia-Molina, H., Kenthapadi, K., Motwani, R., Srivastava, U., Thomas, D., Xu, Y.: Two can keep a secret: a distributed architecture for secure database services. In: Proceedings of CIDR 2005, Asilomar, CA, USA, January 2005
2. Agrawal, R., Kierman, J., Srikant, R., Xu, Y.: Order preserving encryption for numeric data. In: Proceedings of ACM SIGMOD, Paris, France, June 2004
3. Anagnostopoulos, A., Goodrich, M.T., Tamassia, R.: Persistent authenticated dictionaries and their applications. In: Proceedings of ISC 2001, Malaga, Spain, October 2001
4. Anglano, C., Canonico, M., Guazzone, M.: FC2Q: exploiting fuzzy control in server consolidation for cloud applications with SLA constraints. Concurrency Comput. Pract. Experience **22**(6), 4491–4514 (2014)
5. Arasu, A., Blanas, S., Eguro, K., Kaushik, R., Kossmann, D., Ramamurthy, R., Venkatesan, R.: Orthogonal security with cipherbase. In: Proceedigs of CIDR 2013, Asilomar, CA, USA, January 2013
6. Arman, A., Foresti, S., Livraga, G., Samarati, P.: A consensus-based approach for selecting cloud plans. In: Proceedings of IEEE RTSI 2016, Bologna, Italy, September 2016
7. Ateniese, G., Burns, R., Curtmola, R., Herring, J., Kissner, L., Peterson, Z., Song, D.: Provable data possession at untrusted stores. In: Proceedings of ACM CCS 2007, Alexandria, VA, USA, October/November 2007
8. Boneh, D., Gentry, C., Lynn, B., Shacham, H.: Aggregate and verifiably encrypted signatures from bilinear maps. In: Proceedings of EUROCRYPT 2003, Warsaw, Poland, May 2003

 9. Ceselli, A., Damiani, E., De Capitani di Vimercati, S., Jajodia, S., Paraboschi, S., Samarati, P.: Modeling and assessing inference exposure in encrypted databases. ACM TISSEC **8**(1), 119–152 (2005)
10. Ciriani, V., De Capitani di Vimercati, S., Foresti, S., Jajodia, S., Paraboschi, S., Samarati, P.: Combining fragmentation and encryption to protect privacy in data storage. ACM TISSEC **13**(3), 22:1–22:33 (2010)
11. Ciriani, V., De Capitani di Vimercati, S., Foresti, S., Jajodia, S., Paraboschi, S., Samarati, P.: Selective data outsourcing for enforcing privacy. JCS **19**(3), 531–566 (2011)
12. Ciriani, V., De Capitani di Vimercati, S., Foresti, S., Livraga, G., Samarati, P.: An OBDD approach to enforce confidentiality and visibility constraints in data publishing. JCS **20**(5), 463–508 (2012)
13. Cloud Security Alliance: Cloud Control Matrix v3.0.1. https:// cloudsecurityalliance.org/research/ccm/
14. Damiani, E., Capitani, D., di Vimercati, S., Jajodia, S., Paraboschi, S., Samarati, P.: Balancing confidentiality and efficiency in untrusted relational DBMSs. In: Proceedings of CCS 2003, Washington, DC, USA, October 2003
15. Dastjerdi, A.V., Buyya, R.: Compatibility-aware cloud service composition under fuzzy preferences of users. IEEE TCC **2**(1), 1–13 (2014)
16. De Capitani di Vimercati, S., Foresti, S., Jajodia, S., Livraga, G., Paraboschi, S., Samarati, P.: Enforcing dynamic write privileges in data outsourcing. Comput. Secur. **39**, 47–63 (2013)
17. De Capitani di Vimercati, S., Foresti, S., Jajodia, S., Livraga, G., Paraboschi, S., Samarati, P.: Fragmentation in presence of data dependencies. IEEE TDSC **11**(6), 510–523 (2014)
18. De Capitani di Vimercati, S., Foresti, S., Jajodia, S., Livraga, G., Paraboschi, S., Samarati, P.: Integrity for distributed queries. In: Proceedings of IEEE CNS 2014, San Francisco, CA, USA, October 2014
19. De Capitani di Vimercati, S., Foresti, S., Jajodia, S., Paraboschi, S., Samarati, P.: Encryption policies for regulating access to outsourced data. ACM TODS **35**(2), 12:1–12:46 (2010)
20. De Capitani di Vimercati, S., Foresti, S., Jajodia, S., Paraboschi, S., Samarati, P.: Integrity for join queries in the cloud. IEEE TCC **1**(2), 187–200 (2013)
21. De Capitani di Vimercati, S., Foresti, S., Jajodia, S., Paraboschi, S., Samarati, P.: Efficient integrity checks for join queries in the cloud. JCS **24**(3), 347–378 (2016)
22. De Capitani di Vimercati, S., Foresti, S., Paraboschi, S., Pelosi, G., Samarati, P.: Efficient and private access to outsourced data. In: Proceedings of ICDCS 2011, Minneapolis, Minnesota, USA, June 2011
23. De Capitani di Vimercati, S., Foresti, S., Paraboschi, S., Pelosi, G., Samarati, P.: Shuffle index: efficient and private access to outsourced data. ACM TOS **11**(4), 1–55 (2015). Article 19
24. De Capitani di Vimercati, S., Foresti, S., Samarati, P.: Managing and accessing data in the cloud: Privacy risks and approaches. In: Proceedings of CRiSIS 2012, Cork, Ireland, October 2012
25. De Capitani di Vimercati, S., Livraga, G., Piuri, V.: Application requirements with preferences in cloud-based information processing. In: Proceedings of IEEE RTSI 2016, Bologna, Italy, September 2016
26. De Capitani di Vimercati, S., Livraga, G., Piuri, V., Samarati, P., Soares, G.: Supporting application requirements in cloud-based IoT information processing. In: Procedings of IoTBD 2016, Rome, Italy, April 2016

27. Di Battista, G., Palazzi, B.: Authenticated relational tables and authenticated skip lists. In: Proceedings of DBSec 2007, Redondo Beach, CA, USA, July 2007
28. Foresti, S., Piuri, V., Soares, G.: On the use of fuzzy logic in dependable cloud management. In: Proceedings of IEEE CNS 2015, Florence, Italy, September 2015
29. Frey, S., Claudia, L., Reich, C., Clarke, N.: Cloud QoS scaling by fuzzy logic. In: IEEE IC2E 2014, Boston, MA, USA, March 2014
30. Garg, S.K., Versteeg, S., Buyya, R.: SMICloud: A framework for comparing and ranking cloud services. In: Proc. of IEEE UCC 2011, Melbourne, Australia, December 2011
31. Garg, S.K., Versteeg, S., Buyya, R.: A framework for ranking of cloud computing services. Future Gener. Comput. Syst. $29(4)$, 1012–1023 (2013)
32. Gentry, C.: Fully homomorphic encryption using ideal lattices. In: Proceedings of STOC 2009, Bethesda, MA, USA, May 2009
33. Goscinski, A., Brock, M.: Toward dynamic and attribute based publication, discovery and selection for cloud computing. Future Gener. Comput. Syst. $26(7)$, 947–970 (2010)
34. Goyal, V., Pandey, O., Sahai, A., Waters, B.: Attribute-based encryption for fine-grained access control of encrypted data. In: Proceedings of ACM CCS 2006, Alexandria, VA, USA, October/November 2006
35. Hacigümüs, H., Iyer, B., Mehrotra, S.: Ensuring integrity of encrypted databases in database as a service model. In: Proceedings of DBSec 2003, Estes Park, CO, USA, August 2003
36. Hacigümüs, H., Iyer, B., Mehrotra, S.: Efficient execution of aggregation queries over encrypted relational database. In: Proceedings of DASFAA 2004, Jeju Island, Korea, March 2004
37. Hacigümüs, H., Iyer, B., Mehrotra, S., Li, C.: Executing SQL over encrypted data in the database-service-provider model. In: Proceedings of SIGMOD 2002, Madison, WI, USA, June 2002
38. Jhawar, R., Piuri, V.: Fault tolerance management in IaaS clouds. In: Proceedings of IEEE-AESS ESTEL 2012, Rome, Italy, October 2012
39. Jhawar, R., Piuri, V., Samarati, P.: Supporting security requirements for resource management in cloud computing. In: Proceedings of IEEE CSE 2012, Paphos, Cyprus, December 2012
40. Jhawar, R., Piuri, V., Santambrogio, M.: Fault tolerance management in cloud computing: a system-level perspective. IEEE Syst. J. $7(2)$, 288–297 (2013)
41. Juels, A., Kaliski Jr., B.S.: PORs: Proofs of retrievability for large files. In: Proceedings of ACM CCS 2007, Alexandria, VA, USA, October/November 2007
42. Liu, R., Wang, H.: Integrity verification of outsourced XML databases. In: Proceedings of CSE 2009, Vancouver, Canada, August 2009
43. Merkle, R.: A certified digital signature. In: Proceedings of CRYPTO 1989, Santa Barbara, CA, USA, August 1989
44. Mykletun, E., Narasimha, M., Tsudik, G.: Authentication and integrity in outsourced databases. ACM TOS $2(2)$, 107–138 (2006)
45. Naveed, M., Kamara, S., Wrigh, C.: Inference attacks on property-preserving encrypted databases. In: Proceedings of CCS 2015, Denver, CO, USA, October 2015
46. Pang, H., Jain, A., Ramamritham, K., Tan, K.: Verifying completeness of relational query results in data publishing. In: Proceedings of SIGMOD 2005, Baltimore, MA, USA, June 2005

47. Pawluk, P., Simmons, B., Smit, M., Litoiu, M., Mankovski, S.: Introducing STRATOS: A cloud broker service. In: Proceedings of IEEE CLOUD 2012, Honolulu, HI, USA, June 2012
48. Popa, R., Redfield, C., Zeldovich, N., Balakrishnan, H.: Cryptdb: Protecting confidentiality with encrypted query processing. In: Proceedings of SOSP, Cascais, Portugal (2011)
49. Qu, L., Wang, Y., Orgun, M.A.: Cloud service selection based on the aggregation of user feedback and quantitative performance assessment. In: Proceedings of IEEE SCC 2013, Santa Clara, CA, USA, June/July 2013
50. Rao, J., Wei, Y., Gong, J., Xu, C.Z.: DynaQoS: Model-free self-tuning fuzzy control of virtualized resources for QoS provisioning. In: Proceedings of IEEE IWQoS 2011, San Jose, CA, USA, June 2011
51. Rehman, Z., Hussain, O., Hussain, F.: IaaS cloud selection using MCDM methods. In: Proceedings of IEEE ICEBE 2012, Hangzhou, China, September 2012
52. Ruiz-Alvarez, A., Humphrey, M.: An automated approach to cloud storage service selection. In: Proceedings of ACM ScienceCloud 2011, San Jose, CA, USA, June 2011
53. Samarati, P., De Capitani di Vimercati, S.: Data protection in outsourcing scenarios: issues and directions. In: Proceedings of ASIACCS 2010, Beijing, China, April 2010
54. Samarati, P., De Capitani di Vimercati, S.: Cloud security: issues and concerns. In: Murugesan, S., Bojanova, I. (eds.) Encyclopedia on Cloud Computing. Wiley, Chichester (2016)
55. Samreen, F., Elkhatib, Y., Rowe, M., Blair, G.S.: Daleel: Simplifying cloud instance selection using machine learning. In: Proceedings of IEEE/IFIP NOMS 2016, Istanbul, Turkey, April 2016
56. van Dijk, M., Juels, A., Oprea, A., Rivest, R., Stefanov, E., Triandopoulos, N.: Hourglass schemes: How to prove that cloud files are encrypted. In: Proceedings of ACM CCS 2012, Raleich, NC, USA, October 2012
57. Wang, H., Lakshmanan, L.: Efficient secure query evaluation over encrypted XML databases. In: Proceedings of VLDB 2006, Seoul, Korea, September 2006
58. Wang, H., Yin, J., Perng, C., Yu, P.: Dual encryption for query integrity assurance. In: Proceedings of CIKM 2008, Napa Valley, CA, USA, October 2008
59. Waters, B.: Ciphertext-policy attribute-based encryption: an expressive, efficient, and provably secure realization. In: Proceedings of PKC 2011, Taormina, Italy, March 2011
60. Xie, M., Wang, H., Yin, J., Meng, X.: Integrity auditing of outsourced data. In: Proceedings of VLDB 2007, Vienna, Austria, September 2007
61. Zheng, Z., Wu, X., Zhang, Y., Lyu, M.R., Wang, J.: QoS ranking prediction for cloud services. IEEE TPDS 24(6), 1213–1222 (2013)

Native Cloud Applications: Why Monolithic Virtualization Is Not Their Foundation

Frank Leymann[✉], Uwe Breitenbücher, Sebastian Wagner,
and Johannes Wettinger

IAAS, University of Stuttgart, Stuttgart, Germany
{leymann,breitenbucher,wagner,wettinger}@iaas.uni-stuttgart.de

Abstract. Due to the current hype around cloud computing, the term 'native cloud application' becomes increasingly popular. It suggests an application to fully benefit from all the advantages of cloud computing. Many users tend to consider their applications as cloud native if the application is just bundled as a monolithic virtual machine or container. Even though virtualization is fundamental for implementing the cloud computing paradigm, a virtualized application does not automatically cover all properties of a native cloud application. In this work, which is an extension of a previous paper, we propose a definition of a native cloud application by specifying the set of characteristic architectural properties, which a native cloud application has to provide. We demonstrate the importance of these properties by introducing a typical scenario from current practice that moves an application to the cloud. The identified properties and the scenario especially show why virtualization alone is insufficient to build native cloud applications. We also outline how native cloud applications respect the core principles of service-oriented architectures, which are currently hyped a lot in the form of microservice architectures. Finally, we discuss the management of native cloud applications using container orchestration approaches as well as the cloud standard TOSCA.

1 Introduction

Cloud service providers of the early days, such as Amazon, started their Infrastructure as a Service (IaaS) cloud business by enabling customers to run virtual machines (VM) on their datacenter infrastructure. Customers were able to create VM images that bundled their application stack along with an operating system and instantiate those images as VMs. In numerous industry collaborations we investigated the migration of existing applications to the cloud and the development of new cloud applications [1–3]. In the investigated use cases we found that virtualization alone is not sufficient for fully taking advantage of the cloud computing paradigm.

In this article, which is an extension of a previous paper [4] presented at the *6th International Conference on Cloud Computing and Services Science (CLOSER)*, we show that although virtualization lays the groundwork for cloud

M. Helfert et al. (Eds.): CLOSER 2016, CCIS 740, pp. 16–40, 2017.
DOI: 10.1007/978-3-319-62594-2_2

computing, additional alterations to the application's architecture are required to make up a 'cloud native application'. We discuss five essential architectural properties we identified during our industry collaborations that have to be implemented by a native cloud application [5]. Based on those properties we explain why an application that was simply migrated to the cloud in the form of a VM image does not comply with these properties and how the application has to be adapted to transform it into a native cloud application. These properties have to be enabled in any application that is built for the cloud. Compared to the original article, we additionally discuss the deployment and management aspects of native cloud applications. Therefore, we describe how Kubernetes[1] can be employed to manage native cloud applications that are deployed using fine-grained Docker containers. We also point out specific limitations of Kubernetes and similar container orchestration approaches and discuss how to overcome those using TOSCA as cloud management standard. Note that we provide a definition of native cloud applications and how to deploy and manage them; we do not aim to establish a migration guide for moving applications to the cloud. Guidelines and best practices on this topic can be found in our previous work [2,6,7].

Section 2 introduces a reference application that reflects the core of the architectures of our industry use cases. Based on the reference application, Sect. 3 focuses on its transformation from a VM-bundled to a native cloud application. We also discuss why virtualization or containerization alone is not sufficient to fully benefit from cloud environments. Therefore, a set of architectural properties are introduced, which a native cloud application has to implement. Section 4 discusses how native cloud applications are related to microservice architectures, SOA, and continuous delivery. Section 5 discusses how the reference application itself can be offered as a cloud service. How to deploy and run a native cloud application using container orchestration is discussed in Sect. 6 by example of Kubernetes. In Sect. 7 a more holistic and technology-agnostic management approach of native cloud application based on TOSCA is described. Finally, Sect. 8 concludes the article.

2   Reference Application

Throughout the article, the application shown in Fig. 1 is used as running example for transforming an existing application into a cloud native application. It offers functionality for accounting, marketing, and other business concerns. The architecture specification of this application and the following transformation uses the concept of layers and tiers [8]: the functionality of an application is provided by separate components that are associated with logical layers. Application components may only interact with other components on the same layer or one layer below. Logical layers are later assigned to physical tiers for application provisioning. In our case, these tiers are constituted by VMs, which may be hosted by a cloud provider.

---

[1] http://kubernetes.io.

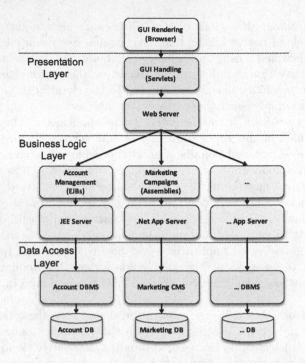

Fig. 1. Reference application to be moved to the cloud.

The reference application is comprised of three layers. Each layer has been built on different technology stacks. The accounting functions are implemented as Enterprise Java Beans[2] (EJB) on a Java Enterprise Edition (JEE) server making use of a Database Management Systems (DBMS); the marketing functions are built in a .Net environment[3] using a Content Management System (CMS). All application functions are integrated into a graphical user interface (GUI), which is realized by servlets hosted on a Web server.

The servlet, EJB and .Net components are *stateless*. In this scope, we differentiate: (i) *session state* - information about the interaction of users with the application. This data is provided with each request to the application and (ii) *application state* - data handled by the application, such as a customer account, billing address, etc. This data is persisted in the databases.

3   Transforming the Reference Application to a Cloud Native Application

When moving the reference application to a cloud environment, the generic properties of this environment can be used to deduct required cloud application

---

[2] https://jcp.org/aboutJava/communityprocess/final/jsr318.
[3] http://www.microsoft.com/net.

properties. The properties of the cloud environment have been defined by the NIST [9]: *On-demand self-service:* the cloud customer can independently sign up to the service and configure it to his demands. *Broad network access:* the cloud is connected to the customer network via a high-speed network. *Resource pooling:* resources required to provide the cloud service are shared among customers. *Rapid elasticity:* resources can be dynamically assigned to customers to handle currently occurring workload. *Measured service:* the use of the cloud by customers is monitored, often, to enable pay-per-use billing models.

To make an application suitable for such a cloud environment, i.e. to utilize the NIST properties, we identified the *IDEAL* cloud application properties [5]: *Isolation of state, Distribution, Elasticity, Automated Management and Loose coupling.* In this section, we discuss why VM-based application virtualization and containerization alone is rather obstructive for realizing them. Based on this discussion, the steps for enabling these properties are described in order to transform a VM-based application towards a native cloud application. As we start our discussion on the level of VMs, we first focus on the Infrastructure as a Service (IaaS) service model. Then we show how it can be extended to use Platform as a Service (PaaS) offerings of a cloud provider.

3.1 Complete Application per Virtual Machine

To provide an application to customers within a cloud environment as quickly as possible, enterprises typically bundle their application into a single virtual machine image (VMI)[4]. Such VMIs are usually self-contained and include all components necessary for running the application. Considering the reference application, the data access layer, the business logic layer, and the presentation layer would be included in that VMI. Figure 2 shows an overview of that package.

Customers now start using the application through their Web browsers. As shown in Fig. 2, all requests are handled by the same VM. Consequently, the more customers are using the application, the more resources are required. At some point in time, considering an increasing amount of customer requests, the available resources will not be able to serve all customer requests any more. Thus, the application needs to be scaled in order to serve all customers adequately.

The first approach to achieve scalability is to instantiate another VM containing a copy of your application stack as shown in Fig. 3. This allows you to serve more customers without running into any bottleneck. However, the operation of multiple VMs also has significant downsides. You typically have to pay for licenses, e.g. for the database server, the application server, and the content management system, on a per VM basis. If customers use the account management features mostly, why should you also replicate the marketing campaigns stack and pay for the corresponding licenses? Next, what about your databases

---

[4] From here on we do not mention containerization explicitly by considering them as similar to virtual machine images - well recognizing the differences. But for the purpose of our discussion they are very similar.

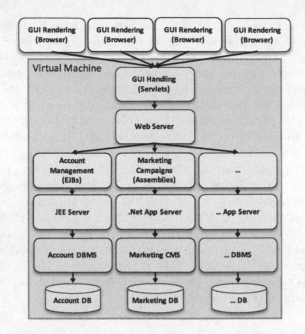

Fig. 2. Packaging the application into one VM.

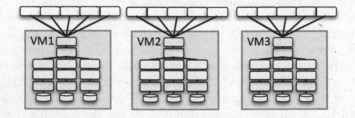

Fig. 3. Scaling based on complete VMs.

that are getting out of sync because separate databases are maintained in each of the VMs? This may happen because storage is associated to a single VM but updates need to be synchronized across those VMs to result in consistent data.

Therefore, it should be possible to scale the application at a finer granular, to ensure that its individual functions can be scaled independently instead of scaling the application as a whole. This can be achieved by following the *distribution property* in the application architecture. This property requires the application functionality to be distributed among different components to exploit the measured service property and the associated pay-per-use pricing models more efficiently. Due to its modularized architecture comprising of logical layers and components, the distribution property is met by the reference application. However, by summarizing the components into one single VM, i.e. in one tier, the modularized architecture of the application gets lost.

Moreover, this leads to the violation of the *isolated state* property, which is relevant for the application to benefit from the resource pooling and elasticity property. This property demands that session and application state must be confined to a small set of well-known stateful components, ideally, the storage offerings and communication offerings of the cloud providers. It ensures that stateless components can be scaled more easily, as during the addition and removal of application component instances, no state information has to be synchronized or migrated, respectively.

Another IDEAL property that is just partly supported in case the application is bundled as a single VM is the *elasticity property*. The property requires that instances of application components can be added and removed flexibly and quickly in order to adjust the performance to the currently experienced workload. If the load on the components increases, new resources are provisioned to handle the increased load. If, in turn, the load on the resources decreases, under-utilized components are decommissioned. This *scaling out* (increasing the number of resources to adapt to workload) as opposed to *scaling up* (increasing the capabilities of a single cloud resource) is predominantly used by cloud applications as it is also required to react to component failures by replacing failed components with fresh ones. Since the distribution is lost, scaling up the application by assigning more resources to the VM (e.g. CPU, memory, etc.) is fully supported, but not *scaling out* individual components. Hence, the elasticity property is just partly met if the application is bundled as a single VM. The incomplete support of the elasticity property also hinders the full exploitation of the cloud resource pooling property, as the elasticity property enables unused application resources to be decommissioned and returned to the resource pool of the cloud if they are not needed anymore. These resources can then be used by other customers or applications.

3.2   Stack-Based Virtual Machines with Storage Offerings

Because of the drawbacks of a single VM image containing the complete application, a suitable next step is to extract the different application stacks to separate virtual machines. Moreover, data can be externalized to storage offerings in the cloud ('Data as a Service'), which are often associated to the IaaS service model. Such services are used similar to hard drives by the VMs, but they are stored in a provider-managed scalable storage offering. Especially the stored data can be shared among multiple VMs when they are being scaled out, thus, avoiding the consistency problems indicated before and hence fostering the isolated state property. Figure 4 shows the resulting deployment topology of the application, where each stack and the Web GUI is placed into a different virtual machine that accesses a Data as a Service cloud offering.

When a particular stack is under high request load, it can be scaled out by starting multiple instances of the corresponding VM. For example, in Fig. 5 another VM instance of the accounting stack is created to handle higher load. However, when another instance of a VM is created the DBMS is still replicated which results in increased license costs.

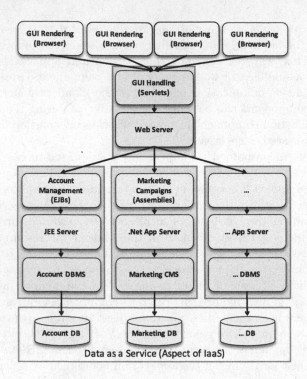

Fig. 4. Packaging stacks into VMs.

3.3   Using Middleware Virtual Machines for Scaling

The replication of middleware components such as a DBMS can be avoided by placing these components again into separate VMs that can be scaled out independently from the rest of the application stack on demand. The middleware component is then able to serve multiple other components. In case of the reference application the DBMS associated with the account management is moved to a new VM (Fig. 6), which can be accessed by different instances of the JEE server. Of course, also the JEE server or the .Net server could be moved into separate VMs. By doing so, the distribution property is increased and elasticity can be realized at a finer granular.

Even though the single components are now able to scale independently from each other, the problem of updating the application components and especially the middleware installed on VMs still remains. Especially, in large applications involving a variety of heterogeneous interdependent components this can become a very time- and resource-intensive task. For example, a new release of the JEE application server may also require your DBMS to be updated. But the new versions of the DBMS may not be compatible with the utilized .Net application server. This, in turn, makes it necessary to run two different versions of the same DBMS. However, this violates an aspect of the *automated management property* demanding that required human interactions to handle management tasks are

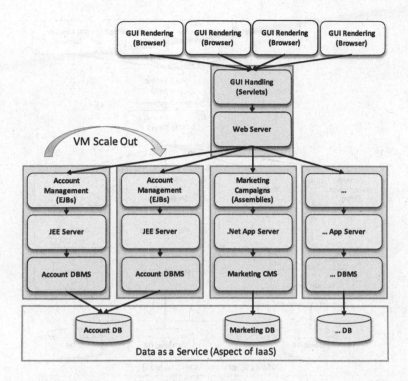

Fig. 5. Scaling stacked VMs.

reduced as much as possible in order to increase the availability and reactiveness of the application.

3.4   Resolving Maintenance Problems

To reduce management efforts, we can substitute components and middleware with IaaS, PaaS, or SaaS offerings from cloud providers. In Fig. 7, the VMs providing the Web server and application server middleware are replaced with corresponding PaaS offerings. Now, it is the cloud provider's responsibility to keep the components updated and to rollout new releases that contain the latest fixes, e.g. to avoid security vulnerabilities.

In case of the reference application, most components can be replaced by cloud offerings. The first step already replaced physical machines, hosting the application components with VMs that may be hosted on IaaS cloud environments. Instead of application servers, one may use PaaS offerings to host the application components of the business logic layer. The DBMS could be substituted by PaaS offerings such as Amazon SimpleDB; marketing campaign .Net assemblies could be hosted on Microsoft Azure, as an example.

To offload the management (and even development) of your .Net assemblies one could even decide to substitute the whole marketing stack by a SaaS offering

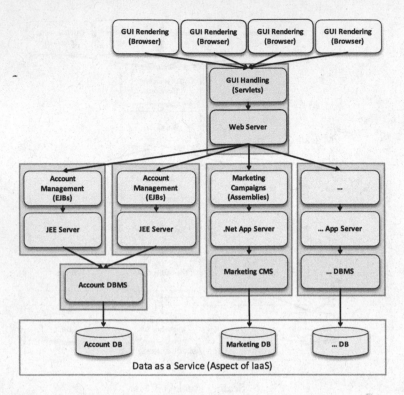

Fig. 6. Middleware-VMs for scaling.

that provides the required marketing functionality. In this case, the Web GUI is integrated with the SaaS offering by using the APIs provided by the offering.

Of course, before replacing a component with an *aaS offering, it should be carefully considered how the dependent components are affected [6]: adjustments to components may be required to respect the runtime environment and APIs of the used *aaS offering.

3.5   The Final Steps Towards a Cloud Native Application

The reference application is now decomposed into multiple VMs that can be scaled individually to fulfill the distribution and elasticity property. Isolation of state has been enabled by relying on cloud provider storage offerings. The software update management has been addressed partially.

However, the addition and removal of virtual machine instances can still be hindered by dependencies among application components: if a VM is decommissioned while an application component hosted on it interacts with another component, errors may occur. The dependencies between application components meaning the assumptions that communicating components make about each other can be reduced by following the *loose coupling property*. This property is implemented

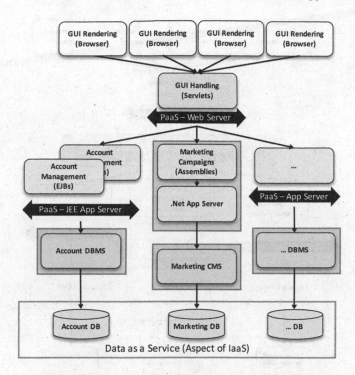

Fig. 7. Making use of cloud resources and features.

by using cloud communication offerings enabling asynchronous communication between components through a messaging intermediate as shown in Fig. 8. This separation of concerns ensures that communication complexity regarding routing, data formats, communication speed etc. is handled in the messaging middleware and not in components, effectively reducing the dependencies among communication partners. Now, the application can scale individual components easier as components do not have to be notified in case other components are provisioned and decommissioned.

To make elastic scaling more efficient, it should be automated. Thus, again the automated management property is respected. This enables the application to add and remove resources without human intervention. It can cope with failures more quickly and exploits pay-per-use pricing schemes more efficiently: resources that are no longer needed should be automatically decommissioned. Consequently, the resource demand has to be constantly monitored and corresponding actions have to be triggered without human interactions. This is done by a separate *watchdog* component [8,10] and elasticity management components [11]. After this step, the reference application became cloud native, thus, supporting the IDEAL cloud application properties: **I**solation of state, **D**istribution, **E**lasticity, **A**utomated Management and **L**oose coupling [5].

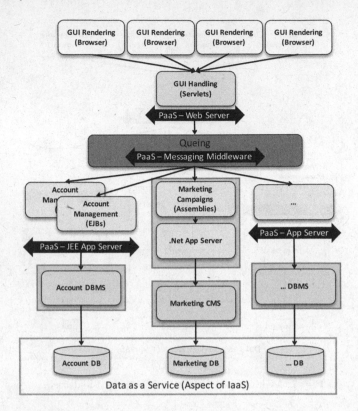

Fig. 8. Making use of cloud communication features.

In terms of virtualization techniques and technologies, fully fledged VMs with their dedicated guest operation system could also be replaced by more lightweight virtualization approaches such as containers, which recently became popular with Docker [12]. However, such approaches may not provide the same degree of isolation, so depending on the specific requirements of an application, the one or the other virtualization approach fits better.

4   Microservices and Continuous Delivery

Microservice architectures provide an emerging software architecture style, which is currently discussed and hyped a lot. While there is no clear definition of what a microservice actually is, some common characteristics have been established [13,14]. Microservice architectures are contrary to monolithic architectures. Consequently, a specific application such as a Web application (e.g. the reference application presented in this paper) or a back-end system for mobile apps is not developed and maintained as a huge single building block, but as a set of 'small' and independent services, i.e. microservices. As of today, there is no common sense how 'small' a microservice should be. To make them meaningful, these services are typically built around business capabilities such as account management and marketing campaigns as outlined by the reference application. Their independence is implemented by running each service in its own process or container [12]. This is a key difference to other component-based architecture styles, where the entire application shares a process, but is internally modularized, for instance, using Java libraries.

The higher degree of independence in case of microservices enables them to be independently deployable from each other, i.e. specific parts of an application can be updated and redeployed without touching other parts. For non-trivial and more complex applications, the number of services involved quickly increases. Consequently, manual deployment processes definitely do not scale anymore for such architectures, because deployment happens much more often and independently. Therefore, fully automated deployment machinery such as continuous delivery pipelines are required [15].

As a side effect of the services' independence, the underlying technologies and utilized programming languages can be extremely diverse. While one service may be implemented using Java EE, another one could be implemented using .Net, Ruby, or Node.js. This enables the usage of 'the best tool for the job', because different technology stacks and programming languages are optimized for different sets of problems. The interface, however, which is exposed by a particular service must be technology-agnostic, e.g. based on REST over HTTP, so different services can be integrated without considering their specific implementation details. Consequently, the underlying storage technologies can also differ, because 'decentralized data management' [13] is another core principle of microservice architectures. As outlined by the reference application, each service has its own data storage, so the data storage technology (relational, key-value, document-oriented, graph- based, etc.) can be chosen according to the specific storage requirements of a particular service implementation.

In addition, microservice architectures follow the principle of 'smart endpoints and dumb pipes' [13], implying the usage of lightweight and minimal middleware components such as messaging systems ('dumb pipes'), while moving the intelligence to the services themselves ('smart endpoints'). This is confirmed by reports and surveys such as carried out by Schermann et al.: REST in conjunction with HTTP as transport protocol is used by many companies today. JSON and XML are both common data exchange formats. There is a trend to minimize the usage of complex middleware towards a more choreography-style coordination of services [16]. Finally, the architectural paradigm of self-contained systems [17] can help to treat an application, which is made of a set of microservices, in a self-contained manner.

In this context, an important fact needs to be emphasized: most of the core principles of microservice architectures are not new at all. Service-oriented architectures (SOA) are established in practice for some time already, sharing many of the previously discussed core principles with microservice architectures. Thus, we see microservice architectures as one possible opinionated approach to realize SOA, while making each service independently deployable. This idea of establishing independently deployable units is a focus of microservice architectures, which was not explicitly a core principle in most SOA-related works and efforts. Therefore, continuous delivery [15] can now be implemented individually per service to completely decouple their deployment.

Our previously presented approach to transition the reference application's architecture towards a native cloud application is based on applying the IDEAL properties. The resulting architecture owns the previously discussed characteristics of microservice architectures and SOA. Each part of the reference application (account management, marketing campaigns, etc.) now represents an independently (re-)deployable unit. Consequently, if an existing application is transitioned towards a native cloud application architecture by applying the IDEAL properties, the result typically is a microservice architecture. To go even further and also consider development as part of the entire DevOps lifecycle, a separate continuous delivery pipeline can be implemented for each service to perform their automated deployment when a bug fix or new feature is committed by a developer. Such pipelines combined with Cloud-based development environments, such as Cloud9 [18], also make the associated application development processes cloud-native in addition to deploying and running the application in a cloud-native way.

5   Moving Towards a SaaS Application

While the IDEAL properties enable an application to benefit from cloud environments and (micro)service-oriented architectures, additional properties have to be considered in case the application shall be offered *as a Service* to a large number of customers [11,19]: Such applications should own the properties *clusterability, elasticity, multi-tenancy, pay-per-use,* and *self-service*. Clusterability summarizes the above- mentioned isolation of state, distribution, and loose coupling. The elasticity discussed by Freemantle and Badger et al. [11,19] is identical to the elasticity mentioned above. The remaining properties have to be enabled in an application-specific manner as follows.

5.1   Multi-tenancy

The application should be able to support multiple tenants, i.e. defined groups of users, where each group is isolated from the others. Multi-tenancy does not mean isolation by associating each tenant with a separate copy of the application stack in one or more dedicated VMs. Instead, the application is adapted to have a notion of tenants to ensure isolation. The application could also exploit multi-tenant aware middleware [20] which is capable to assign tenant requests to the corresponding instance of a component.

In scope of the reference application, the decomposition of the application into loosely coupled components enables the identification of components that can easily be shared among multiple tenants. Other components, which are more critical, for example, those sharing customer data likely have to be adjusted in order to ensure tenant isolation. In previous work, we discussed how such *shared components* and *tenant-isolated components* may be implemented [5]. Whether an application component may be shared among customers or not may also affect the distribution of application components to VMs.

5.2 Pay-Per-Use

Pay-per-use is a property that fundamentally distinguishes cloud applications from applications hosted in traditional datacenters. It ensures that tenants do only pay when they are actually using an application function, but not for the provisioning or reservation of application resources. Pay-per-use is enabled by fine-grained metering and billing of the components of an application stack. Consequently, the actual usage of each individual component within the application stack must be able to be monitored, tracked, and metered. Depending on the metered amount of resource usage, the tenant is billed. What kind of resources are metered and billed depends on the specific application and the underlying business model. Monitoring and metering can also be supported by the underlying middleware if it is capable to relate the requests made to the application components with concrete tenants.

In scope of the reference application, sharing application component instances ensures that the overall workload experienced by all instances is leveled out as workload peaks of one customer happen at the same time where another customer experiences a workload low. This sharing, thus, enables flexible pricing models, i.e. charging on a per-access basis rather than on a monthly basis. For instance, the reference application may meter and bill a tenant for the number of marketing campaigns he persists in the CMS. Other applications may meter a tenant based on the number requests or the number of CPUs he is using. Amazon, for instance, provides a highly sophisticated billing model for their EC2 instances [21].

5.3 Self-service

The application has to ensure that each tenant can provision and manage his subscription to the application on his own, whenever he decides to do so. Especially, no separate administrative staff is needed for provisioning, configuring, and managing the application. Self-service capability applies to each component of the application (including platform, infrastructure, etc.). Otherwise, there would not be real improvements in time-to-market. The self-service functionality can be provided by user interfaces, command line interfaces, and APIs to facilitate the automated management of the cloud application [11].

In scope of the reference application, automated provisioning and decommissioning of application component instances is enabled by the used cloud environment. Therefore, customers may be empowered to sign up and adjust subscriptions to the cloud-native application in a self-service manner, as no human management tasks are required on the application provider side anymore.

6   Native Cloud Applications and Container Orchestration

The previous sections of this paper aim to point out the key concepts of native cloud applications. Applications that are simply packaged as a virtual machine image (VMI) or monolithic container do not benefit from the cloud properties discussed previously. Therefore, such applications do not represent native cloud applications. However, instead of packaging existing legacy applications as monolithic virtual machines, this paper presented an approach to systematically split such an application into fine-grained building blocks (Sect. 3). This approach is compatible with the emerging microservice architecture style (Sect. 4), which essentially is a modernized flavor and opinionated implementation of service-oriented architecture (SOA). Container virtualization (containerization) approaches such as Docker can be utilized to package and deploy applications comprising of such fine-grained building blocks, i.e. components or microservices: each component runs inside a separate container instead of hosting the entire application inside a monolithic container. The latter option would be conceptually the same as putting everything inside a monolithic virtual machine. This is technically different from using virtual machines because the overhead of VMs is significantly larger: each VM runs a completely dedicated instance of an operating system. While this approach leads to strong isolation, it implies the overhead of running each component on top of a dedicated operating system. Containers share the underlying operating system of their host and are, therefore, often referred to as 'lightweight' virtualization. Docker[5] and Rocket[6] are two popular container virtualization approaches. The Open Container Initiative (OCI)[7] aims to standardize a packaging format for containers.

While containers help to package, deploy, and manage loosely coupled components such as microservices that make up a specific application, their orchestration is a major challenge on its own. Container orchestration approaches such as Docker Swarm[8], Kubernetes, and Apache Mesos[9] are emerging to address this issue and simplify the process of combining and scaling containers. The Cloud Native Computing Foundation (CNCF)[10] is an emerging standard in this field, which promotes Kubernetes as one of the most actively developed container orchestration frameworks. Therefore, the remainder of this section discusses how the split application outlined in Fig. 7 can be deployed based on containers that are managed using Kubernetes.

Figure 9 outlines how the application topology can be distributed across a diverse set of containers, e.g. Docker containers. Persistent data is stored in container volumes[11], which are connected to the corresponding containers that run

---

[5] http://www.docker.com.

[6] http://coreos.com/rkt.

[7] http://www.opencontainers.org.

[8] http://docs.docker.com/engine/swarm.

[9] http://mesos.apache.org.

[10] http://cncf.io.

[11] http://docs.docker.com/engine/tutorials/dockervolumes.

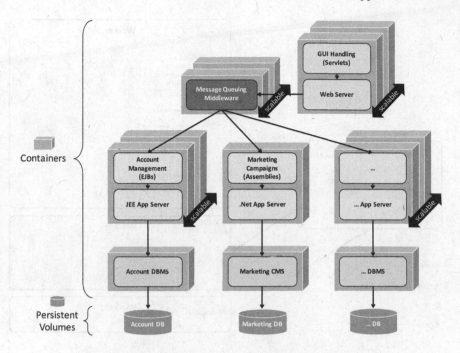

Fig. 9. Container-based application topology.

DBMS and CMS components. This approach makes data persistence more reliable because if a container crashes, a new one can be started and wired with the persistent volume. Some of the outlined containers such as the account management container can be scaled independently, i.e. container instances can be added and removed depending on the current load of this part of the application. This mechanism is the foundation for providing elasticity, an essential property of native cloud applications.

While Docker and similar solutions provide the basic mechanisms, tooling, and APIs to manage and wire containers, the orchestration and management of containers at scale add further challenges that have to be addressed. This is where large-scale container orchestration frameworks such as Kubernetes come into play. They provide key features that would have to be implemented using custom solutions on top of container virtualization solutions such as Docker. These features include the optimized distribution of containers inside a cluster, self-healing mechanisms to ensure that crashed containers are replaced, automated scaling mechanisms based on current load to ensure elasticity, and load balancing of incoming requests among container instances. Figure 10 sketches the simplified architecture using Kubernetes by example of the split application discussed in this paper. A cluster consists of *worker nodes* and at least one *master node*. Technically, a node is either a physical or virtual machine running on-premises or in the cloud. Each worker node runs Docker to host and execute containers. *Pods* are groups of containers that inherently belong together

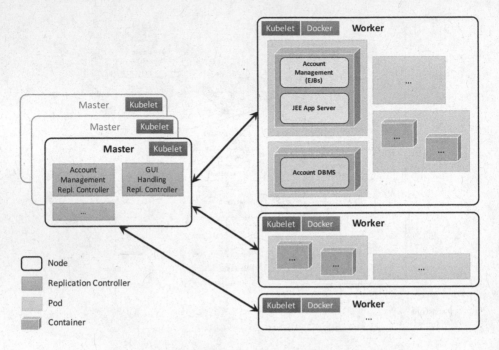

Fig. 10. Simplified Kubernetes architecture overview.

and cannot be deployed and scaled individually. Each pod consists of at least one container. Consequently, pods are the smallest units in Kubernetes that can be managed individually. Each node runs a *kubelet*, which acts as Kubernetes agent for cluster management purposes. The primary master node connects to these kubelets to coordinate and manage the other nodes, i.e. monitoring and scheduling containers, removing crashed ones, etc. Pods can either be maintained manually through the Kubernetes API of a cluster or their management is automated through application-specific *replication controllers* running on the master nodes. A replication controller such as the *account management replication controller* shown in Fig. 10 ensures that the defined number of replicas of a specific pod are up and running as part of the cluster. For example, the previously mentioned replication controller is associated with the pod comprising the container that runs the account management EJBs on top of the JEE application server. If, for instance, a replica of this pod fails on a particular node, the replication controller immediately schedules a new replica on any node in the cluster. While a single master node is technically sufficient, additional master nodes can be added to a cluster in order to ensure high availability: in case the primary master node fails, another master node can immediately replace it.

Technically, JSON or YAML files are used to define pods and replication controllers. The following YAML listing provides an example how the *account management pod* can be specified as outlined in Fig. 10 by the pod that consists of a single container (containing account management EJBs and JEE application server), which runs on the upper worker:

```
1   apiVersion: v1
2   kind: Pod
3   metadata:
4     # Labels describe properties of this pod
5     labels:
6       name: account-management-pod
7   spec:
8     containers:
9     - name: account-management
10      # Docker image running in this pod
11      # Contains EJBs and JEE application server
12      image: account-mgmt-jee:stable
13      ports:
14      - containerPort: 8080
```

In order to maintain pods in an automated manner by Kubernetes, a replication controller can be created instead of defining a manually managed pod. The *account management replication controller* denoted in Fig. 10 can be specified, for example, using the following YAML listing:

```
1   apiVersion: v1
2   kind: ReplicationController
3   metadata:
4     name: account-management-controller
5   spec:
6     # Three replicas of this pod must run in the cluster
7     replicas: 3
8     # Specifies the label key and value on the pod that
9     # this replication controller is responsible
10    selector:
11      app: account-management-pod
12    # Information required for creating pods in the cluster
13    template:
14      metadata:
15        labels:
16          app: account-management-pod
17      spec:
18        containers:
19        - name: account-management
20          image: account-mgmt-jee:stable
21          ports:
22          - containerPort: 8080
```

Each pod dynamically gets an IP address assigned. Especially when dealing with multiple replicas that are managed through a replication controller, it is a major challenge to keep track of the IP addresses. Moreover, the transparent load balancing of requests between existing replicas is not covered. Kubernetes tackles these issues by allowing the definition of *services*, which can be specified as outlined by the following YAML listing:

```
1   apiVersion: v1
2   kind: Service
3   metadata:
4     name: account-management-service
5   spec:
6     ports:
7     - port: 80 # port exposed by the service
8       targetPort: 8080 # port pointing to the pod
9     selector:
10      app: account-management-pod
```

Each service provides a single and stable endpoint that transparently distributes requests between running replicas of the pod specified by the given selector, for example, the previously defined `account-management-pod`. In addition to pods, replication controllers, and services, Kubernetes allows to attach *volumes* to pods. For example, an Amazon Elastic Block Store volume[12] can be mounted into a pod to provide persistent storage. This is key for database back-ends of applications as outlined in Fig. 9.

Kubernetes provides a broad variety of additional features such as liveness and readiness probes[13], also known as health checks. These checks are not only important for monitoring purposes, e.g. to find out whether a particular container is up and running, they are also key to wire different containers. For example, container 1 connects to an endpoint exposed by container 2. However, the component started in container 2 that exposes the required endpoint takes some time until the connection can be established. Therefore, container 1 uses a readiness probe (e.g. an HTTP endpoint that must return the success response code 200) to find out when container 2 is ready for establishing the actual connection between the two components. Moreover, Kubernetes provides mechanisms to perform rolling updates[14] on a cluster, which is a key enabler for continuously delivering updated application instances into production as discussed in Sect. 4.

Although container orchestration such as Kubernetes provide a comprehensive framework to deploy and manage native cloud applications at scale, there are several constraints and assumptions made by such container orchestration approaches. (i) It is typically assumed that each application component runs inside a stateless container, i.e. a container can be terminated and replaced by a new instance of it at any point in time. Many existing legacy applications were not designed this way, so significant effort is required to wrap them as stateless containers. Moreover, the scaling mechanisms of some legacy application components rely on specific features provided by application servers such as Java EE servers. Migrating them toward a container orchestration is a non-trivial challenge. (ii) A container is considered immutable, i.e. container instances are not patched or updated (security fixes, etc.), but instances are dropped and replaced by updated containers. (iii) A Kubernetes cluster is based on a pretty homogenous infrastructure, e.g. a set of virtual servers running at a single cloud provider. Consequently, multi-cloud or hybrid cloud deployments are hard to implement. However, the Kubernetes community is working on cluster federation solutions[15]. (iv) Logic cannot be added to relationships between containers, i.e. the relations between containers are dumb (e.g. sharing TCP ports), so the containers must be smart enough to establish corresponding connections between them. This is achieved by utilizing readiness and liveness probes as discussed previously in this section. (v) Application components cannot be split into arbitrarily fine-grained

---

[12] http://kubernetes.io/docs/user-guide/volumes/#awselasticblockstore.
[13] http://kubernetes.io/docs/user-guide/production-pods.
[14] http://kubernetes.io/docs/user-guide/rolling-updates.
[15] http://github.com/kubernetes/kubernetes/blob/master/docs/proposals/federation.md.

containers. For example, the account management EJBs outlined in Fig. 9 must run in the same container as the JEE application server because the application server exposes the port that is used by other components to utilize the functionality provided by those EJBs. An EJB itself would not be capable of providing and exposing such a port through a separate container. (vi) Finally, external services such as platform-as-a-service offerings cannot be immediately managed as part of a container orchestration solution such as Kubernetes. Although containerized application components running inside the cluster can connect to external services such as a hosted database instance, their configuration and management is another challenge that needs to be tackled in addition to managing the cluster.

7   Standards-Based Modeling, Orchestration, and Management of Native Cloud Applications

In the previous sections, we discussed how native cloud applications can be developed, how the approach is compatible with the microservice architecture style, and how container virtualization approaches can be utilized to package and deploy such applications. However, we have seen that there are several constraints in terms of orchestrating containers, for example, application components must be stateless, strong assumptions about the communication in and between containers, and the difficulties regarding multi- and hybrid cloud applications. Moreover, the automation of holistic management processes, such as migrating the whole application including its data to another cloud provider, is an unsupported challenge. Therefore, in this section, we describe a solution to tackle these issues by discussing how the TOSCA [22–25] standard supports modeling, orchestrating, and managing various kinds of applications components.

The *Topology and Orchestration Specification for Cloud Applications (TOSCA)* is an OASIS standard released in 2013 that enables modeling cloud

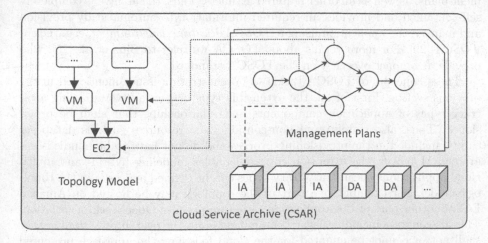

Fig. 11. The TOSCA concepts: topology model and management plans.

applications and their management in a portable manner. Beside portability, the automation of application provisioning and management is one of the major goals of this standard. TOSCA models conceptually consist of two parts as depicted in Fig. 11: (i) application topology model and (ii) management plans. An *application topology model* is a directed, possibly cyclic graph that describes the structure of the application to be provisioned. The nodes of this graph represent the application components (virtual machines, etc.), edges represent relationships between those. The topology model also specifies information about components and their relationships in terms of types, properties, available management interfaces, etc. Especially, TOSCA provides an extensible type system for describing the types of components and relationships. Thus, the standard allows modeling arbitrary components and relationship types, even the definition of proprietary types for supporting legacy applications—a specific feature that gets important in the following discussion.

The second part of TOSCA models are *management plans*, which are executable process models implementing management functionality for the described application, for example, to backup all application data or to migrate the application from one provider to another. Management plans are typically implemented using the workflow technology [26] in order to provide a reliable execution environment. In addition, standardized workflow languages such as BPEL [27] and BPMN [28] support this choice. To implement their functionality, plans invoke operations provided by the nodes of the topology model. Therefore, additional artifacts are required: (i) *deployment artifacts (DAs)* implement business logic whereas (ii) *implementation artifacts (IAs)* implement management operations provided by the components of the topology model. Typical deployment artifacts are, for example, the binaries of an application server whereas a corresponding implementation artifact may be a shell script that starts the application server. Moreover, TOSCA defines a portable packaging format, namely *Cloud Service Archive (CSAR)*, that contains the topology model, all management plans, as well as all other required artifacts. Thus, the archive is completely self-contained and provides all required information to automatically provision and manage the application using a *TOSCA Runtime Environment*, e. g., *Open-TOSCA* [29]. For more details about TOSCA we refer to Binz et al. [25], who provide a compact overview on the TOSCA standard.

These concepts of TOSCA tackle the orchestration issues mentioned in the previous section. First of all, the extensible type system allows modeling arbitrary types of application components and relationships that shall be provisioned. Thus, also stateful and non-containerized resources such as databases can be included in the provisioning process and do not have to be handled separately. Moreover, the extension concept enables modeling hybrid- and multi-cloud applications because arbitrary types can be defined and used in the topology model. For example, one part of the topology may be hosted on Amazon EC2, another part of the application on an on-premise OpenStack installation. This also enables including legacy components into the overall provisioning that shall not or cannot be adapted for the cloud following the approach presented in Sect. 3. To specify non-functional requirements regarding the provisioning or management of the application, TOSCA employs the concept of *policies* [22] that may be used, for example, to specify the regions of a cloud provider in which an application is allowed to be deployed [30].

However, it is important to emphasize that the concept of topology models does not compete with containerization and container orchestration: containerized components, e. g., in the form of Docker containers, can be modeled as part of a topology model. Furthermore, container orchestration technologies such as Kubernetes can be modeled in the form of a component that hosts other components, in this case, Docker containers. Thus, arbitrary technologies can be included in these topology models since the type system is extensible. As a result, container technologies may be used as building blocks for TOSCA models. Moreover, using this modeling concept, they can be combined with other technologies that are not supported natively by container technologies: for example, a web application that is hosted on a web server contained in a Docker container may be connected to a legacy, non-containerized application that gets provisioned, too. In addition, containers hosted in different clouds may be wired with each other using this higher-level modeling concept. Therefore, TOSCA can be seen as an orchestration language on top of existing provisioning, virtualization, and management technologies.

To automate the provisioning of such *complex composite cloud applications*, TOSCA employs two different approaches: (i) declarative and (ii) imperative [23]. Using the declarative approach, the employed TOSCA Runtime Environment interprets the topology model and derives the required provisioning logic by itself. Based on well-defined interfaces, such as the *TOSCA lifecycle interface* [23] which defines operations to install, configure, start, stop, and terminate components, runtimes are enabled to provision arbitrary types of components as long as the Cloud Service Archive provides implementations for the required operations. Moreover, there are approaches that are capable of automatically generating provisioning plans out of topology models [31–33]. As the semantics of component types and relationship types—or at least of their management interfaces—must be defined and understood by the employed runtime, the declarative approach is limited to provisioning scenarios that consist of common components such as web servers, web applications, and databases. However, if the provisioning of a complex application requires custom provisioning logic for individual components, for example, legacy components, and an application-specific orchestration of those components, automatically deriving all required provisioning steps is often not feasible.

TOSCA tackles these issues by the imperative approach, which employs management plans to model all activities that have to be executed, including their control and data flow. As management plans, in this case *Build Plans* that automate the provisioning of the application [22], describe all steps to be executed explicitly, this concept enables customizing the provisioning for arbitrary applications including custom, application-specific logic. Thus, also complex applications that possibly employ containerization technologies hosted on different clouds connected to stateful non-containerized legacy components can be provisioned automatically using this approach as any required activity can be described explicitly. The TOSCA-based management plan modeling language BPMN4TOSCA [34,35] additionally supports creating management plans based on the management operations provided by the components in the topology model. In addition, combined with the aforementioned plan generation, Build Plans may be generated even for complex applications and refined afterwards for custom, application-specific needs.

8 Summary

Based on the IDEAL cloud application properties we have shown how an existing application can be transformed to a native cloud application. Moreover, we discussed the relation of native cloud applications to (micro)service-oriented architectures and continuous delivery. Additional properties defined by Freemantle and Badger et al. – multi-tenancy, pay-per-use, and self-service – enabling a native cloud application to be offered as a service require significant adjustments of the application functionality. Multi-tenancy commonly requires adaptation of application interfaces and storage structures to ensure the isolation of tenants. Functionality to support pay-per-use billing and self-service commonly has to be newly created using application-specific knowledge.

Based on the transformation of the reference application we have shown that virtualization is a mandatory prerequisite for building a native cloud application, but just virtualizing an application in a monolithic manner does not satisfy all cloud application properties. Hence, it is insufficient to simply move an application into a monolithic virtual machine and call it a native cloud application. Furthermore, we outlined how emerging container orchestration approaches such as Kubernetes and Docker Swarm can be utilized to manage native cloud applications. Since these container orchestration approaches make specific assumptions about containerized application components, we further discussed how TOSCA can be used as standards-based modeling, orchestration, and management approach for native cloud applications. More specifically, TOSCA provides overarching and extensible mechanisms to integrate and orchestrate legacy, typically stateful, and non-containerized components with stateless and containerized application components.

References

1. Fehling, C., Leymann, F., Schumm, D., Konrad, R., Mietzner, R., Pauly, M.: Flexible process-based applications in hybrid clouds. In: Liu, L., Parashar, M., (eds.) IEEE International Conference on Cloud Computing, CLOUD 2011, Washington, DC, USA, 4–9 July 2011, pp. 81–88. IEEE (2011)
2. Fehling, C., Leymann, F., Ruehl, S.T., Rudek, M., Verclas, S.A.W.: Service migration patterns - decision support and best practices for the migration of existing service-based applications to cloud environments. In: 2013 IEEE 6th International Conference on Service-Oriented Computing and Applications, Koloa, HI, USA, 16–18 December 2013, pp. 9–16. IEEE Computer Society (2013)

3. Brandic, I., Dustdar, S., Anstett, T., Schumm, D., Leymann, F., Konrad, R.: Compliant cloud computing (C3): architecture and language support for user-driven compliance management in clouds. In: IEEE International Conference on Cloud Computing, CLOUD 2010, Miami, FL, USA, 5–10 July 2010, pp. 244–251 (2010)
4. Leymann, F., Fehling, C., Wagner, S., Wettinger, J.: Native cloud applications: why virtual machines, images and containers miss the point! In: Proceedings of the 6th International Conference on Cloud Computing and Service Science (CLOSER 2016), Rome, pp. 7–15. SciTePress (2016)
5. Fehling, C., Leymann, F., Retter, R., Schupeck, W., Arbitter, P.: Cloud Computing Patterns - Fundamentals to Design, Build, and Manage Cloud Applications. Springer, Heidelberg (2014)
6. Andrikopoulos, V., Binz, T., Leymann, F., Strauch, S.: How to adapt applications for the cloud environment - challenges and solutions in migrating applications to the cloud. Computing **95**, 493–535 (2013)
7. Binz, T., Breitenbücher, U., Kopp, O., Leymann, F.: Migration of enterprise applications to the cloud. IT - Information Technology, Special Issue: Architecture of Web Application, vol. 56, pp. 106–111 (2014)
8. Fowler, M.: Patterns of Enterprise Application Architecture. Addison-Wesley Longman Publishing Co., Inc., Boston (2002)
9. Mell, P.M., Grance, T.: The NIST definition of cloud computing. Technical report, Gaithersburg, MD, United States (2011)
10. Ornstein, S.M., Crowther, W.R., Kraley, M.F., Bressler, R.D., Michel, A., Heart, F.E.: Pluribus: a reliable multiprocessor. In: AFIPS Conference Proceedings of American Federation of Information Processing Societies: 1975 National Computer Conference, Anaheim, CA, USA, 19–22 May 1975, vol. 44, pp. 551–559. AFIPS Press (1975)
11. Freemantle, P.: Cloud Native (2010). http://pzf.fremantle.org/2010/05/cloud-native.html
12. Mouat, A.: Using Docker: Developing and Deploying Software with Containers. O'Reilly Media, Sebastopol (2015)
13. Fowler, M.: Microservices Resource Guide (2016). http://martinfowler.com/microservices
14. Newman, S.: Building Microservices: Designing Fine-Grained Systems, 1st edn. O'Reilly Media, Sebastopol (2015)
15. Humble, J., Farley, D.: Continuous Delivery: Reliable Software Releases Through Build, Test, and Deployment Automation, 1st edn. Addison-Wesley Professional, Upper Saddle River (2010)
16. Schermann, G., Cito, J., Leitner, P.: All the services large and micro: revisiting industrial practices in services computing. PeerJ PrePrints **3**, 36–47 (2015)
17. SCS: Self-contained System (SCS) Assembling Software from Independent Systems (2016). http://scsarchitecture.org
18. Cloud9 IDE Inc.: Cloud9 website (2016). https://c9.io
19. Badger, M.L., Grance, T., Patt-Corner, R., Voas, J.M.: Cloud computing synopsis and recommendations. Technical report, Gaithersburg, MD, USA (2012)
20. Azeez, A., Perera, S., Gamage, D., Linton, R., Siriwardana, P., Leelaratne, D., Weerawarana, S., Fremantle, P.: Multi-tenant SOA middleware for cloud computing. In: IEEE International Conference on Cloud Computing, CLOUD 2010, Miami, FL, USA, 5–10 July 2010, pp. 458–465 (2010)
21. Amazon: Amazon Elastic Compute Cloud (EC2) Pricing (2016). http://aws.amazon.com/ec2/pricing

22. OASIS: Topology and Orchestration Specification for Cloud Applications (TOSCA) Version 1.0. Organization for the Advancement of Structured Information Standards (OASIS) (2013)
23. OASIS: Topology and Orchestration Specification for Cloud Applications (TOSCA) Primer Version 1.0. (2013)
24. OASIS: TOSCA Simple Profile in YAML Version 1.0 - Committee Specification (2016)
25. Binz, T., Breitenbücher, U., Kopp, O., Leymann, F. Advanced web services. In: Bouguettaya, A., Sheng, Q.Z., Daniel, F. (eds.) TOSCA: Portable Automated Deployment and Management of Cloud Applications, pp. 527–549. Springer, New York (2014)
26. Leymann, F., Roller, D.: Production Workflow: Concepts and Techniques. Prentice Hall PTR, Upper Saddle River (2000)
27. OASIS: Web Services Business Process Execution Language (WS-BPEL) Version 2.0. Organization for the Advancement of Structured Information Standards (OASIS) (2007)
28. OMG: Business Process Model and Notation (BPMN) Version 2.0. Object Management Group (OMG) (2011)
29. Binz, T., Breitenbücher, U., Haupt, F., Kopp, O., Leymann, F., Nowak, A., Wagner, S.: OpenTOSCA – a runtime for TOSCA-based cloud applications. In: Basu, S., Pautasso, C., Zhang, L., Fu, X. (eds.) ICSOC 2013. LNCS, vol. 8274, pp. 692–695. Springer, Heidelberg (2013). doi:10.1007/978-3-642-45005-1_62
30. Waizenegger, T., et al.: Policy4TOSCA: a policy-aware cloud service provisioning approach to enable secure cloud computing. In: Meersman, R., Panetto, H., Dillon, T., Eder, J., Bellahsene, Z., Ritter, N., Leenheer, P., Dou, D. (eds.) OTM 2013. LNCS, vol. 8185, pp. 360–376. Springer, Heidelberg (2013). doi:10.1007/978-3-642-41030-7_26
31. EI Maghraoui, K., Meghranjani, A., Eilam, T., Kalantar, M., Konstantinou, A.V.: Model driven provisioning: bridging the gap between declarative object models and procedural provisioning tools. In: Steen, M., Henning, M. (eds.) Middleware 2006. LNCS, vol. 4290, pp. 404–423. Springer, Heidelberg (2006). doi:10.1007/11925071_21
32. Mietzner, R.: A method and implementation to define and provision variable composite applications, and its usage in cloud computing. Dissertation, Universitt Stuttgart, Fakult ät Informatik, Elektrotechnik und Informationstechnik (2010)
33. Breitenbücher, U., Binz, T., Képes, K., Kopp, O., Leymann, F., Wettinger, J.: Combining declarative and imperative cloud application provisioning based on TOSCA. In: International Conference on Cloud Engineering (IC2E 2014), pp. 87–96. IEEE (2014)
34. Kopp, O., Binz, T., Breitenbücher, U., Leymann, F.: BPMN4TOSCA: a domain-specific language to model management plans for composite applications. In: Mendling, J., Weidlich, M. (eds.) BPMN 2012. LNBIP, vol. 125, pp. 38–52. Springer, Heidelberg (2012). doi:10.1007/978-3-642-33155-8_4
35. Kopp, O., Binz, T., Breitenbücher, U., Leymann, F., Michelbach, T.: A domain-specific modeling tool to model management plans for composite applications. In: Proceedings of the 7th Central European Workshop on Services and their Composition (ZEUS 2015), CEUR Workshop Proceedings, pp. 51–54 (2015)

Papers

Delegated Audit of Cloud Provider Chains Using Provider Provisioned Mobile Evidence Collection

Christoph Reich[✉] and Thomas Rübsamen

Institute for Cloud Computing and IT Security,
Furtwangen University of a Applied Science, Furtwangen, Germany
{christoph.reich,thomas.ruebsamen}@hs-furtwangen.de
http://wolke.hs-furtwangen.de

Abstract. Businesses, especially SMEs, increasingly integrate cloud services in their IT infrastructure. The assurance of the correct and effective implementation of security controls is required by businesses to attenuate the loss of control that is inherently associated with using cloud services. Giving this kind of assurance, is traditionally the task of audits and certification done by auditors. Cloud auditing becomes increasingly challenging for the auditor, if you be aware, that today cloud services are often distributed across many cloud providers. There are Software as a Service (SaaS) providers that do not own dedicated hardware anymore for operating their services, but rely solely on other cloud providers of the lower layers, such Infrastructure as a Service (IaaS) providers. Cloud audit of provider chains, that is cloud auditing of cloud service provisioned across different providers, is challenging and complex for the auditor.

The main contributions of this paper are: An approach to automated auditing of cloud provider chains with the goal of providing evidence-based assurance about the correct handling of data according to predefined policies. A concepts of individual and delegated audits, discuss policy distribution and applicability aspects and propose a lifecycle model. The delegated auditing of cloud provider chains using a provider provisioned platform for mobile evidence collection is the policy to collect evidence data on demand. Further, the extension of Cloud Security Alliance's (CSA) CloudTrust Protocol form the basis for the proposed system for provider chain auditing.

1 Introduction

As cloud computing becomes more accepted by mainstream businesses and replaces more and more on-premise IT installations, compliance with regulation, industry best-practices and customer requirements becomes increasingly important. The main inhibitor for even more widespread adoption of cloud services still remain security and privacy concerns of cloud customers [3]. In Germany, a preference for cloud providers that fall under German jurisdiction and also run

© Springer International Publishing AG 2017
M. Helfert et al. (Eds.): CLOSER 2016, CCIS 740, pp. 43–64, 2017.
DOI: 10.1007/978-3-319-62594-2_3

their own data centers in Germany or at least inside the European Union can be observed recently [2]. This comes as no surprise when privacy violations that have become known to the general population in recent years are considered (e.g., NSA and Snowden revelations). A feasible way to assess and ensure compliance of cloud services regularly is by using audits. For any technical audit, information has to be collected in order to assess compliance. This automated process is called evidence collection in our system. In our previous work on cloud auditing, the focus was put on automating the three major parts of an audit system, (i) evidence collection and handling, (ii) evaluation against machine-readable policies and (iii) presentation of audit results [15, 17, 18].

Today, it is common to not only have a single cloud provider to provision a service to its customers, but multiple. The composition of multiple services provided by different providers can already be observed where Software as a Service (SaaS) providers host their offering on top of the computing resources provided by an Infrastructure as a Service (IaaS) provider. For instance, Dropbox and Netflix both host their services using Amazon's infrastructure. These composed services - they can be considered to form a chain of cloud providers, therefore cloud provider chain - can become very complex and opaque with respect to the flow of data between providers. Several new challenges for the auditing of such cloud provider chains can be identified, which will be discussed in this paper. The other major contribution of this paper is a proposed solution to auditing of cloud provider chains, which is an extension to our previous work in this area.

This paper is structured as follows: in Sect. 2, related research projects and industrial approaches are discussed. In Sect. 3 the authors elaborate on the definition and properties of cloud provider chains and auditing. Afterwards, a discussion of three different approaches of auditing provider chains in Sect. 4 is presented. In Sect. 5, the lifecycle of the delegated audit of cloud providers is described. How the evidence data is collected in a provider chain, how it is integrated, and automated is shown in Sect. 6. To realize the delegated audit by CloudTrust Protocol (CTP) it has to be extended (see Sect. 7). Section 8 evaluates the proposed approaches and a conclusion of the paper is drawn in Sect. 9.

2 Related Work

Standards and catalogues such as ISO27001, Control Objectives for Information and Related Technology (COBIT) or NIST 800-53 define information security controls. There are some extensions to the previous frameworks such as the Cloud Controls Matrix [4]. However, conducting audits based on these standards is mostly a manual process, still. Our proposed approach supports the automatic collection and evaluation of evidence based on policies that may stem from these frameworks and therefore could enable continuous certification.

Auditing and monitoring in cloud computing has gained more momentum in recent years and a growing number of research projects is addressing their unique challenges. Povedano et al. [13] propose Distributed Architecture for Resource manaGement and mOnitoring in cloudS (DARGOS) that enables efficient distributed monitoring of virtual resources based on the publish/subscribe paradigm. While they introduce isolation of tenants in cloud environments, they do not at this stage show how their system would work in a multi-provider scenario.

Massonet et al. [10] propose an approach to monitoring data location compliance in federated cloud scenarios, where an infrastructure provider is chained with a service provider (i.e., the service provider uses resource provided by the infrastructure provider). A key requirement of their approach is the collaboration of both providers with respect to collecting monitoring data. Infrastructure monitoring data (from the IaaS provider) is shared with the service provider (SaaS provider) that uses it to generate audit trails. However, their main focus is to monitor virtual execution environments (VEE) that "are fully isolated runtime modules that abstract away the physical characteristics of the resource", which roughly translates to virtual machines.

Reference [8] follow the idea of tightly integrating monitoring into their management system for federated clouds, inorder to facilitate provider selection on the basis of availability and reliability metrics. They introduce service monitoring by reusing SALMonADA [12]. Their approach is geared towards provider decision making for stateless services based on performance metrics and does neither include protection mechanisms and dynamic collector distribution that are required in a system for evidence collection in the cloud.

Security and privacy auditing are increasingly important topics in cloud auditing. They demonstrate that security controls are put in place by the provider and also that they are functioning correctly (i.e., data protection mechanisms are working correctly and effectively). There are some projects working on the architectural and interface level regarding the automation of security audits, such as the Security Audit as a Service (SAaaS) project [7]. SAaaS is specifically designed to detect incidents in the cloud and thereby consider the dynamic nature of such ecosystems, where resources are rapidly provisioned and removed. However, SAaaS does not address provider chain setups or treat gathered data as evidence.

ABTiCI (Agent-Based Trust in Cloud Infrastructure) describes a system used for monitoring [19]. All relevant parts of a cloud infrastructure are monitored to be able to detect and verify unauthorized access. Integrity checks are done at boot-time, using Trusted Platform Module (TPM) boot or at runtime. Monitoring hardware and software configurations allow the system to detect changes at runtime. The aforementioned system is similar to our approach. Instead of using agents we utilize CTP. Furthermore, our approach relies on evidence collection through audits with pull and trigger mechanisms.

A centralized trust model is introduced by Rizvi et al. [14]. Trust between consumer and provider is established by using an independent third-party auditor. With the adoption of a third-party auditing system, consumers are able to create baseline evaluation for providers they have never worked with to generate initial trust. The model acts as a feedback mechanism providing valuable insight into the providers processes. After initial trust was generated the third-party auditor continues to obtain trust values for the consumer. We see initial trust in the provider as a given factor and focus on obtaining trust values based on evidence within a multi-provider scenario.

The DMTF is also working on cloud auditing with the Cloud Audit Data Federation (CADF) working group. They focus on developing standardized interfaces and data formats to enable cloud security auditing [6]. A similar project is the Cloud Security Alliance's Cloud Trust Protocol (CTP), which defines an interface for enabling cloud users to "generate confidence that everything that is claimed to be happening in the cloud is indeed happening as described, ..., and nothing else" [5], which indicates an additional focus on providing additional transparency of cloud services. The latter two projects, however, do not elaborate on how the interfaces should be implemented nor do they describe explicitly focus on privacy and accountability. We use CTP as a basis and propose its extension and use in our proposed auditing system to enable automated auditing of cloud provider chains.

3   Complex Cloud Provider Chains for Service Provision

While a lot of today's cloud use cases only involve one service provider for service provision, there are also many cases where multiple providers are involved. A prominent example is Dropbox that heavily uses Amazon's S3 and EC2 services to provide its own SaaS offering.

There are several terms for the concept of provider chains such as *federated cloud*, *inter-cloud* and *cloud service composition*. In this work these terms are used synonymously. The concept of a provider chain is defined as follows:

1. At least two cloud providers (either IaaS, PaaS or SaaS) are involved in the provision of a service to a consumer (who can be an individual or business).
2. One cloud provider acts as a primary service provider to the cloud consumer.
3. Subsequent cloud providers do not have a direct relationship with the consumer.
4. The primary provider *must* be and the subsequent providers *can* be cloud consumers themselves, if they use services provided by other clouds.
5. The data handling policies agreed between the cloud consumer and the primary service provider must not be relaxed if data is processed by a subsequent provider.

The terms *cloud consumer* and *cloud customer* are used synonymously as well, while relying on the definition of a cloud consumer and auditor provided by NIST [9].

Figure 1 depicts a simplified scenario where three cloud service providers are involved in provisioning of a seemingly single service to a cloud consumer. The SaaS provider acts as the primary service provider, while it uses the PaaS provider's platform for hosting its service. The PaaS provider in turn does not have its own data center but uses resources provided by an IaaS provider.

The data handling policy applies to the whole chain (depicted by the dashed rectangle in Fig. 1). Data handling policies thereby govern the treatment of data such as data retention (the deletion of data after a certain time), location (geographical restrictions) and security requirements (access control rules and protection of systems that handle the data).

All cloud provider produce evidence of their cloud operations.

3.1 Auditing Cloud Provider Chains

According to NIST, a cloud auditor is defined as "A party that can conduct independent assessment of cloud services, information system operations, performance and security of the cloud implementation" [9]. In our proposed system, the auditor is supported by a system for automated evidence collection and assessment. Evidence in the audit system is any kind of information that is indicative of compliance with policies or a violation of

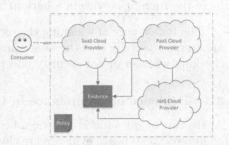

Fig. 1. Cloud provider chains.

those. Typically, evidence is collected at the auditee. In general, an auditee is an organization that is being audited, which in this paper, is always a cloud provider.

Complex cloud service provision scenarios introduce new challenges with respect to auditing. While in a typical scenario, where there is one cloud provider and one cloud consumer, policies can be agreed upon relatively easily between the two, this is not as easy in a provider chain. In fact, the cloud consumer might not be aware of or even interested in the fact that there is an unknown third-party that might have access to his data as long as his expectations regarding the protection and processing of his data are fulfilled. However, to assert compliance, the whole chain of providers, including data flows that are governed by the previously mentioned policies, have to be considered. This means that an audit with respect to a single policy rule may need to be split into several smaller evidence collection and evaluation tasks that are distributed among the providers.

For instance: assuming there is a restriction on data retention put in place that states that certain types of data (e.g., Personal Identifiable Information PII) has to deleted by the provider after a certain fixed period of time and no copies may be left over. This restriction can stem from regulatory framework such as the European Data Protection Directive or simply preferences that were stated by the data subject, whose data is being processed in the cloud. Such requirements can be formulated and enforced in for example the Accountability PrimeLife Policy Language (A-PPL) and its enforcement engine (A-PPL-E) [1].

Auditing for compliance with such a policy requires, on a higher level, the check for the implementation of appropriate mechanisms and controls at each provider where the data itself or a copy thereof could have been stored. On a lower-level, the correct enforcement of the data retention rule could be evaluated in an audit by using evidence of data deletion that is being collected from all the cloud providers. In the overview depicted in Fig. 1, that evidence could comprise of:

- Data deletion enforcement events generated by the service at the primary service provider as a reaction to the retention period being reached,
- Database delete log events produced by a database management system at the PaaS provider,
- And scan results on the IaaS level for data that may be still available outside of the running service in a backup subsystem provided by the IaaS provider.

The importance of widening the scope of audits in such dynamic scenarios is apparent, especially if at the same time the depth of analysis is widened beyond checking whether or not security and privacy controls are put in place.

3.2   Evidence of Compliance in Cloud Provider Chains

At the core of any audit is evidence of compliance or non-compliance that is being analyzed. The types of evidence are closely linked to the type of audit (e.g., security audit, financial audit etc.) and are - from a technological perspective - especially diverse in the cloud due to the heterogeneity of its subsystems, architectures, layers and services. The notion of evidence for cloud audits was discussed in our previous work in more detail [15].

In general, we follow the definition of digital evidence that is "information of probative value that is stored or transmitted in binary form" [20]. This means, that the types of evidence are diverse and include for example logs, traces, files, monitoring and history data from cloud management system like OpenStack's Nova service.

Evidence collection at a single cloud provider is already a complex task due to the diverse types of evidence sources and sheer amount of potentially required data that is being produced continuously. In a provider chain, these problems are intensified by the introduction of administrative domains and the lack of transparency regarding the number of involved providers and their relationships.

Another problem that is introduced with the concept of provider chains are changing regulatory domains. In a single-provider scenario, there are typically only two regulatory domains to be considered: (i) the one that applies to the cloud consumer and (ii) the one that applies to the cloud provider. With the addition of more cloud providers, the complexity of achieving regulatory compliance increases tremendously.

A simple example for such a case is the recent decision of the European Court in 2015 to declare Safe Harbor invalid, which leads to data transfers outside the European Union that are only governed by Safe Harbor to be invalid. In a provider chain, where a European Cloud provider transfers data about European individuals to another provider in the US, regulatory compliance could have been lost overnight. Here, it can be seen that regulatory domains can have a tremendous impact on how a compliance audit may have to look like, and on the type of evidence that may need to be collected at the different providers.

As previously suggested, the third major challenge for evidence collection in cloud provider chains is their inherent technological heterogeneity. APIs, protocols and data formats differ by provider and typically cannot be integrated easily (e.g., providers offering proprietary APIs). There are some approaches to homogenize some of the technologies, such as for example CSA CloudTrust Protocol [5] that aims to provide a well-defined API that enables cloud providers to export transparency-enhancing information to auditors and cloud consumers.

4   Audit Frameworks for Cloud Provider Chains

In this chapter, we illustrate different variations of auditing cloud provider chains. We thereby focus on traditional individual audits and delegated provider audits.

4.1   Audit Frameworks and Audit Automation

Policy compliance assessment and validation is the main goal of our audit system. Policies can be of various kinds, for instance, a data protection policy is a typical tool used by cloud providers to frame their data protection and handling practices. In such policies, limits and obligations that a provider has to fulfill are defined in well-known standards, frameworks and industry best practices, like CSA's Cloud Controls Matrix (CCM) [4] mentioned in "related work" (Sect. 2). Typically, these documents are not machine-readable and are geared towards limiting liability of the provider, but for automation this is necessary precondition. The nature of the obligations defined in standards are general and the expertise of an auditor is then to design machine-readable policies, which can be monitored.

4.2   Individual Provider Audits

Figure 2 describes the process of auditing individual providers in a service provision chain. All policies will be distributed to each provider (as seen in Fig. 2). Policy distribution can either be:

1. Manual policy evaluation: This approach is based on the specified policy documents (e.g., terms of service in human-readable form) given by the provider. The auditor manually maps statements of such documents to information requests for the providers (e.g., asking for specific process documentation or monitoring data and audit logs).
2. Deploying machine-readable policies: In this approach the auditor deploys a machine-readable policy document (XML, JSON) onto the provider. The provider will then automatically audit the tasks specified within the document. The auditor can request the results for the audited policy rules to verify if everything is fulfilled. The policy needs to be deployed to each involved provider. Within this approach new policies can easily be added and deployed for automated auditing.

The audit results are used to assure the consumer that policy and rule compliance is given or not. As previously described, a service provision chain contains at least one provider. In this case, two providers - a primary and a 2nd-level provider. To audit the service as a whole, it is necessary to audit each provider separately and then aggregate the results to form a complete picture of the service from an audit perspective. This means, that regarding data handling policies (e.g., location restric-

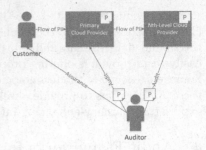

Fig. 2. Individual audit.

tions, access control etc.), each provider that holds data is audited. The same is true for the auditing of security and privacy controls that are put in place at the providers. Obviously, the consumer-facing provider has to transparently disclose all his sub-providers and notify auditors about every sub-provider his data was stored at and where his data is currently stored. Even though not every provider will get the consumer's data, the auditing process gains more complexity with an increasing number of Nth-level providers. Requests must be sent to each provider separately and each provider will deliver audit reports to the auditor.

4.3 Delegated Provider Audits

An alternative to individual audits are what we call delegated audits, where the auditor only interfaces with the primary service provider that in turn takes over the auditing of its sub-provider(s). Therefore the auditor only has to audit the primary service provider to obtain policy compliance results of all involved service providers.

This allows less influential stakehold-ers such as the cloud consumer to act as an auditor towards the primary provider while not having the same rights towards the Nth-level provider(s). Whereas the individual audit scenario is an example of how chain audits could be performed with more influential stakeholders, such as data protection authorities. Figure 3 depicts the delegated provider audit sce-nario. Every audit request is sent to the

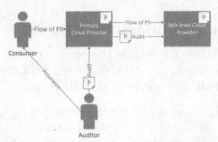

Fig. 3. Delegated audit.

primary provider who will then extract CTP calls from a previously deployed policy document (machine-readable document deployed by the auditor). Since the primary provider is acting as a mediator he has to delegate requests and communication. Existing problems regarding policy compliance is of major con-cern for the primary provider because complaints will always be addressed to him, even if he is not responsible for a failed audit. For the case a given audit response did not satisfy policy compliance the consumer will contact the primary provider with a complaint (e.g., data was transferred outside valid location).

On the other hand the consumer's payoff can be much higher due to the central-ized structure using a mediator ensuing low complexity for the auditor. There-fore, he can always rely on the data controller to forward his request to the data holding sub-provider without the need of adaption (send requests to different entities, use different API-calls).

5   Audit Lifecycle in Delegated Provider Audits

In the following section the audit system lifecycle in delegated provider audits is described in three phases: **(i)** *Preparation*, **(ii)** *Processing* and **(iii)** *Presentation*. In the following, we describe each phase in more detail:

(i) Preparation Phase: The most important task during the *Preparation Phase* is resource identifier distribution, which is required for request handling. Request handling is done using unique resource identifiers (URI), which are used to iden-tify any kind of resource that is part of an audit. A URI unambiguously identifies an object within a provider's domain. In our approach, each provider has its own namespace in which identifiers can be assigned arbitrarily.

The preparation process *Policy Adding* allows the auditor to create new machine-readable rules based on already existing policies, like specifying a new data location rule to ensure that his data will not leave his countries jurisdic-tion. From the new rule, auditable elements are derived, that an automated audit process provides all necessary information to enable the possibility of pol-icy compliance assessing. Auditable elements include for example the location of data, logs and configurations.

The *Policy Mapping* process, maps each new added rule or policy to trans-parency elements and associated requests. If a newly added rule cannot be mapped to an already existing transparency element a new element needs to be created. The mapping is done based on the specified policies. For this reason the policy adding process is limited to already defined policies and the associated rules within the contract. During the mapping each non-standard policy (i.e., a policy that requires a transparency element that is not part of CTP) will receive an URI and all necessary data sources needed to answer a request. The mapping process generates URIs and defines all auditable attributes for an element.

The preparation process *Policy Distribution* propagates the resource identi-fiers throughout the system. Each sub-provider sends his resource identifiers to the primary provider. Afterwards, when all identifiers are known by the primary provider, they will be forwarded to the auditor.

(ii) Processing Phase: The information defined in *Preparation Phase* will be col-lected and stored in the evidence store. Is the primary cloud provider requesting evidence using the CTP, the secondary cloud provider retrieves the information from its evidence store and replies to the CTP response. A policy evaluation is done to determine the policy compliance.

(iii) Presentation: Within this phase the results are presented to the auditor. Thereby, each requested element will be presented containing the name of the policy rule as well as its achieved compliance state.

Depending on the auditor, the lifecycle can continue with *Preparation Phase* again. Returning to the *Preparation Phase* is necessary if new policies/rules were added or in a continuous auditing scenario, where policy compliance is audited in short intervals or event-driven (e.g., on new or changed policy, on infrastructure change or on custom triggers defined by the auditor). During the new cycle only newly generated URIs will be distributed.

6   Approaches for Collecting Evidence in Cloud Provider Chain Audits

In provider chains the problem of collecting evidence for audit purposes in a service provider chain differ in the following aspects:

1. The level of control an auditor has over the extent of the data that is being published, i.e. whether the auditor is limited to information that a provider is already providing or if he has more fine-grained control and access to a provider's infrastructure.
2. Technical limitations imposed by the technological environment, i.e. the extent to which cloud providers have to implement additional evidence collection mechanisms.
3. The expected willingness or acceptance to provide such mechanisms by the publishing service provider, i.e. the potential disclosure of confidential provider information and required level of access to the provider's systems.

In the remainder of this chapter, three approaches are described and rated by the above-mentioned factors.

The focus lies thereby on inspecting common components at two exemplary cloud providers that form a provider chain for the provision of a service. These components are:

AuditSys. An audit system that enables automated, policy-based collection of evidence as well as the continuous and periodic evaluation of said evidence during audits.

Collector. A component that enables the collection of evidentiary data such as logs at various architectural layers of the cloud, while addressing the heterogeneous nature of said evidence sources by acting as an adapter.

Source. A location where evidence of cloud operations is generated.

Now we consider three approaches how to collect the evidence. The first approach focuses on reusing already existing evidence sources by collecting via remote APIs of a cloud system. The second approach uses provider-provisioned evidence collectors and the third approach leverages the mobility of software agents (as used in the prototype implementation of our system) for evidence collection.

6.1   Remote API Evidence Collector

The first approach for collecting evidence that is relevant to automated auditing, leverages already existing APIs in cloud ecosystems. Several cloud providers, such as Amazon (CloudWatch) or Rackspace (Monitoring), already provide improved transparency over their cloud operations by providing their customers with access to proprietary monitoring and logging APIs. The extent to which data is shared is typically limited to information that is already produced by the cloud provider's system (e.g., events in the cloud management system) and restricted to information that immediately affects the cloud customer (e.g., events that are directly linked to a tenant).

Data such as logs that are generated by the underlying systems are very important sources of evidence, since they expose a lot of information about the operation of cloud services. A specific example of such evidence are for instance: VM lifecycle events (created, suspended, snapshotted etc.) including timestamp of the oper-

Fig. 4. Remote API evidence collector.

ation and who performed. This can be requested from OpenStack's Nova service via its REST interface. The type of information is highly dependent on the actual system, the granularity of the produced logs and the scope of the provided APIs. For instance, on the infrastructure level, there are log events produced and shared that provide insight on virtual resource lifecycle (e.g., start/stop events of virtual machine).

Figure 4 depicts such a scenario. The AuditSys at Cloud A operates a collector that implements the API of the remote data source at Cloud B. It is configured with the access credentials of Cloud A, thus enabling the collector to request evidence from Cloud B. Furthermore, since different services may provide different APIs (e.g., OpenStack vs. OpenNebula API), the collector is service-specific. For instance, a collector implements the data formats and protocols as defined in the OpenStack Nova API to collect evidence about the images that are owned or otherwise associated with Cloud A as a customer of Cloud B.

Level of Auditor Control: The amount of evidence that can be collected is severely limited by the actual APIs that are provided by a cloud provider. It is either: (i) the evidence that an auditor is looking for is immediately available because the provider already monitors all relevant data sources and makes that data accessible via the API or (ii) the data is not available. Since a lot of the cloud provider's systems expose remote APIs anyway, they have to be considered. However, the completeness of the exposed APIs and therefore the completeness of the collected evidence is questionable due to the aforementioned reasons.

If an auditee for some reason does not implement or provide access to the audit system, an auditor may still collect evidence to a limited degree using this approach.

Technical limitations: If lower-level access to the providers infrastructure is required to collect evidence (e.g., log events generated on the network layer or block storage-level access to data), an auditor might not be able to gain access to that information.

Acceptance: This approach poses some challenges with respect to security, privacy and trust required by the auditee. Since the auditee is already exposing the APIs publicly, it can be expected that they will be used for auditing and monitoring purposes. The implementation of security and privacy-preserving mechanisms on the API-level is therefore assumed. However, the extent to which such mechanisms are implemented highly depends on the actual implementation of the APIs on the provider side.

While this way of providing evidence to auditors is likely to be accepted by cloud providers, it may be too limited with respect to the extent to which evidence can be collected at lower architectural levels.

6.2   Provider Provisioned Evidence Collector

In this approach, the audit system still is the main component for evidence collection. All cloud providers that are part of the service provision chain are running a dedicated system for auditing. However, the instantiation and configuration of the collector is delegated to the auditee. The auditee assumes full control over the collector and merely grants the audi-

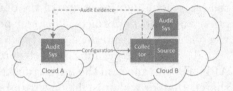

Fig. 5. Provider-provisioned evidence collector.

tor access to interact with the collector for evidence collection. The auditee (see Cloud B in Fig. 5) provisions evidence collectors and provides access to them to the auditor. The auditor (who is using AuditSys at Cloud A) configures evidence collection for the audit to connect to the collectors at Cloud B.

Level of Auditor Control: The configuration of the evidence collector can be adjusted by the auditor to a degree that is controlled by the auditee (e.g., applying filters to logs). He is provided limited means to configure a collector but no direct, low-level access such as freely migrating the collector in the auditee's infrastructure. At any time, the auditee can disconnect, change or otherwise control the collector. An auditor may be put off by the limitations posed by this approach, since he is effectively giving up control over the central part of evidence collection and is relying solely on the cooperation of the auditee. For instance, simple tasks such as reconfiguring or restarting a collector may require extensive interaction between the two audit systems and potentially intervention by a human (e.g., an administrator).

Technical limitations: This approach is only limited by the availability of collectors for evidence sources.

Acceptance: In this approach, the auditee retains full control over the collector and the potential evidence that can be collected by it. The auditor can take some influence on the filtering of data that is collected from the evidence source and on general parameters, such as whether evidence is pushed by or pulled from the collector. Most of the baseline configuration though, is performed by the auditee (such as access restrictions and deployment of the collector). The auditor's ability to influence the collector is severely limited by the restriction of interactions to a well-defined set of configuration parameters and the evidence exchange protocol. This level of control that the auditee has over the evidence collection process may have positive influence on provider acceptance.

6.3   Provider Provisioned Platform for Mobile Evidence Collectors

This approach is specific to a central characteristic of software agent systems, which is the ability to migrate over a network between runtime environments. In this approach, the migration of evidence collectors between separate instances of the audit system running at both Cloud A and B is proposed.

In our implementation, we opted for the well-known Java Agent Development framework (JADE) for implementing collectors. The migration of collectors between providers is thereby performed by using JADE's mobile agent capabilities.

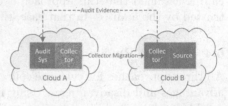

Fig. 6. Mobile evidence collector.

As depicted in Fig. 6, the auditor prepares the required collector fully (i.e., agent instantiation and configuration) and then migrates the collector (shaded box named *Collector*) to the auditee (*Collector'*). There, the collector gathers evidence that is sent back to the auditor for evaluation. Generally however, agents do not cross from one particular administrative domain to another, but remain at one. In this case, the collector crosses from Cloud A's administrative domain to Cloud B. This may have significant impact on the acceptance of the approach.

Level of Auditor Control: The auditor retains full control over the type of collector and its configuration. The auditee may not in any way change or otherwise influence the collector since this could be deemed a potentially malicious manipulation.

Technical limitations: Since the auditor knows most about the actual configuration required for a collector, it is logical to take this approach and simply hand-over a fully prepared collector to the auditee. However, this only works if both run the same audit system, or the auditee at the very least provides a runtime environment for the collector. In any case, this approach offers the most complete and most flexible way of collecting evidence at an auditee due to the comprehensive evidence collection capabilities.

Acceptance: The main problem with this approach is required trust by the auditee. Since the collector that is being handed over to him by the auditor is in fact software that the auditee is supposed to run on its infrastructure, several security, privacy and trust-related issues associated with such cross-domain agent mobility need to be addressed. Several security controls need to be implemented in order to make cloud providers consider the implementation of an audit system including the proposed approach of using mobile collectors.

The main security concerns of this approach stem from the fact that the auditee is expected to execute software on his infrastructure over which he does not have any control. He cannot tell for certain whether or not the agent is accessing only those evidence sources which he expects it to.

Without any additional security measures, it cannot be expected that any cloud provider is willing to accept this approach. However, with the addition of security measures such as ensuring authenticity of the collector (e.g., using collector code reviews and code signing) this approach becomes more feasible. The discussion of such measures depends on the technology used by the implementation and is out of scope of this paper. Without any additional measures, it can be assumed that this approach is only feasible, if the auditor is completely trusted by the auditee. In that case, this approach is very powerful and flexible.

6.4   Critical Review

All three approaches for evidence collection in provider chains have their distinct advantages and disadvantages. Using remote API evidence collectors is simple, quickly implemented, securely and readily available, but severely limited regarding access to evidence sources. Using provider-provisioned evidence collectors is more powerful with respect to access to evidence sources, but requires more effort in the configuration phase and leaves full control to the auditee. Using mobile evidence collectors is the most flexible approach that allows broad access to evidence sources at the auditee's infrastructure and leaves full control over the evidence collection to the auditor. Therefore, a balance has to be struck between broad access to evidence sources when using mobile collectors (effectively having low-level access to logs and other files for evidence collection) and more limited access when using remote APIs (evidence limited to what the system that exposes the API provides).

In the audit system, the use of remote APIs is integrated due to its simplicity and mobile collectors due to their flexibility and powerfulness as the main approaches to evidence collection.

7   Extending CloudTrust Protocol (CTP) for Provider Chain Auditing

So far there is no intention of permit other providers a more flexible evidence collection by defining audit evidence collectors on demand as described in "Provider Provisioned Platform for Mobile Evidence Collectors" (Sect. 6.3). Therefore auditors depend on standards like the CloudTrust Protocol (CTP).

7.1   CloudTrust Protocol (CTP)

The CloudTrust Protocol [5] establishes a mechanism that allows users to audit a CSP. An auditor can choose from a set of transparency elements for instance: geographic location of data objects and affirmation or results of latest vulnerability assessments. CTP has 23 pre-defined transparency elements and supports user-specified elements on which cloud consumer and provider agreed on. The purpose of the CTP is to transparently provide the user with important information about the cloud to show that processing is done as promised. By providing information about the inner-workings of the cloud service (with respect to the transparency elements), trust between the cloud provider and the consumer is supposed to be strengthened. For automation purpose of the audit the CTP had to be extended.

7.2   CloudTrust Protocol (CTP) Extended

In our approach, we leverage CTP as a means for evidence exchange between cloud providers in complex cloud auditing scenarios [16]. Additional functions and components are located above the protocol (as seen in Fig. 7) and are responsible to exchange requests and responses with the CTP. This structure enables us to utilize the benefits of CTP out of the protocol's operational area without changing the protocol itself. Although the operational structure of CTP remains unchanged some optimisations for audit reports are required to be able to transfer additional information e.g. more detailed user access lists. In this case, the additional information would give the auditor not only authorized users but also since when they have authorization and who authorized user permissions.

Figure 7 illustrates the systems architecture in a two provider scenario. Within the figure it is assumed, that the Preparation Phase did end and all for the audit request necessary policies and rules were already mapped and distributed. Incoming transparency element requests will directly go to the Remote Evidence Collection component. In the following paragraph the system components of our proposed approach are described:

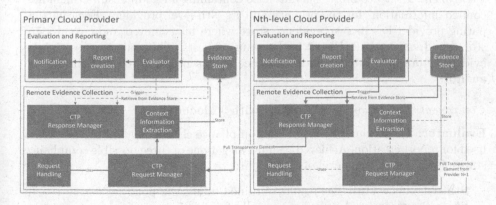

Fig. 7. Multi-provider audit system architecture.

Component: Remote Evidence Collection Consists of 4 Modules (see Fig. 7):

Request handling: Every incoming audit request will arrive at the Request handler of the primary provider. A decision is made which resource identifier should be used based on current data location (Nth-level provider, primary provider). The resource identifiers are used to set up CTP-calls. Therefore, it will forward each request to the CTP Request Manager. Each request is processed separately to guarantee that context information or states do not get mixed up.

CTP Request Manager: The Request Manager, sends each given request to the CTP Response Manager of a Nth-level provider (solid line in Fig. 7 between both providers) using a pull pattern (see [16]). Inter-provider communication is initiated by the Request Manager.

Context information extraction: An incoming CTP-response contains the general response (specified in [5]) as well as the corresponding context information and the compliance state for the requested element. The context information are extracted from the response and securely stored inside the evidence store. The remaining information which are used for report creation are stored as well for the audit report creation.

CTP Response Manager: After receiving a request the Response Manger pulls data from the evidence store if the Nth-level provider is not able to determine the compliance state of his own or receives the data from the Evaluator. Obtained results are packed into a CTP-response and sent back to the primary providers CTP Request Manager. In case a trigger is fired the Response Manager will push the response immediately to the primary providers CTP Request Manager even in the absence of an audit request for the triggered element. A primary provider might be a Nth-level provider of another provider and thus needs a Response Manager. Requests for context information are send from the auditor to the primary providers Response Manager. Like a normal response the Manager pulls the context information for the requested object from the evidence store and writes them into a CTP-Response.

Component: Evidence Store is a database containing all audit results (including context information) for the primary and its Nth-level providers. Each participating provider has its own evidence store where his achieved audit results are stored and can be pulled from by pending requests. The main purpose of the evidence store is to provide audit evidence and to make them accessible to the auditor.

Component: Evaluation and Reporting with 3 Modules:

Evaluator: The Evaluator runs policy compliance checks on all obtained results used for report creation. Achieved results can get one of three possible compliance states depending on the level of fulfillment:

– **State 1 successful:** The results obtained fulfills the policy.
– **State 2 partially:** A policy is partially fulfilled.

– **State 3 failed:** No records for this policy were found or the given results were unsatisfying.

Configured triggers (see Rübsamen [16]) are fired if a compliance check for a request failed or a deviation from the trigger specification is identified.

Report creation: The stored content (state, CTP transparency elements results) is used to create the final report. The report can be of different types, for instance a representation of the results on a web dashboard or in a auto-generated document.

Notification: An audit can take some time to finish. This largely depends on the size and scope of the audit. Therefore, asynchronous mechanisms are required to present audit results. An auditor can be notified via mail when his audit is finished and his audit report is available.

Implementation of each above described part is mandatory for every provider. It is possible that a primary provider is a Nth-level provider in a different audit-chain, whereas a Nth-level provider might be a primary provider in another audit chain.

A functional scenario based evaluation, a STRIDE [11] analysis, and a general security analysis can be found in the paper Rübsamen [16].

8 Scenario-Based Provider Chain Auditing Evaluation

In the previous Sect. 6, the approaches that can be taken when collecting evidence for auditing purposes in cloud provider chains were described. In this section, it is demonstrated how to incorporate the feasible approaches into an extension of the proposed audit system to enable automated, policy-driven auditing of cloud provider chains. The focus is put on the *Remote API Evidence Collector* and *Provider Provisioned Platform for Mobile Evidence Collectors* approaches (see Sects. 6.1 and 6.3 respectively). The approach is validated by discussing a fictitious use case.

8.1 Audit Agent System

In Fig. 8, an example deployment of the automated audit system is depicted. This deployment is not necessarily representative of real-world cloud environments but used to highlight possible combinations of services and data flows that can happen in a multi-cloud scenario. There are four cloud providers, which are directly or indirectly involved in the service provision. The SaaS provider A1 uses the platform provided by a PaaS provider B1, who does not have its own data center but uses computing resources provided by yet another IaaS provider C1. The IaaS provider C2 provides a low-level backup as a service solution that is used by provider C1. To enable auditing of the whole provider chain, each provider is running its own instance of the audit system (AuditSys, see Sect. 6).

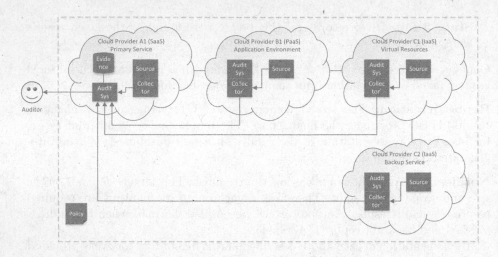

Fig. 8. Provider chain auditing architecture.

8.2   Provider Chain Auditing Extension

The auditor that uses AuditSys at the primary service provider A1 defines and configures continuous audits based on data protection and handling policy statements. Since these policy statements do not include any information about the service architecture, the auditor introduces his knowledge about the cloud deployment into the audit task, by defining evidence collection tasks that gather data on the PaaS and IaaS layer and also at the primary service provider. An audit task consists of collector, evaluator and notification agents. The type of evidence collection approach that has to be taken (as described in Sect. 6) is also defined by the auditor.

In this scenario it is assumed that all providers allow the auditor at A1 to collect evidence using the mobile evidence collectors and that the infrastructure providers also provide the auditor with access to their management system's APIs. As previously mentioned, the auditor is assumed trustworthy by all parties, which enables broad access to all cloud providers. Additionally, it is assumed that all cloud providers are acting in good faith and see the audit process as an opportunity to transparently demonstrate that they are acting in compliance with data handling policies.

As depicted in Fig. 8, the auditor uses A1's AuditSys to define and audit task based on the data handling policy that is in effect. That task refers to the data retention obligation that was described earlier in Sect. 3.1. The retention time is defined as 6 months for every PII data record that is gathered about the users of provider A1. If the retention time is reached, the following delete process is executed as part of the normal operation of the service A1 provides:

1. The delete event fires at A1 due to max retention time being reached and the event is propagated to B1.

2. The data record is deleted from the database at B1.
3. The database is hosted on virtual machines provided by C1 and therefore does not require any delete actions.
4. A backup of the B1's database is available in C2's backup system and the delete action was not triggered in C2.

As part of the delete event, the following evidences are collected by the mobile evidence collectors to build an evidence trail for compliance evaluation at A1.

1. The data retention event is recorded by the collector running at A1.
2. The delete action of the database is recorded by the collector running at B1.
3. No evidence is recorded by the collector at C1 since there are no leftover copies such as virtual machine snapshots available.
4. The backup's meta-data such as creation timestamps are recorded as evidence by the collector at C2.

The evidence from all collectors (A1, B1, C2) is sent to the AuditSys at A1, where it is evaluated and a policy compliance statement is generated for the auditor. In this particular case, a policy violation is detected, because the audit trail shows that the record that should have been deleted is still available in a backup at C2. Provider A1, and B1 acted compliant by deleting the data, whereas C1 never stored a copy outside of B1's database.

8.3   Pre-processing and Intermediate Results of Audit Evidence Evaluation

The audit system uses a component at the AuditSys that is responsible for storing evidence records that are collected by the collector agents. Externally collecting evidence and merging it at a central evidence store that is only reachable via the network, can easily become a bottle-neck in audit scenarios where either a lot of evidence records are produced externally or where the record size is big. This obviously has significant impact on the scalability of the whole system.

   The problem can be addressed by making the evidence store (which is just a specialized form of an agent with a secure storage mechanism) distributable and also by de-centralizing parts of the evidence evaluation process. There are generally two concepts:

Pre-processing allows the evidence collector agent to apply filtering and other types of evidence pre-processing. The goal is to reduce the amount of collected evidence to a manageable degree (without negatively impacting the completeness of the audit trail) and to reasonably reduce the amount of network operations by grouping evidence records and storing them in bulk. For example, by filtering the raw data at the evidence source for certain operations, subjects, tenants, or time frames. Data that is not immediately required for the audit is filtered out.

Intermediate Result Production: A second pre-processing strategy is to move (parts of) the evaluation process near the collector. This means that the collected evidence is already reduced to the significant portions that indicate partial compliance or violation of policies. However, this strategy requires specific audit task types (e.g. audit result, that can be produced by combining intermediate results).

The three concepts bring several implications with them with respect to privacy and security. Pre-processing can be considered a manipulation of evidence. Therefore, the unaltered source upon which the pre-processing happened should be protected to later be able to trace pre-processed evidence back to its unaltered form. Immediate result production effectively moves the evaluation of evidence step of the audit into the domain of the auditee, where it would be easy for him to manipulate the result. However, the same applies to the collection of evidence as well where an auditor can intentionally manipulate the evidence source or the collector.

This case is not considered in the current iteration of the system but it is assumed that cloud providers (auditees) are acting in good faith. This assumption can be justified by the potential increase in transparency and the associated strengthening of trust in the cloud provider that can mean a competitive advantage. On the other hand, intentional manipulation of evidence or intermediate results can have disastrous impact on a provider's credibility, reputation and trustworthiness upon detection.

9   Conclusions

Cloud auditing is becoming increasingly important as cloud adoption increases and compliance of data processing is put into focus of the cloud consumer. While there are many systems for monitoring cloud providers (with varying level of completeness), there are fewer systems that automate audit tasks and even fewer still that enable continuous auditing, which is a key enabler of continuous certification. In this paper we discussed two different types of chain audits: (i) individual provider audits where the auditor has to audit each Nth-level provider separately and (ii) delegated provider audits where the primary cloud provider acts as an mediator. A powerful approach, that provides a provider prvisioned platform for mobile evidence collection allows on demand and policy-driven evidence collection, has been shown. We also discussed the applicability of data handling and privacy policies and how they apply in complex scenarios where multiple providers share a cloud consumer's data. In the latter part of this paper, we focused on the architectural integration of the CloudTrust Protocol in the evidence collection and transport of our audit system. Finally, their implementation in an audit system was presented and validated using a scenario-based approach. It was shown how automated cloud audits can be extended to scenarios, where more than one cloud provider is involved in the service provision. We evaluated the functional soundness by demonstrating an audit scenario that involves a cloud consumer using a service that is intransparently provided by two different cloud providers. Additionally, we evaluated our approach by defining a threat model using threat scenarios and addressing those threats.

In our future work, we focus on even more complex service provision scenarios, where even more layers of service providers are involved. We will also put special focus on ensuring the scalability of our approach. Another interesting topic emerges, when any of the cloud providers is considered untrustworthy. This can be the case when a malicious insider tries to intrude in our system. We consider ensuring the integrity of evidence in the whole chain of providers to be a major challenge.

Acknowledgements. This work has been partly funded: EC:FP7/2007-2013, 317550, A4Cloud.

References

1. Azraoui, M., Elkhiyaoui, K., Önen, M., Bernsmed, K., Oliveira, A.S., Sendor, J.: A-PPL: an accountability policy language. In: Garcia-Alfaro, J., Herrera-Joancomartí, J., Lupu, E., Posegga, J., Aldini, A., Martinelli, F., Suri, N. (eds.) DPM/QASA/SETOP -2014. LNCS, vol. 8872, pp. 319–326. Springer, Cham (2015). doi:10.1007/978-3-319-17016-9_21
2. Bitkom Research GmbH: Cloud Monitor 2015 (2015). https://www.kpmg.com/DE/de/Documents/cloudmonitor%202015_copyright%20_sec_neu.pdf
3. Cloud Security Alliance: Top threats to cloud computing survey results update 2012 (2013). https://downloads.cloudsecurityalliance.org/initiatives/top_threats/Top_Threats_Cloud_Computing_Survey_2012.pdf
4. Cloud Security Alliance: Cloud Controls Matrix (2014). https://cloudsecurityalliance.org/research/ccm/
5. Cloud Security Alliance: CloudTrust Protocol (2016). https://cloudsecurityalliance.org/research/ctp
6. Distributed Management Task Force, Inc. (DMTF): Cloud auditing data federation (CADF) - data format and interface definitions specification (2014). http://www.dmtf.org/sites/default/files/standards/documents/DSP0262_1.0.0.pdf
7. Doelitzscher, F., Rübsamen, T., Karbe, T., Reich, C., Clarke, N.: Sun behind clouds - on automatic cloud security audits and a cloud audit policy language. Int. J. Adv. Netw. Serv. **6**(1&2) (2013)
8. Kertesz, A., Kecskemeti, G., Oriol, M., Kotcauer, P., Acs, S., Rodríguez, M., Mercè, O., Marosi, A., Marco, J., Franch, X.: Enhancing federated cloud management with an integrated service monitoring approach. J. Grid Comput. **11**(4), 699–720 (2013). http://dx.doi.org/10.1007/s10723-013-9269-0
9. Liu, F., Tong, J., Mao, J., Bohn, R., Messina, J., Badger, L., Leaf, D.: Nist cloud computing reference architecture (2011). http://www.nist.gov/customcf/get_pdf.cfm?pub_id=909505
10. Massonet, P., Naqvi, S., Ponsard, C., Latanicki, J., Rochwerger, B., Villari, M.: A monitoring and audit logging architecture for data location compliance in federated cloud infrastructures. In: 2011 IEEE International Symposium on Parallel and Distributed Processing Workshops and Phd Forum (IPDPSW), pp. 1510–1517, May 2011
11. Microsoft Developer Network: The Stride Threat Model (2014). https://msdn.microsoft.com/en-US/library/ee823878(v=cs.20).aspx

12. Muller, C., Oriol, M., Rodriguez, M., Franch, X., Marco, J., Resinas, M., Ruiz-Cortes, A.: Salmonada: a platform for monitoring and explaining violations of WS-agreement-compliant documents. In: 2012 ICSE Workshop on Principles of Engineering Service Oriented Systems (PESOS), pp. 43–49, June 2012

13. Povedano-Molina, J.; Lopez-Vega, J.M., Lopez-Soler, J.M., Corradi, A., Foschini, L.: dargos: a highly adaptable and scalable monitoring architecture for multi-tenant clouds. Future Gener. Comput. Syst. **29**(8), 2041–2056 (2013). http://www.sciencedirect.com/science/article/pii/S0167739X13000824

14. Rizvi, S., Ryoo, J., Liu, Y., Zazworsky, D., Cappeta, A.: A centralized trust model approach for cloud computing. In: 2014 23rd Wireless and Optical Communication Conference (WOCC), pp. 1–6, May 2014

15. Rübsamen, T., Reich, C.: Supporting cloud accountability by collecting evidence using audit agents. In: 2013 IEEE 5th International Conference on Cloud Computing Technology and Science (CloudCom), vol. 1, pp. 185–190, December 2013

16. Rübsamen, T., Hölscher, D., Reich, C.: Towards auditing of cloud provider chains using cloudtrust protocol. In: Proceedings of the 6th International Conference on Cloud Computing and Service Science (CLOSER 2016), pp. 83–94. SciTePress (2016)

17. Rübsamen, T., Pulls, T., Reich, C.: Secure evidence collection and storage for cloud accountability audits. In: CLOSER 2015 - Proceedings of the 5th International Conference on Cloud Computing and Services Science, Lisbon, Portugal, 20–22 May 2015. SciTePress (2015)

18. Rübsamen, T., Reich, C.: An architecture for cloud accountability audits. In: Baden-Württemberg Center of Applied Research Symposium on Information and Communication Systems, SInCom 2014 (2014)

19. Saleh, M.: Construction of agent-based trust in cloud infrastructure. In: 2014 IEEE/ACM 7th International Conference on Utility and Cloud Computing (UCC), pp. 941–946, December 2014

20. Scientific Working Groups on Digital Evidence, Imaging Technology: SWGDE and SWGIT Digital and Multimedia Evidence Glossary (2015). https://www.swgde.org/documents/Current%20Documents/2015-05-27%20SWGDE-SWGIT%20Glossary%20v2.8

Trade-offs Based Decision System for Adoption of Cloud-Based Services

Radhika Garg[✉], Marc Heimgartner, and Burkhard Stiller

Communication Systems Group CSG@IfI, University of Zürich, UZH,
Binzmühlestrasse 14, 8050 Zürich, Switzerland
{garg,stiller}@ifi.uzh.ch, marc.heimgartner@uzh.ch

Abstract. Decision of adopting any new technology in an organization is a crucial decision as it can have impact at the technical, economical, and organizational level. Therefore, the decision to adopt Cloud-based services has to be based on a methodology that supports a wide array of criteria for evaluating available alternatives. Also, as these criteria or factors can be mutually interdependent and conflicting, a trade-offs-based methodology is needed to take a decisions.

This paper discusses the design, implementation, and evaluation of the prototype developed for automating the extended theoretical methodology of Trade-offs based Methodology for Adoption of Cloud-based Services (TrAdeCIS) developed in [5]. This system is based on Multi-attribute Decision Algorithms (MADA), which selects the best alternative, based on priorities of criteria of a decision maker. The applicability of this methodology to the adoption of cloud-based services in an organization is validated with a use-case and is even extended to other domains, especially for Train Operating Companies.

Keywords: Cloud computing · Cloud adoption · Decision support system · Multi-attribute Decision Algorithms

1 Introduction

Traditional IT (Information Technology) aligns resources according to applications in order to fulfill their business requirements. Each application has its own dedicated infrastructure and data storage [11]. For data protection and continuity of business operations, dedicated backup and recovery solutions are also deployed. As an alternative, Cloud Computing (CC) has recently emerged as a paradigm that offers its users the flexibility of scaling their computing resource usage without the concern of over or under-provisioning. CC is the result of evolution and embracement of various technologies as that of Virtualization (separating physical devises into one or more virtual devices), Service-oriented Architecture (based on loosely coupled independent services), and Utility Computing (which charges the user based on the usage instead of a fixed rate). The major benefits of cloud-based services include pay-as-you-go model, business agility and flexibility, increase in economies of scale. However, there also exist disadvantages

© Springer International Publishing AG 2017
M. Helfert et al. (Eds.): CLOSER 2016, CCIS 740, pp. 65–87, 2017.
DOI: 10.1007/978-3-319-62594-2_4

in terms of security, privacy risk, or vendor-lock in [1]. CC has four deployment models (1) Private Cloud, (2) Public Cloud, (3) Hybrid Cloud, and (4) Community Model [1]. CC today can be delivered as XaaS (Anything-as-Service), which includes the fundamental service models of Software-as-a-Service (SaaS), Platform-as-a-Service (PaaS), and Infrastructure-as-a-Service (IaaS) and can be extended to anything such as Network-as-a-Service, Database-as-a-Service, or Communication-as-a-Service, Business-as-a-Service [5]. Owing to several available options an organization has to decide various following aspects:

- **Selection of Deployment Model:** Each deployment model has its advantage and disadvantages; therefore, several factors have to be considered while making a decision.
- **Selection of Service Model:** Each service model consists of various requirements to be fulfilled both from the side of Cloud Service Provider (CSP) and the organization that plans to adopt the solution. For example, in case of PaaS, CSP provides both hardware and software on which applications run, whereas, in IaaS a virtual machine is provided by CSP. For OS and middleware, organization is responsible. Therefore, here again the decision of which service model can be adopted depends on various requirements.
- **Selection of Appropriate Service Package:** Also, there is a variation in terms of capabilities CSP provider in numerous different packages. These packages can have different benefits or drawbacks. For example, some CSPs might offer services at low cost, however, they might then not offer backups or redundant storage of data at multiple locations. This implies that the factors influencing the decision can be dependent and mutually contradictory. Therefore, organization has to make a trade-off and make the selection based on the best match to its requirement.

Due to this wider range of decisions to be taken and selections to be made, an automated Decision Support System with industrial strength will have to make trade-off decisions, which need to show a respective detailed evaluation of alternative options. Thus, the research questions to be answered are the following:

- How can a quantified trade-off based strategy be established?
- How can such a strategy evaluate several alternatives with respect to numerous interdependent and contradictory requirements?

To address this problem of decision making while adopting Cloud-based services in an organization, the methodology TrAdeCIS was introduced [5]. TrAdeCIS automates the decision process and the paper evaluates its applicability and validity not only in the context of Cloud Computing but also in the decision of adopting any new technology in an organization.

The remainder of this paper is structured as follows. Section 2 discusses related work in the field of the decision analysis for adopting any technology in an organization. It also highlights existing gaps and how TrADeCIS bridges them. Section 3 presents the architecture of the prototype for implementing TrAdeCIS and discusses the applicability and relevance of the algorithms used for making such a decision. Section 4 presents key functionality and tests as well as evaluates the methodology with respect use cases from the domain of cloud computing and Internet on train. Furthermore, this section also evaluates the performance of the implementation of the prototype. Finally, Sect. 5 concludes the paper.

2 Related Work

Spokesperson of Gartner stated that customers should be very careful while selecting the correct service provider, and ask them detailed questions about contractual terms [10]. Therefore, the decision maker has to be aware of complete requirements, their interdependencies, and conflicts in order to evaluate different CSPs. This was performed in [6]. The second challenge is to develop a quantitative approach to make decision of adopting best alternative that encompasses all requirements (criteria) and their interrelations. There have been efforts in the past to make a decision whether to move the legacy infrastructure into cloud or not. [1] and [14] propose two different approaches. While [1] compares the cost of using a cloud-based service with the costs of a datacenter on an hourly basis, [14] presents an approach to compare the costs of leasing and purchasing a CPU (Central Processing Unit) over several years. Both of these approaches only consider cost as a factor, when there are multiple conflicting factors that must be considered. Also, this approach is not open to an extension to multiple quantitative factors (that can have different measurement units) and to factors that are of qualitative nature [9]. Therefore, there is a need for a methodology encompassing multiple factors for evaluating several alternatives.

In the past, Multi-attribute Decision Algorithms (MADA) have been used for the decision on outsourcing [15] that supports multiple factors. MADAs include a finite set of alternatives, and their performance in multiple criteria is identified in the beginning of the analysis. These methods can either be used to sort or classify available alternatives. However, the current research is restricted to a number of predefined factors for taking a decision. Research so far on a cloud adoption decision process also suggests approaches such as Goal-oriented Requirements Engineering (GRE) ([2, 16]) and a quantified method using MADA [9, 13]. GRE-based approaches are based on a step-by-step process of fulfilling requirements of the cloud user and are qualitative in nature. MADA-based approaches are quantitative in nature, however, fail to evaluate impact such an adoption will have on an organization and do not incorporate business or organizational aspects in the decision. They also do not consider the influence of one attribute over another. In addition, they do not establish a trade-off strategy, where conflicting factors are involved. A trade-off strategy refers to the technique of reducing or forgoing one or more desirable parameters in exchange of increasing or obtaining other desirable outcomes in order to maximize the total return.

As shown in Table 1, a gap still exists in terms of not only developing a trade-offs-based methodology for decision making while adopting CC, but also in automating it. The comparison of related work to the work done in this paper is based on four key features, "√" describing the presence and "×" denoting

Table 1. Comparison of related work with respect to main characteristics.

Features	Cost-based approaches	MADA-based approaches	TrAdeCIS
Interrelations of factors	Partially	Partially	√
Trade-offs based quantified methodology	×	×	√
Automated decision support system	×	×	√
Applicability to other domains	×	Partially	√

the lack of that feature. This paper, therefore, fills this gap by (a) automating trade-off based quantified methodology and (b) studying its applicability for a CC use-cases, which models all relevant factors and their interrelations.

3   Research Methodology and Architecture of the System

The methodology followed to establish trade-offs-based decision of selecting the best alternative is based on algorithms of MADA- The Technique for Order of Preference by Similarity to Ideal Solution (TOPSIS) and Analytic Network Process (ANP). The methodology of TrAdeCIS is shown in detail with the flow diagram in Fig. 1. Both of these algorithms of ANP and TOPSIS require multiple alternatives and criteria as inputs. TOPSIS is used to rank the alternatives from the technical perspective. ANP is used to rank the same alternatives from economical and business perspective. The relevant criteria from the domain of CC, has already been identified in [6]. A user can either select the relevant criteria from this list, or enter their own requirements. The details of these algorithms and their implementations are described in the following sections. Furthermore, the architecture as implemented for the prototype of TrAdeCIS and the database model of the system developed is also discussed below.

3.1   TOPSIS

TOPSIS is based on the concept that the optimal solution is the one, which has geometrically the shortest distance from the best possible solution and the longest distance from the worst possible solution [8]. TOPSIS does pair-wise comparisons of all alternatives across all the criteria and facilitates trading-off a poor performance of an alternative in one factor by a good performance in another factor. Pair-wise comparison is a process of comparing alternatives in pairs to judge which of the two alternatives is preferred, has better performance with respect to a factor, or whether or not the two alternatives are performing at the same level with respect to a factor. Listing 1, expects three inputs:

- a N × M matrix of the values with N criteria as columns and M alternatives as rows.

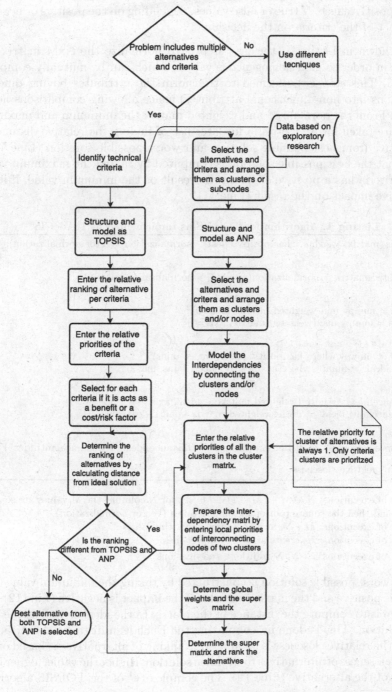

Fig. 1. Flow diagram for TraAdeCIS.

– weights, N priority values in order to prioritize the criteria.
– has_positiv_effect, N true or false values, depending on the positive or negative impact of the criteria on the decision.

As shown in Listing 1, the first step is to normalize the NxM matrix and weights in order to gain homogeneous values, which can be mutually compared (Line 4). This step is performed to transform the attributes having different dimensions into non-dimensional attributes, hence allowing comparisons across criteria. From the normalized and weighted matrix the minimum and maximum values are taken for each criteria for later use of finding the relative distance of alternative from best possible solution and worst possible solution (Line 7, 8). After that the best possible solution is computed by taking the maximum value, if the criteria has a positive effect on the result or the minimum value, if it has a negative impact on the result (Line 11).

Listing 1. Algorithm for TOPSIS as Implemented in TrAdeCIS.

```
def topsis(matrix, weights,  has_positiv_effect , normalization=vector_normalization):

    # normalize and apply weights
    weighted_matrix = normalization(weights) * normalization(matrix)¹

    # extract min and max values for each column
    mins = numpy.min(weighted_matrix, axis=0)²
    maxs = numpy.max(weighted_matrix, axis=0)²

    # create ideal and anti ideal arrays
    ideal  = numpy.where(has_positiv_effect, maxs, mins)³
    anti_ideal  = numpy.where(has_positiv_effect, mins, maxs)³

    # calculate distances to the ideal and anti ideal arrays
    distance_ideal  = norm(weighted_matrix − ideal, axis=1)¹
    distance_anti_ideal  = norm(weighted_matrix − anti_ideal, axis=1)¹

    # compute relative closeness
    relative_closeness  = distance_anti_ideal / ( distance_ideal + distance_anti_ideal )⁴

    return  relative_closeness
```

[1] Number of executions: $N * (M + (M − 1) + M^{1/2})$, for N columns the M values are squared, summed and then the square root of the sum is taken (vector normalization)
[2] Number of executions: $M * N$, for N columns check the values
[3] Number of executions: N, create arrays with N values
[4] Number of executions: $2 * N$, N additions and divisions

The worst possible solution is constructed by taking the minimum value if the impact is positive and the maximum value if the impact is negative (Line 12). The next step is to compute the distance of the matrix to the ideal as well as the anti-ideal solution. This is done by computing the Euclidean distance (Line 15, 16). Finally the relative closeness is computed. Ranking of alternatives is based on the relative closeness of alternatives to the ideal solution. Higher the value, higher is the ranking of the alternative (Line 19). The complexity[1] of the TOPSIS algorithm,

---

[1] Complexity: $N * (M + (M − 1) + M^{1/2}) + 2 * M * N + 2 * N + 2 * (N * (M + (M − 1) + M^{1/2}) + 2 * N = 8 * M * N + N + 3 * M^{1/2} = O(M * N)$.

with respect to implementation shown above is $O(N * M)$ where N is the number of alternatives and M the number of criteria.

3.2 ANP

ANP is the generalization of the Analytic Hierarchy Process (AHP) [8], and is a method where dependencies can be modeled between any of the elements. In ANP, all criteria are represented as a cluster, and their sub-criteria (if any) are modeled as elements or nodes within that cluster. Also, all available alternatives constitute an additional cluster. Each connection symbolizes the interdependency between the 2 connected nodes or clusters. On one hand, it results in the modeling of more accurate models, but on other hand it increases the complexity of the required input. This is because pairwise comparison matrices where criteria or alternatives are compared with respect to every other interconnected element in the network(see use cases for an example) has to be constructed. As shown in Listing 2 the next step is to compute the super matrix. Super matrix is generated with the eigenvectors of all the possible pairwise comparison matrices.

Listing 2. Generation of the Super Matrix as Implemented in TrAdeCIS.

```
supermatrix: function (clusterNodes) {
    var children = graph.findChildren(clusterNodes);
    var matrix = utils.matrix(children.length, 0);

    children.each(function (column, sourceNode) {
        children.each(function (row, targetNode) {
            matrix[row][column] = graph.getValue(sourceNode, targetNode) || 0;
        });
    });

    return matrix
}
```

As the super matrix represents all interrelations between any two nodes in the network, the alternatives are listed always as last n elements of the super matrix, where n is the number of alternatives. The ranking of alternatives in obtained by computing the limit matrix, and transforming it into an array. In order to compute the limit matrix, the super matrix has to be raised to high odd powers until it converges. It can be shown that the limit exists if the matrix is column stochastic [12].

As shown in Listing 3, the computation of limit matrix is an iterative process where the matrix is raised by the power of 3 and then again normalized in order to keep the matrix column stochastic (Line 7). Then the result is checked if it is equal up to the 8th decimal precision with the previous result (Line 12). If so the process is ended. Usually this takes around 3 iterations to find a result. The number of iterations depends on the limit of the super matrix (the power at which the matrix converges), and therefore on the values in the super matrix, which consist of the global cluster comparison, the criteria comparison of the cluster, and the criteria value.

Listing 3. Generation of Limit Matrix as Implemented in TrAdeCIS.

```
def limit_matrix(matrix):
    result = matrix
    previous_matrix = result

    while True:
        result = linear_normalization(numpy.linalg.matrix_power(result, 3))¹

        if numpy.isnan(numpy.sum(result))²:
            raise ArithmeticError('received not a number')

        if numpy.allclose(previous_matrix, result)³:
            break

        previous_matrix = result

    return result
```

[1] Number of executions: $2 * N^3$, two matrix multiplications are done, matrix multiplication has a complexity of $O(N^3)$
[2] Number of executions: $M * N$, all the values are summed up
[3] Number of executions: $M * N$, all the values are compared

The reason for raising the super matrix by the power of 3 is that odd numbers have the advantage of preserving the structure of the matrix (in matrix multiplication, depending on where a zero is the other values might switch places with the zeros). When the limit is found, the values for the whole row are the same. The advantage, however, is that if is raised by an odd number the first column will certainly have non-zero values. These values denote the ranking of the alternatives. The value of 3 is chosen so as to maintain a balance between the rising complexity of the computation with higher values, and the number of iterations needed to compute the limit matrix. The complexity[2] of the algorithm is $O(N^3)$, where N is the dimension of the super matrix.

3.3 Trade-off Based Decision

A trade-offs-based strategy is required, if the ranking of alternatives obtained by using TOPSIS (from technical perspective) and ANP (from business – economical and organizational – perspective) is different. TrAdeCIS therefore, as shown in Fig. 1, compares the rankings obtained and gives the option to a decision maker to select the best technical solution at a trade-off of business value. Trade-offs are achieved by altering the priorities of criteria. There are essentially three possible dimensions at which priorities can be adjusted in ANP.

1. At the global cluster level, prioritizing an entire cluster compared to others. The alternatives cluster can also be compared as an exception to other clusters if needed.

---

[2] Complexity: $2 * N^3 + M * N + M * N = O(N^3)$.

2. At the cluster level, comparing the importance of criteria in a cluster.
3. At the criteria level, changing the values of the comparison matrix.

In order to assist a decision maker in deciding which criteria or alternative should be adjusted, TrAdeCIS provides a basic approach to calculate the impact of changing any of these values. A column of the super matrix shows the influence that a node has on other criteria and alternatives (outgoing influence). Therefore, by increasing and decreasing the row values in a super matrix the influence of a node to the final rankings as per ANP can be evaluated. Hence, the approach developed and followed here for calculating trade-offs is outlined in Algorithm 1. The increase (*inc*) and decrease (*dec*) for each element of original super matrix as shown in Algorithm 1 is calculated using on the current normalized value and half of the minimum element of the original super matrix (*relative_change*). Choosing *relative_change* to be less than the minimum element of the super matrix removes the possibility of division by zero error in *inc*. Also, *dec* will never be zero or negative, because the *relative_change* is smaller than the lowest value in the super matrix.

Table 2. Original super and limit matrices for trade-offs.

(a) Super matrix

	C1	C2	A1	A2
C1	0	0	0.5	0.5
C2	0	0	0.5	0.5
A1	0.75	0.3333	0	0
A2	0.25	0.6667	0	0

(b) Limit matrix

	C1	C2	A1	A2
C1	0	0	0.5	0.5
C2	0	0	0.5	0.5
A1	0.5417	0.5417	0	0
A2	0.4583	0.4583	0	0

The above developed methodology is illustrated here with an example to understand the application of this trade-offs process. In this example the result of TOPSIS favors A2, while ANP favors A1. A user now has to make a trade-off decision by adjusting ANP values so that the result will also be A2. By analyzing the super matrix in Table 2(a) and comparing it with the model structure shown in Fig. 2 the meaning of values in the super matrix can be analyzed.

Each row of the super matrix represents how much a node is influenced by other criteria or alternatives (incoming influence), *e.g.*, the first row of Table 2(a)

Algorithm 1. Algorithm for establishing Trade-offs.

▷ % S_x denotes the super matrix x
▷ %: L_x denotes the limit matrix of super matrix x
▷ %: *Row* is the set of elements in a row of a super matrix or limit matrix
▷ %: *Node* denotes to a criteria or alternative
▷ %: *inc* denotes the value by which values of super matrix has to be increased
▷ %: *dec* denotes the value by which values of super matrix has to be decreased
▷ %: *current_value* denotes an element of original super matrix
▷ %: *relative_change* = 0.5 * (minimum element of the original super matrix)

Compute the limit matrix L_c from the current super matrix S_c

for each row in the super matrix **do**
    $inc = relative\_change/(1 - (current\_value + relative\_change))$
    $dec = relative\_change/(1 + (current\_value - relative\_change))$
    Construct super matrix S_{row+} by increasing each element in *Row* by *inc*
    Construct super matrix S_{row-} by decreasing each element in *Row* by *dec*
end for

for each constructed super matrix $\in \{S_{row+}, S_{row-}\}$ **do**
    Compute the limit matrix L_{row+}
    Compute the limit matrix L_{row-}
end for

for each computed limit matrix $\in \{L_{row+}, L_{row-}\}$ **do**
    Compute the differences to the original limit matrix L_c, $L_{row+} - L_c$ and $L_{row-} - L_c$
end for
Sort the the differences of the limit matrix in the previous step to obtain a list of the nodes which have the highest impact

for each positive difference of the limit matrix **do**
    Repeat the process but only specific to one element at a time
end for

---

shows that C1 is equally influenced by A1 and A2 and not influenced by C1 and C2 at all. In Table 2(a) the minimum value is 0.25, therefore the *relative_change* is 0.125. The algorithm is applied here only specific to row 3, for illustration purposes.

Tables 3(a) and 4(a) show the super matrices after increasing and decreasing the non-zero values of row 3 by *inc* and *dec* respectively. By computing the limit

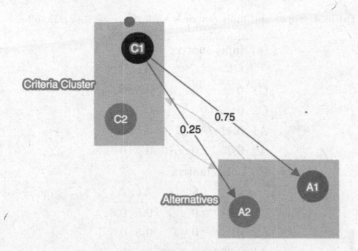

Fig. 2. ANP model for trade-off.

Table 3. Super and limit matrices with increase for trade-offs.

(a) Super matrix				
	C1	C2	A1	A2
C1	0	0	0.35	0.35
C2	0	0	0.25	0.25
A1	1.75	0.5641	0	0
A2	0.25	0.6666	0	0

(b) Limit matrix				
	C1	C2	A1	A2
C1	0	0	0.5	0.5
C2	0	0	0.5	0.5
A1	0.7163	0.7163	0	0
A2	0.2837	0.2837	0	0

matrix for these super matrices (cf. Tables 3(b) and 4(b)), and subsequently calculating the difference to the original limit matrix (cf. Tables 5(a) and (b)) the influence of A2 on the result of ANP can be identified.

This process is repeated for each row and the values corresponding to A2 (as this was the highest ranked alternative according to TOPSIS) from limit matrices are shown in Table 6(a). Table 6(a) shows that increasing A2 in terms of its interconnected node is the most beneficial (highest value corresponds to A2) for ranking A2 as the highest ranked alternative per ANP. Table 6(b) shows for every positive value in Table 6(a) those values obtained by increasing or decreasing only one element at time in the original super matrix. For example,

Table 4. Super and limit matrices with decrease for trade-offs.

(a) Super matrix

	C1	C2	A1	A2
C1	0	0	0.5	0.5
C2	0	0	0.5	0.5
A1	0.4167	0.1754	0	0
A2	0.25	0.6666	0	0

(b) Limit matrix

	C1	C2	A1	A2
C1	0	0	0.5	0.5
C2	0	0	0.5	0.5
A1	0.3924	0.3924	0	0
A2	0.6076	0.6076	0	0

Table 5. Differences of limit matrices.

(a) Difference of increased matrix

	C1	C2	A1	A2
C1	0	0	0.5	0.5
C2	0	0	0.5	0.5
A1	0.1746	0.1746	0	0
A2	−0.1746	−0.1746	0	0

(b) Difference of decreased limit matrix

	C1	C2	A1	A2
C1	0	0	0.5	0.5
C2	0	0	0.5	0.5
A1	−0.1492	−0.1492	0	0
A2	0.1492	0.1492	0	0

as C2 has positive value in Table 6(a), two values are obtained in Table 6(b) corresponding to the interrelation of C2 to A1 and C2 to A2. This leads to the identification of the node that is interrelated to A2: when changed in terms of its associated priority it will make A2 the highest ranked alternative. In this example, as A2 associated with C2 has highest positive value, change in priority of C2 will lead to the desired ranking.

Table 6. Trade-offs result.

(a) Row specific limit matrix values

Node	Increase	Decrease
C1	−0.0520	0.0520
C2	0.0520	−0.0520
A1	−0.1746	0.1492
A2	0.1548	−0.1421

(b) Element specific limit matrix values

Node	Connected node	Increase	Decrease
C1	A1	-	0.0246
C1	A2	-	0.0226
C2	A1	0.0297	-
C2	A2	0.0285	-
A1	C1	-	0.0917
A1	C2	-	0.0392
A2	C1	0.0492	-
A2	C2	0.1250	-

3.4   Implementation Architecture

TrAdeCIS 1.0 was implemented as a traditional client-server architecture, where for each action a new request is made and the server answers with the corresponding markup. This architectural implementation lead to duplication of code, because some parts of the webpage had to update asynchronous. The further details of TrAdeCIS 1.0 are available in [4,7]. For TrAdeCIS 2.0 the implemented architecture was designed based on the Single-Page-Application (SPA) architecture. A SPA is defined as web application or web site that fits on a single web page. This can be either achieved by initially loading all necessary code for the representation (HTML, JavaScript and CSS) or resources are dynamically loaded and added to the page as necessary. This design has been implemented in order to improve performance as well as avoid potential duplication of code. A cycle of requests in a tradition client-server architecture with asynchronous elements (*e.g., TrAdeCIS 1.0*) can be outlined like the following:

1. Client visits the webpage
2. Backend processes the request, renders the webpage via a template language
3. Client receives the webpage
4. Client changes data on the previously received webpage
5. Frontend issues an AJAX request and updates the user interface

This approach shows overlapping logic at step 2 and 5 just in different environments (frontend, backend) and programming languages. A SPA architecture has the advantage to overcome those issues by separating all logic for the representation on the frontend and all the data manipulation logic on the backend. The added advantages are (1) the clear separation of concerns for the frontend and backend, (2) that user no longer experiences browser reloads (everything will now be loaded asynchronously by the client) and that (3) the data flow is minimized (only the necessary data is loaded, no markup). Additionally future implementations of native clients would be simplified as they have already an API to address. However this approach has also disadvantages, indexing will no longer work, *e.g.,* from Google and also because routing has to be done in the frontend, therefore a request on a route in a web browser will not exist. There is one way to solve both problems by using the concept of "universal javascript". Briefly said it means that javascript code will also run in the backend. In order that this is possible a node.js instance has to process the JavaScript code and then return a html page. However this will raise the complexity of the backend because the communication between the two instances has to be managed. Due to this added complexity TrAdeCIS does currently not implement a "universal javascript" approach. In TrAdeCIS 2.0 the implemented architecture reads as follows:

- The backend is built with Django and Django REST framework and exposes REST endpoints for all the necessary tasks.
- The frontend is built with React.js, consumes the data from the REST endpoints and visualizes them on the client side.

Backend. The architecture of the system follows the community standards with Django projects. Django is fullstack web framework for Python. In Django coherent logic is bundled in a so-called "app". TrAdeCIS is built with two

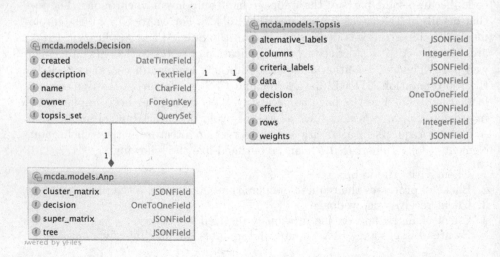

Fig. 3. Database interrelations.

apps: (1) "mcda", for storing, computing and visualizing TOPSIS and ANP, and (2) "account", to manage the different access levels which TrAdeCIS provides. The app "mcda" consists of three database models namely, Decision, TOPSIS and ANP (cf. Fig. 3). Decision model denotes one use-case, which consists of a name, optional description and the data for TOPSIS and ANP, which are stored in their respective tables. The access as well as the modification of the data is exposed via REST endpoints by utilizing the "Django REST Framework", which simplifies the creation of REST APIs with Django.

Frontend. The SPA frontend is built with several new technologies, which are briefly introduced: (1) "React.js", is a Frontend-Framework developed and maintained by Facebook. It challenges current standards by writing everything in Javascript. Rather than DOM changes it utilizes a virtual DOM which is performs faster on changes and computes the minimal changes which have to be done to the actual DOM. The key feature is that a webpage is built with reusable components, which are defined by programmer. There are many other Frontend-Frameworks which could be used and there is no specific reason to favor React.js. (2) "Webpack", is a module bundler, which allows to build browser javascript which can be modularized. Apart from that there are many additional features which make Webpack valuable, for example minification of code, import of other files (e.g. css), removing unused code, etc. (3) "Redux", is a state management library which makes the state immutable at all time and therefore leads to clearer, as well as better testable state. Additional new technologies and concepts like "Css Modules", "PostCss", "ES6", "JSX" were used as well as the following libraries "React-Router", "Cytoscape" and "Chartjs".

4 Testing and Evaluation

This section tests the developed system with the objective of evaluating its applicability, usability in various use-cases of making such decisions. The performance values of all alternatives with respect to criteria is taken from [3], which is platform to measure and monitor these values. In addition, performance of the system is also evaluated, which includes calculation of load-up and processing time of the system, depending of number of alternatives and criteria.

4.1 Decision of Adopting PaaS (Use Case 1)

For Use Case 1 (UC1) the scenario of adopting PaaS is considered, with alternative providers and criteria as shown in Table 7. For TOPSIS the criteria of "Runtimes" denotes the number of supported programming languages. "Services" are additional services that are supported (for example databases), and "Add-ons" are additional other programs which can be used. Also, "uptime" of the service in the past 30 days for all the providers is included. In this scenario as well, all these criteria have a positive impact and are weighted equally. The result of TOPSIS is computed with code snippet 1 and is "Heroku".

Table 7. Decision of adopting PaaS input for TOPSIS.

Alternatives	Uptime	RAM (MB)	Runtimes	Services	Add-ons
Heroku	99.91	512	9	2	17
dotcloud	99.95	32	5	1	7
AppHarbor	99.99	512	1	3	33

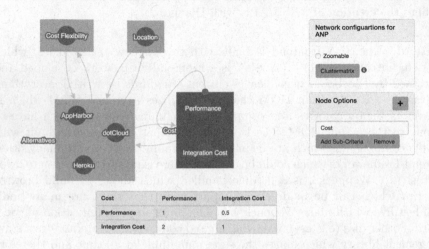

Fig. 4. Decision of adopting PaaS ANP model.

Table 8. Decision of adopting PaaS resulting super matrix.

	Location	Performance cost	Integration cost	Cost flexibility	Heroku	dotcloud	AppHarbor
Location	0	0	0	0	0.083	0.083	0.083
Performance cost	0	0	0	0	0.021	0.056	0.028
Integration cost	0	0	0	0	0.062	0.028	0.056
Cost flexibility	0	0	0	0	0.083	0.083	0.083
Heroku	0.100	0.037	0.104	0.035	0	0	0
dotcloud	0.050	0.025	0.022	0.144	0	0	0
AppHarbor	0.100	0.022	0.040	0.071	0	0	0

For ranking the alternatives from the business perspective the model in Fig. 4 for ANP is constructed. In this case there is a self-loop on the cost cluster, which allows to give relative priority to each criteria in a cluster. Here the criteria of Integration Cost is considered 2 times more important than that of Performance Cost. For this scenario, the resulting super matrix is shown in Table 8 and not every pairwise comparison matrix.

Again by applying Listing 3 the limit matrix is found, shown in Table 9, which ranks dotCloud the highest.

Table 9. Decision of adopting PaaS resulting limit matrix.

	Location	Performance cost	Integration cost	Cost flexibility	Heroku	dotcloud	AppHarbor
Location	0	0	0	0	0.333	0.333	0.333
Performance cost	0	0	0	0	0.140	0.140	0.140
Integration cost	0	0	0	0	0.193	0.193	0.193
Cost flexibility	0	0	0	0	0.333	0.333	0.333
Heroku	0.335	0.335	0.335	0.335	0	0	0
dotcloud	0.343	0.343	0.343	0.343	0	0	0
AppHarbor	0.322	0.322	0.322	0.322	0	0	0

Table 10. Decision of adopting PaaS tradeoff.

Node	Connected node	Increase	Decrease
Location	Heroku	0.00087	-
Location	dotcloud	0.00087	-
Location	AppHarbor	0.00079	-
Performance cost	Heroku	0.00131	-
Performance cost	dotcloud	0.00146	-
Performance cost	AppHarbor	0.00118	-
Integration cost	Heroku	0.00560	-
Integration cost	dotcloud	0.00418	-
Integration cost	AppHarbor	0.00458	-
Cost flexibility	Heroku	-	0.00514
Cost flexibility	dotcloud	-	0.00437
Cost flexibility	AppHarbor	-	0.00427
Heroku	Location	0.01561	-
Heroku	Performance cost	0.00704	-
Heroku	Integration cost	0.01535	-
Heroku	Cost flexibility	0.01089	-
dotcloud	Location	-	0.00643
dotcloud	Performance cost	-	0.00294
dotcloud	Integration cost	-	0.00353
dotcloud	Cost flexibility	-	0.01148
AppHarbor	Location	-	0.00799
AppHarbor	Performance cost	-	0.00264
AppHarbor	Integration cost	-	0.00376
AppHarbor	Cost flexibility	-	0.00656

However, now the results of TOPSIS and ANP do not match and therefore a trade-off is necessary to match the results of TOPSIS and ANP. Table 10 shows the trade-offs result based on the approach shown in Sect. 3.3. These values conclude that the highest change towards "Heroku" can be achieved by increasing the priority of the criteria Location (highest positive value is 0.01561 in Table 10).

4.2   Applicability of TrAdeCIS in Other Domains (Use Case 2)

TrAdeCIS was developed to support organizations in the adoption of cloud-based services. However, organizations may also utilize TrADeCIS to improve their understanding of the value of technologies from other domains than cloud-based services. This is illustrated by applying TrAdeCIS to Train Operating Companies (TOC), who need to take a decision of choosing the best technology, to improve both voice- and data coverage on-board trains (UC2). This decision takes the perspective of the TOC, who is hoping to sell more tickets by providing the service. For the train-to-wayside connection, all on-board solutions are assumed to use the same technology, especially a connection to mobile base stations by 3G or beyond. The following alternatives are considered (cf. Table 11) to be installed on-board of trains:

- Option 1: Wireless Access Point (WAP)
- Option 2: Analog repeater
- Option 3: Femtocells

Technical requirements from these and their relative priorities read:

- Internet has to be available to all passengers with a mobile device (Priority 1)
- Quality of voice calls needs to be improved for all passengers with a phone (Priority 2)
- Internet speed should be as high as possible (Priority 3)

Therefore, after applying TOPSIS to these technical requirements, installation of WAPs is ranked the highest. From the financial/economic requirements perspective, ANP is used to model it as shown in Fig. 5. The factor of Net Present Value, which should be positive as soon as possible, is of highest priority. However, it is broken into sub-factors as that of low deployment time, high revenue,

Table 11. Applicability of TrAdeCIS in other domains input for TOPSIS.

Alternatives	Internet availability	Voice coverage	Internet speed
Option 1	3	1	3
Option	2	2	2
Option 3	2	2	2

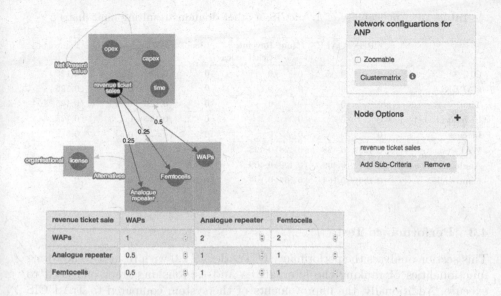

Fig. 5. Applicability of TrAdeCIS in other domains ANP model.

Table 12. Applicability of TrAdeCIS in other domains resulting super matrix.

	OPEX	CAPEX	Time	Revenue ticket sales	License	WAPs	Analogue repeater	Femtocells
OPEX	0	0	0	0	0	0.083	0.083	0.083
CAPEX	0	0	0	0	0	0.083	0.083	0.083
Time	0	0	0	0	0	0.083	0.083	0.083
Revenue ticket sales	0	0	0	0	0	0.083	0.083	0.083
License	0	0	0	0	0	0.333	0.333	0.333
WAPs	0.05	0.041	0.021	0.308	0.425	0	0	0
Analogue repeater	0.05	0.081	0.041	0.154	0.425	0	0	0
Femtocells	0.05	0.027	0.014	0.154	0.142	0	0	0

low capital expenditure, and low operational expenditure. In addition, all these factors contribute differently to the factor of Net Present Value. This is represented with a self-loop and the respect weighing or priorities of these factors are entered in the corresponding comparison matrix. Also in terms of organizational requirements, the TOC prefers to avoid the use of licensed spectrum (medium importance). The resulting super matrix, which is constructed from all comparison matrices, is shown in Table 12. The highest ranked alternative from ANP is also WAPs, as calculated in the limit matrix (cf. Table 13). Therefore, as the ranking obtained from both TOPSIS and ANP is the same, for the scenario of providing internet and voice call connectivity on-board train, WAP is the best alternative.

Table 13. Applicability of TrAdeCIS in other domains resulting limit matrix.

	OPEX	CAPEX	Time	Revenue ticket sales	License	WAPs	Analogue repeater	Femtocells
OPEX	0	0	0	0	0	0.125	0.125	0.125
CAPEX	0	0	0	0	0	0.125	0.125	0.125
Time	0	0	0	0	0	0.125	0.125	0.125
Revenue ticket sales	0	0	0	0	0	0.125	0.125	0.125
License	0	0	0	0	0	0.5	0.5	0.5
WAPs	0.428	0.428	0.428	0.428	0.428	0	0	0
Analogue repeater	0.409	0.409	0.409	0.409	0.409	0	0	0
Femtocells	0.164	0.164	0.164	0.164	0.164	0	0	0

4.3   Performance Test

This section analyses the performance of TrAdeCIS 2.0 with respect to how long functionalities of ranking the alternatives and establishing trade-offs need to execute. Additionally, the improvements of the system compared to TrAdeCIS 1.0, where the implemented architecture was based on a traditional client-server architecture are discussed. All the performance tests are executed on a system with a 2.6 GHz Intel Core i5 CPU, 8 GB 1600 MHz DDR3 RAM and a Intel HD Graphics 4000 1536 MB graphics card. The implemented architecture introduced in Sect. 3.4 has a major influence on the overall performance as well as the user experience. This is due to the fact that all needed files for the representation are loaded on the first connection, and only the necessary data has to be retrieved in subsequent requests. This minimizes data flow and results in load-up time for pages in around 100–300 ms. In TrAdeCIS 1.0 depending on the complexity of the ANP model a user had to wait up to 3 s until the page was fully loaded. It can be concluded that based on these implementation specific changes the execution time has improved by a factor of 5–10.

Another crucial performance impact is on the visualization of TOPSIS and ANP model on GUI. While TOPSIS scales well even with growing number of alternatives and criteria, ANP does not. TOPSIS only needs tabular data as input. ANP, however, has a complex model of interrelations which is created by a html canvas, as shown Fig. 5. The highly complex model of ANP, and its corresponding input value limits the size at which it is user friendly to work with. By trial and error it could be evaluated that the number of nodes in the visualization should not be higher than 100, because otherwise the user has to wait longer intervals to be able to interact with the model. Because of changes in the interaction with the canvas the performance has been increased by a factor of 3 compared to TrAdeCIS 1.0. Figure 6 shows average execution time for 1000 executions of ranking alternatives with TOPSIS and ANP with respect to different number of criteria and alternatives. In case of ANP the values are obtained by an average over 1000 computation of random super matrices. The random super matrices are generated by random values in the interval of 0 and 1 with the given dimension from the number of criteria and alternatives. This can

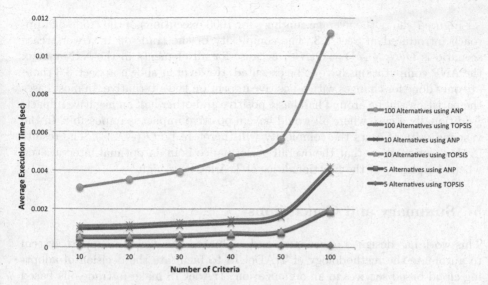

Fig. 6. Execution time measurements of TOPSIS and ANP.

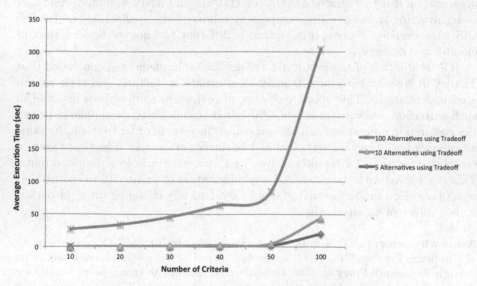

Fig. 7. Execution time measurements of the tradeoff approach.

be justified by the fact that the values are normalized, therefore a higher interval is not necessary. Also when considering the distribution of the random values, zero will occur sparsely, this implies that the resulting random super matrix will be very well connected and therefore computationally more expensive. In general, the execution time should be slightly below the given values since such well connected models are not frequently constructed.

Figure 7 shows the average runtimes for 1000 executions of the trade-off approach introduced in Sect. 3.3. The complexity of the trade-off in a worst case scenario is $O(N^2 * N^3) = O(N^5)$, because for all elements in the NxN matrix the ANP computations have to be executed. However as shown in Sect. 3.3 those are only done for changes with a positive impact on the alternative. In most cases the partition will be around half has a positive and other half an negative impact. Additionally a case where all would have a positive impact is impossible. In the vast majority of cases the complexity will therefore be $O((N/2)^5)$. The results in this section show that the overall performance is in an optimal interval even when complexity of the functionalities of TrAdeCIS is high.

5 Summary and Conclusions

This work has designed, developed, and evaluated the decision support system to automate the methodology of TrADeCIS to facilitate the decision of adopting cloud-based services in an organization. TrAdeCIS makes a trade-offs based quantified decision of selecting the best alternative as per requirements of the organization using integrated MADAs of TOPSIS and ANP. An appropriate use-case (involving train operating companies) validated the applicability of TrADeCIS in a decision process of adopting a different technology besides that of cloud-based services.

The evaluation of the prototype TrAdeCIS 2.0 implemented concluded that TrAdeCIS is scalable to large number of alternatives, factors, and their associated interrelations. This allows modeling of real-world complexities involved in such a decision making. Trade-offs established are measured by change in priorities required in economical and organizational factors, in order to reach the same alternative both from the technical and business perspective. The time taken to establish trade-offs for 50 alternatives and 100 criteria is less than a minute. This is achieved with the optimized implementation of TrAdeCIS to ensure that results obtained are not outdated due to dynamically changing input in performance values of an alternative.

Acknowledgements. This work was partly funded by FLAMINGO, the Network of Excellence Project ICT-318488, supported by the European Commission under its Seventh Framework Program. The authors would also like to thank Bram Naudts for discussions and excellent input with respect to applying TrAdeCIS to Train Operating Companies.

References

1. Armburst, M., Fox, A., Griffith, R., Anthony, J.D., Katz, R., Konwinski, A., Lee, G., Patterson, D., Rabkin, A., Stoica, I., Zaharia, M.: A view of cloud computing. Commun. ACM **53**(4), 50–58 (2010)

2. Beserra, P.V., Camara, A., Ximenes, R., Albuquerque, A.B., Mendonça, N.C.: Cloudstep: a step-by-step decision process to support legacy application migration to the cloud. In: IEEE 6th International Workshop on the Maintenance and Evolution of Service-Oriented and Cloud-Based Systems (MESOCA), Trento, Italy, pp. 7–16 (2012)

3. Cloud Harmony Inc.: Cloud Harmony (2015). https://cloudharmony.com/. Accessed October 2015

4. Garg, R., Heimgartner, M., Stiller, B.: Decision support system for adoption of cloud-based services. In: 6th International Conference on Cloud Computing and Services Science, Rome, Italy, pp. 71–82, April 2016

5. Garg, R., Stiller, B.: Trade-off-based adoption methodology for cloud-based infrastructures and services. In: Sperotto, A., Doyen, G., Latré, S., Charalambides, M., Stiller, B. (eds.) AIMS 2014. LNCS, vol. 8508, pp. 1–14. Springer, Heidelberg (2014). doi:10.1007/978-3-662-43862-6_1

6. Garg, R., Stiller, B.: Factors affecting cloud adoption and their interrelations. In: Proceedings of the 5th International Conference on Cloud Computing and Services Science (CLOSER), SCITEPRESS (Science and Technology Publications, Lda), Lisbon, Portugal, pp. 87–94, May 2015

7. Heimgartner, M.: Design and implementation of prototype for TrAdeCIS. Department of Informatics, University of Zurich, Zurich, Switzerland. Communication Systems Group, December 2015

8. Ishizaka, A., Nemery, P.: Multi-criteria Decision Analysis: Methods and Software. Wiley, Chichester (2013)

9. Menzel, M., Schönherr, M., Tai, S.: $(MC^2)^2$: criteria, requirements and a software prototype for cloud infrastructure decisions. Softw. Pract. Experience 43(11), 1283–1297 (2013)

10. Moore, S.: Gartner says worldwide cloud Infrastructure-as-a-Service spending to grow 32.8% in 2015e. http://www.gartner.com/newsroom/id/3055225. Accessed October 2015

11. NetApp: The journey from traditional IT to the Cloud-Net App (2015). http://webobjects.cdw.com/webobjects/media/pdf/netapp/NetApp-Virtualization-To-Cloud-Brochure-1.pdf?cm_sp=NAPShowcase-_-Cat4-_-CloudComputing. Accessed October 2015

12. Saaty, T.L., Vargas, L.G.: Decision Making with the Analytic Network Process. Springer, New York (2006)

13. Saripalli, P., Pingali, G.: MADMAC: Multiple Attribute Decision Methodology for Adoption of Clouds. In: IEEE 4th International Conference on Cloud Computing (CLOUD), Washington D.C., USA, pp. 316–323 (2011)

14. Walker, E.: The real cost of a CPU hour. IEEE Comput. 42(4), 35–41 (2009)

15. Wang, J.J., Yang, D.L.: Using a hybrid multi-criteria decision aid method for information systems outsourcing. Comput. Oper. Res. 34(12), 3691–3700 (2007)

16. Zardari, S., Bahsoon, R.: Cloud adoption: a goal-oriented requirements engineering approach. In: ACM 2nd International Workshop on Software Engineering for Cloud Computing, Honolulu, Hawaii, USA, pp. 29–35 (2011)

Toward Proactive Learning of Multi-layered Cloud Service Based Application

Ameni Meskini[1(✉)], Yehia Taher[2], Amal El Gammal[3],
Béatrice Finance[2], and Yahya Slimani[1]

[1] INSAT, LISI Research Laboratory, University of Carthage, Tunis, Tunisia
ameni.meskini@gmail.com
[2] Laboratoire PRiSM, Universite de Versailles/Saint-Quentin-en-Yvelines,
Versailles, France
yehia.taher@gmail.com, beatrice.finance@gmail.com
[3] Faculty of Computers and Information, Cairo University, Cairo, Egypt
a.elgammal@gmail.com
http://prism.uvsq.fr,
http://fci-cu.edu.eg

Abstract. Cloud computing is becoming a popular platform to deliver service-based applications (SBAs) based on service oriented architecture (SOA) principles. Monitoring the performance and functionality in all the layers which affects the final step of adaptations of SBAs deployed on multiple Cloud providers and adapting them to variations/events produced by several layers (infrastructure, platform, application, service, etc.) are challenges for the research community, and the major challenge is handling the impact of the adaptation operations. A crucial dimension in industrial practice is the non-functional service aspects, which are related to Quality-of-Service (QoS) aspects. Service Level Agreements (SLAs) define quantitative QoS objectives and is a part of a contract between the service provider and the service consumer. Although significant work exists on how SLA may be specified, monitored and enforced, few efforts have considered the problem of SLA monitoring in the context of Cloud Service-Based Application (CSBA), which caters for tailoring of services using a mixture of Software-as-a-Service (SaaS), Platform-as-a-Service (PaaS) and Infrastructure-as-a-Service (IaaS) solutions. With a preventive focus, the main contribution of this paper is a novel learning and prediction approach for SLA violations, which generates models that are capable of proactively predicting upcoming SLAs violations, and suggesting recovery actions to react to such SLA violations before their occurrence. A prototype has been developed as a Proof-Of-Concept (POC) to ascertain the feasibility and applicability of the proposed approach.

Keywords: Cloud Service Based Application · SLA violations prevention · Cloud environments · Decision tree

© Springer International Publishing AG 2017
M. Helfert et al. (Eds.): CLOSER 2016, CCIS 740, pp. 88–108, 2017.
DOI: 10.1007/978-3-319-62594-2_5

1 Introduction

Cloud computing is a model for enabling ubiquitous, convenient, [2] on-demand network access to a shared pool of configurable computing resources (e.g., networks, servers, storage, applications, and services) that can be rapidly provisioned and released with minimal management effort or service provider interaction. There are three-service models in the cloud, Software as a Service (SaaS), Platform as a Service (PaaS) and Infrastructure as a Service (IaaS). An application built or composed of a set of these services is called Cloud Service Based Application (CSBA).

In CSBA, a client can rent a cloud service or a set of cloud services from single or multi providers to make up an application. The provisioning of these services relies on Service Level Agreement (SLA), [3] which is a contract between the client and the service provider including a set of non-functional requirements called Quality of Service (QoS). In the case of SLA violations, costly penalty payments and adjustments to the contract or the underlying system may occur as a consequence. For this specific reason monitoring the SLA and detecting, predicting and resisting violations appears to be important, and it is more challenging in CSBAs since various providers are involved.

Although in the body of literature there is a lot of work on monitoring, detecting and predicting SLA violations. Most of the available systems either rely on grid or service oriented infrastructures, which are not directly compatible to Clouds due to the difference of resource usage model. Considerable efforts has been made lately to monitor and detect SLA violations in CSBA, but these approaches are reactive, i.e., they can only detect a violation and take the necessary action only when the violation has already happened.

Today, Elasticity in Cloud Computing allows the Cloud environment to -ideally automatically- assign a dynamic number of resources to a task, aiming to ensure that the amount of resources needed for its execution is indeed provided to the respective task. The argue that elasticity in cloud computing is a replacement for monitoring SLA violations is a controversial research problem, that is out of the scope of our research. But from our perspective systems providing elasticity today like AmazonWatch and AmazonScale are specific for single provider and don't fit to the CSBAs that is formed of cloud services from various providers. Nevertheless these tools reacts to the symptoms of resource deficiency in the system rather than being proactive and pre-dicting violations in the system. Moreover what we propose is a leading paradigm in the area of proactive learning from SLA violations in CSBAs.

Learning from history and the failures is in the instinct of human being. We as human beings learn from our historical events and failures that we face in our life to make up some rules in our mind that adapts to new context and proactively anticipate future events and avoid the failure. This idea is the main core of proactive event processing and artificial intelligence researches. For example if our mind learns that an event **A** followed by an event **B** leads to a failure **C**, the next time we face **A** followed by **B** we proactively act to the failure and anticipate its consequences.

The question that arises here are, can we adapt the concept of learning from history and making up proactive rules in the context of SLA violations in CSBAs?

In CSBAs a failure in an SLA might lead to a failure in another SLA or in the global SLA, can we correlate or associate these precedencies? The most important question is what is the mechanism to figure out these precedencies and make the proactive rules?

The contributions of this paper are twofold: (i) we present a novel monitoring framework for CSBA, namely Proactive learning from SLA violation, based on MAPE-K[1] adaptation loop, and (ii) we concretely address the 'Analysis' component of the proposed framework. This novel proactive learning approach takes advantage of the massive amount of past process execution data in order to predict potential violations. It identifies the best counter measures that need to be applied. As a proof-of-concept of the proposed approach, a prototype has been developed that ascertains the applicability and feasibility of the proposed solution.

The rest of this paper is structured as follows. Section 2 discusses related work. Section 3 introduces a running scenario that will be used throughout the paper. Section 4 lays the background needed to understand the work proposed in this paper. The proposed predictive monitoring framework is presented in Sect. 5. Section 6 presents our proposed SLA violations learning approach. Section 7 discusses the implementation of the proposed approach on a real-life log. Finally, Sect. 8 draws conclusions and perspectives.

2 Literature Review

There is a massive amount of work in the literature related to cloud-based environment, covering various aspects of this multi-disciplinary domain. In the next discussion, we are focusing on prominent efforts in the area of SLA monitoring and management in CBSA, which is appraised against the work proposed in this paper.

In principle, approaches for monitoring and detecting SLA violations with respect to QoS constraints are mainly based on techniques and strategies to adapt QoS settings according to changes and violations detected during execution of CSBA. In this case QoS parameters are generally used to repair and optimize a web service. Generally, these adaptive approaches are based on the ability to select and replace the failed services dynamically at runtime or during deployment. The selection is governed not only by the need to substitute services but to optimize the requirements of QoS of the system. Accordingly, the system must autonomously adapt itself in order to improve the quality of service of the process. In [11] proposed a novel hybrid and adaptive multi learners approach for online QoS modeling in the cloud; they described an adaptive solution that dynamically selects the best learning algorithm for prediction.

The proposals in [6, 10] address the problem of violation detection and adaptation of SLA contracts between several layers. For example, [6] proposed a methodology to create, monitor and adapt the inter-layer SLA contracts. The SLA model includes parameters such as KPI (key performance indicators), KGI (indicators key objectives), and metrics infrastructure. [10] proposed a solution to avoid SLA violations by

---

[1] MAPE-K (Monitoring, Analysis, Planning Execution-Knowledge).

applying inter-layer techniques. The proposed approach uses three layers for the prediction of SLA violations. The identification of adaptation needs is based on the prediction of QoS, which uses assumptions about the characteristics of the execution context. In [11], the authors introduced a Cloud Application and SLA monitoring architecture, and proposed two methods for determining the frequency these applications need to monitor, they also identified the challenges in regard with application provisioning and SLA enforcement, especially in multi-tenant Cloud services.

2.1 Monitoring and Detecting SLA Violations in CSBA Related to QoS Management

Considerable efforts has been made lately to monitoring and detecting SLA violations in CSBA [8–10] but all proposed approaches are reactive, i.e., they can only detect a violation and take the necessary action only when the violation has already happened.

These approaches are based on the techniques and strategies to adapt settings QoS according to changes and violations detected during execution of CSBA (Cloud Service Based Application). In this case QoS parameters are generally used to repair and optimize a web service. Generally, these adaptive approaches are based on the ability to select and replace the failed services dynamically at runtime or during deployment. The selection is governed not only by the need to substitute services and/or optimize the requirements of QoS of the system. Accordingly, the system must autonomously adapt it self in order to improve the quality of service process. The objective of [8–10], is to select the best set of services available at the time of execution, taking into account the constraints of the process but also the preferences of the user and the execution context.

Reference [11] proposed a novel hybrid and adaptive multilearners approach for online QoS modeling in the cloud; they described an adaptive solution that dynamically selects the best learning algorithm for prediction.

To determine the inputs of QoS model at runtime, they partition the inputs space into two sub-spaces, each of which applies different symmetric uncertainty based selection techniques, and then they combine the sub-spaces results.

In [12] the PREVENT approach is described to support the prediction and prevention of SLA violation in service compositions, based on the monitoring of events and the machine learning techniques. The prediction of violations is calculated only for predefined controls in the composition, based on classifiers prediction models regression. However, this approach does not support the changes in the composition according to the problems that may appear in any portion of the composition.

In contrast, [3] addressed the QoS degradation problem that can create SLA violations of a compound service and suggest that management while performing service composition requires consideration of the structure composition and the dependencies between the participating services. They proposed an approach that determines the impact of each service on the overall performance of a composition. This approach aims to estimate the impact factor of the QoS of each service involved in the composition. The shaft concept composition is proposed to characterize the dependency relationships between a service composition and SLA when analyzing the components of the impact factor services.

2.2 SLA Management Including Violation Detection SLA Violations

The works [6, 11] addressee the problem of violation detection and adaptation of SLA contracts between several layers. For example, [6] proposed a methodology to create, monitor and adapt the inter-layer SLA contracts. The SLA model includes parameters proposed KPI (key performance indicators) and KGI (indicators key objectives) and metrics infrastructure. In a second work, [11] proposed a solution to avoid SLA violations by applying the inter-layer techniques. The proposed approach uses three layers for the prediction of SLA violations. The identification of adaptation needs is based on the prediction of QoS, which uses assumptions about the characteristics of the execution context.

Reference [3] proposed a methodology and a tool for learning adaptative strategies of web services to automatically select the optimal repair actions. The proposed methodology is able to learn its repair knowledge incrementally once a detected fault is previously repaired. It is therefore possible to ensure adaptability, at runtime, according current characteristics of the faults and in the history of repair actions performed previously. So this methodology includes the ability to learn autonomously, the two model parameters, which are useful to determine the type of fault and repair strategies that are successful and suitable for particular fault.

Reference [13] introduced a Cloud Application and SLA monitoring architecture, and proposed two methods for determining the frequency that applications needs to be monitored, they identified the challenges facing application provisioning and SLA enforcement especially in multi-tenant Cloud services. [5] discuss autonomous QoS management using a proxy-like approach, SLAs can be exploited to define certain QoS parameters that a service has to maintain during its interaction with a specific customer. However, their approach is limited to Web services and does not consider Cloud.

2.3 SLA Violation Prediction

Reference [8] introduced a general approach to prediction of SLA violations for composite services, taking into account both QoS and process instance data, and using estimates to approximate not yet available data, in contrary to our work their work is to alert the provider to potential quality problems, it clearly cannot directly help preventing them or suggesting proactive actions, moreover in their approach the system introduces the notions of checkpoints (points in the execution of the composition where prediction can be done).

Reference [4] introduced some concepts, such as the basic idea of using prediction models based on machine learning techniques, the trade-off between early prediction and prediction accuracy. However, the authors do not discuss important issues such as the integration of instance and QoS data, or strategies for updating prediction models. Additionally, this work does not take estimates into account, and relatively little technical information about their implementation is publicly available.

Reference [11] proposed a new methodology that predicts any deviation in SLA thresholds, for a response time metric in a SOA-based system, using stochastic models. They created an analytic model that implements different SOA features and then accompanied it with a failure model that is able to figure out if the result from the analytic model fails to comply with a predefined SLA. The model in this work cannot predict any QoS metric in an SLA; rather, it is mainly used for predicting response time compliance according to changes in workload.

Discussion: The main limitation of the aforementioned approaches is that they only consider certain services regions (execution points) of the composition and do not consider all process tasks. Most of the works targeting SLA violations prediction is addressing grid environments or service-oriented infrastructure that differs from cloud infrastructure, therefore the applicability of these approaches on CSBA is limited.

To the best of our knowledge, the approach proposed in this paper is the first that uses data mining techniques to learn from SLA violations in order to correlate between multiple violated SLAs. It recommends actions for automatically reconfiguring the CSBA to avoid the predicted violation before its occurrence.

2.4 Discussion

Although the adaptation approaches presented above address the problem of SLA violations and service execution failures. In particular, these approaches aim at establishing a connection only with new services that comply with functional and non-functional requirements of adaptation and ignore generally similar services that do not meet the requirements for interoperability and interaction (e.g., services with incompatible communication protocols or different interfaces). This may neglect candidate services with best characteristics and affects, as a result. The main limitation of these approaches is that they only consider certain services regions (execution points) of the composition and do not consider all process tasks. Most of the works targeting SLA violations prediction is considered on grid environments or service-oriented infrastructure that differs from cloud infrastructure, there for the applicability of these approaches on CSBA is limited.

To the best of our knowledge, the approach proposed in this paper is the first that uses data mining techniques to learn from SLA violations in order to correlate between violated SLAs and recommends actions for configuring the CSBA to avoid the predicted violation.

3 Motivation Scenario

To illustrate the ideas presented in this paper we will use a simple travel agency scenario, which is composed of three services: (i) Reserve Flight, (ii) Payment Service, and (iii) Reserve Hotel Service.

We summarize in Table 1 the Service Level Objectives (SLOs) for our CSBA. It corresponds to the SLA specifications of all the QoS constraints for the whole application. Each cloud service provider involved in the Travel-Agency-CSBA configuration promises to satisfy the stipulated Qualities of Services (QoS) through a Service Level Agreement (SLA) with his consumer.

Table 1. QoS constraints of SLA relevant to the scenario.

SLA parameter	Value
Response time	≤ 20 s
Storage (St)	≥ 2 GB
CPU number	≥ 3

Each of these services is made up of a mixture of rented Cloud Services (SaaS, PaaS and IaaS). This work aims at locating the failure event and determine adaptation actions in order to prevent its spread at the others layers, as soon as possible. The central focus of Travel-Agency CSBA scenario is the SLA between the client Travel-Agency and the cloud service providers offering the Reserve Fly, the Reserve hotel and the payment cloud services. Upon receiving a request placed by the customer 'C', a process instance is created (Fig. 1).

Fig. 1. A simple process describing the travel agency scenario.

For this instance, the process execution starts with the activity 'Reserve Fly' (S1). Then, the SLA monitor is called and the software services are invoked if they are available. The maximum duration of the Response time of the whole process should be less than 20 s, a violation of the respective SLA occurs, as it can be seen in Fig. 2.

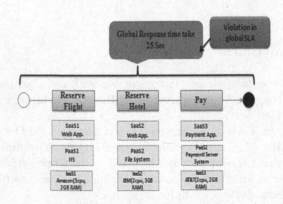

Fig. 2. Travel agency SLA violations.

4 Background

Numerous tasks are reached by data mining. They can be classified in descriptive tasks which are the association rules in our case and predictive tasks which is here the decision Tree used for the prediction from execution logs.

In this section, we first introduce the decision Tree, a commonly used data mining technique in order to build predictions models from execution logs to be able to predict potential violation and react to it proactively. This is followed by an overview of association rules.

4.1 CSBA Layers

CSBA is a set of different services, each one is provided by different suppliers. There are three-service models in the cloud, Software as a Service (SaaS), Platform as a Service (PaaS) and Infrastructure as a Service (IaaS). An application built or composed of a set of these services is called Cloud Service Based Application (CSBA) which is a composition of list of rented service.

In CSBA, a client can rent a cloud service or a set of cloud services from single or multi providers to make up an application. The provisioning of these services relies on Service Level Agreement (SLA), which is a contract between the client and the service provider including a set of non-functional requirements called Quality of Service (QoS) [2].

4.2 Decision Tree

Objective: classification of people or things into groups by recognizing patterns. The user or the expert has always a tendency to structure or classify data into groups of similar objects called classes. For this purpose, he uses distance measurements in order to evaluate the belonging of an object to a class. The most known classification methods are nearest neighbor and decision trees.

For detecting the violation we have used the decision tree learning which uses a decision tree as a model to predict the value of a target variable based on input variables (features), it generate the tree using the tool WEKA (https://weka.wikispaces.com/). The WEKA tool uses the j48 algorithm to generate the decision tree. Additionally, based on monitoring data of historical process instances, we used decision tree learning order to learn the dependencies and to construct classification models which are then used to predict the value of an instance while it is still running.

If a violation is predicted, we identify adaptation requirements and adaptation strategies are extracted from the decision tree in order to prevent the violation. We have decided to use decision trees because of their following advantages in our context: (i) they constitute a white box model as they show explicitly the relationships between explanatory attribute value ranges and categorical target attributes (i.e., KPI classes).

Thus they are easy to understand and interpret for people and enable human support in the learning and adaptation phases. (ii) They support both explanation and prediction. (iii) In particular, they support extraction of adaptation requirements.

A decision tree algorithm works by splitting the instance set into subsets by selecting an explanatory attribute (new node in the tree) and corresponding splitting predicates on the values of that attribute (branches). This process is then repeated on each derived subset in a recursive manner until all instances of the subset at a node have the same value of the target attribute or when splitting does not improve the prediction accuracy.

In order to find out these dependencies, we use classification learning known from machine learning and data mining. In a classification problem, a dataset is given consisting of a set of examples (a.k.a. instances) described in terms of a set of explanatory attributes (a.k.a. predictive variables) and a categorical target attribute. The explanatory attributes may be partly categorical and partly numerical. By using a learning algorithm, based on the example dataset (a.k.a. training set) a classification model is learned (a.k.a. supervised learning), whose purpose is to identify recurring relationships among the explanatory variables which describe the examples belonging to the same class of the target attribute. The so created classification model can be used to explain the dependencies in past instances but in particular also to predict the class of (future) instances for which only the values of the explanatory attributes are known. Each leaf of the decision tree is associated with a learning rule classified by class support (class support) and a probability distribution (class probability). Class support represents the number of examples in the training set, that follow the path from the root to the leaf and that are correctly classified; class probability (prob) is the percentage of examples correctly classified with respect to all the examples following that specific path, as shown in the formula as shown:

$$prob = \frac{\neq (corr\_class\_leaf\_examples)}{\neq (corr\_class\_leaf\_examples + incorr\_class\_leaf\_exapmles)}$$

A rule is always given with two measures (the support (S) and the confidence (C)) describing its strength and its interestingness. The support (Eq. (1)) is the percentage of transactions that satisfy A and B among all the transactions of the transactions base. The confidence (Eq. (2)) is the percentage of transactions that verify the consequent of a rule among those that satisfy the antecedent (premise) data.

$$Support\, A \rightarrow P(A \cup B) \tag{1}$$

$$Confidence\, A \rightarrow B = P(B/A) = support \frac{A \cup B}{support}(A) \tag{2}$$

4.3   Association Rules

Objective: associating what events are likely to occur together.

Association models aim to discover relationships or correlations in a set of items. Association rules is a data mining technique intending to find associated values in a given dataset and serving decision making. It has the following form: A-> B (S %), (C %). This rule means that tuples satisfying conditions in **A** also satisfy conditions in **B**.

A rule is always given with two measures (the support (S) and the confidence (C)) describing its strength and its interestingness. The support (Eq. (1)) is the percentage of transactions that satisfy A and B among all the transactions of the transactions base. The confidence (Eq. (2)) is the percentage of transactions that verify the consequent of a rule among those that satisfy the antecedent (premise) data. A rule is always given with two measures (the support (S) and the confidence (C)) describing its strength and its interestingness. The support (Eq. (1)) is the percentage of transactions that satisfy A and B among all the transactions of the transactions base. The confidence (Eq. (2)) is the percentage of transactions that verify the consequent of a rule among those that satisfy the antecedent (premise) data.

The extraction of association rules is the generation of the interesting rules with support and confidence greater than minimum thresholds of support and confidence. The process of extracting association rules involves two distinct phases. Firstly, the items having a support level that exceeds a certain threshold are segregated. Secondly, the most frequent items are combined in order to generate associations.

5   A Framework for Proactive Learning from SLA Violations

Figure 3 portrays a high-level architectural view of the proposed cross-layer self-adaptation framework. The framework is based on MAPE-K adaptation loop [5], introduced by IBM as an efficient and novel approach for self-adaptation in autonomic computing. As shown in Fig. 3, the MAPE-K adaptation loop comprises of five main components corresponding to its acronym, which will be discussed in the following. The main focus of this paper is on the Analysis component of the MAPE-K loop, where a proactive learning approach is proposed (cf. Sect. 5) to predict potential QoS violations based on historical execution logs and react accordingly to avoid/prevent the predicted violation. Therefore, the upcoming sections will focus on presenting the details of these two main components.

Knowledge: As services of CSBA CSBS rely on third party cloud service providers, a SLA involves a 'consumer' and one to many 'providers'. For example, a client "X" buys an infrastructure services from Amazon, a platform service from Google, and a software service from SalesForce. These providers must satisfy the obligations they promise to the client "X". A violation indicates either a cloud service consumer or a provider fails to satisfy one or more constraints contained in the agreement.

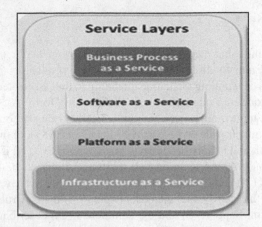

Fig. 3. Architecture of service-based business processes running on the cloud.

The SLA violation patterns are classified into QoS constraint violation patterns, security constraint violation patterns, privacy constraint violation patterns, and regulatory constraint violation patterns. The term 'constraint' refers to obligations related to QoS, privacy, regulatory, and security, which are stipulated and must be satisfied by the involved service providers. The knowledge component is responsible of storing and maintaining the SLAs specified upon a cloud service-based system (Fig. 4).

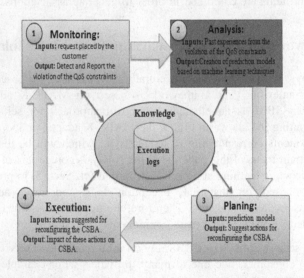

Fig. 4. High-level architectural view of the proposed proactive monitoring framework.

Monitoring: Monitoring of SLA compliance is of crucial importance to the proposed framework. Monitoring is intended to be in a near real-time fashion in order to take corrective actions before it is too late [14]. Our intention also is to predict possible SLA violation and avoid them before they occur. To tackle the monitoring task, we will depend on complex event processing (CEP) technology.

In traditional systems, data are static in the system while the queries used are changing. For example in traditional database systems, the data are stored in tables and users can write different queries that access those tables to process data and get the results. However, when using complex event processing (CEP) technology, the roles of data and queries are reversed; where the queries will be static and data or events will be dynamic based on the input event streams from different sources. Events are heterogeneous and are generated from different sources. Events can be emitted from sources like sensors, software applications, databases, etc. Events generated from such sources are called *raw* events. Raw events have more value when they are combined together.

There are several CEP platforms. For the work proposed here, we will use ESPER (http://www.espertech.com/esper/index_redirected.php), an open source CEP platform. For each of the event data sources, an event type is registered within ESPER. To process these events, ESPER defines the query language Event Processing Language (EPL) which is an SQL-like language with select/from/where/order by/group by clauses in addition to built-in temporal operators to reason about the sequencing of events.

For more details about the monitoring component, we refer the reader to [14].

Analysis: Based on the monitoring results, the analysis component is responsible for performing complex data analysis and reasoning, by the continuous interaction with the knowledge component.

Particularly, the analysis component carries out the processing, the correlation, and the analysis of events stemming from the history of the instances and predicts potential violations for the running instance. Based on that prediction, the planning component will take necessary actions. For this to happen, we will rely on a set of historical traces on the cloud service-based system that will provide insights on how the process was executed in the past. Based on information extracted from the historical traces, predictions and recommendations are provided for running instances. Such predictions and recommendation will be relying on data mining techniques, more specifically, decision trees.

Planning: Once a violation is predicted, the planning component takes the hand over. To deal with such a predicted violation, the planning component starts by searching for alternative solutions in order to avoid the occurrence of the predicted violation. The planning component will attempt to adapt the smallest possible set of services without directly targeting a re-engineering process of the whole system.

Execution: constructed in the previous planning component, the execution component is responsible for selecting the adaptation plan (in the form of recommendations as passed from the planning component) with the highest probability of success.

The plan will involve recommendations of adaptation actions for all directly and indirectly affected layers in the cloud stack. That's the propagation of the adaptation actions from top to down. The selected actions will be executed to avoid anticipated QoS violations, and then the applied actions will be evaluated to check their impact. This involves evaluating whether the applied adaptation plan was actually an efficient plan and prevented anticipated QoS violations, and if it resulted in any other negative impacts. This evaluation will iteratively enhance the quality of the predication models by better learning.

6 The Proposed Learning Approach in Cloud Environments

In this section, we present the details of the proposed learning approach, which combines different existing techniques ranging from learning approaches to decision tree learning, to provide predictions, at runtime, about the achievement of business goals in a Business Process (BP) execution trace. In the following sections, we provide an overview of the approach. Section 7 discusses the implementation of the proposed approach as a Proof-Of-Concept (POC) of the realization of the proposed approach.

6.1 The Proposed Learning Approach

The process of learning from SLA violation and making dependencies precedence between different SLAs violations in CSBAs have been identified as major research challenges in Cloud environments.

SLA does not contain information about the dynamicity of the system. In other words, it is independent of the context of the business process, and it contains information about the service behavior or quality provided by the service which we aim to exploit. SLAs are not mathematically defined. That means that the semantics of the SLA elements and metrics are defined in natural languages, which makes it harder to understand the semantic of QoS, and it is usually dependent on the client and provider contract. Thus, being precise and formal about SLA semantic is necessary. SLAs violations come from different kind of failures, determining the appropriate type of actions to be taken when predicting an SLA violation is equally important.

First Phase: Learning phase: It is a continuous evolving process. The association rules extraction is explained as follows:

> **Given**: a set of historical BP event logs of SLA violations.
> **Find:** Association rules.

Our method of Association Rules (AR) determination goes through three steps:

- The first step is to discover frequent itemsets.
- The second step is dedicated to find out ARs on the basis of the first step outputs.
- The third step is the refinement of the extracted ARs.

These steps are depicted in the diagram shown in Fig. 5. Two steps are combined in order to carry out ARs. They are respectively the generation of frequent itemsets and the extraction of association rules. The frequent itemsets are extracted as defined in Algorithm 1.

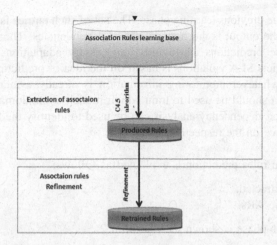

Fig. 5. Proposed SLA violation learning approach: Architecture overview.

Algorithm 1: Extraction and refinement of AR
    Inputs: Execution logs
    Outputs: ARs
        Reject-rule: = Boolean
    Begin
        Reject-rule: =False
    Do
    (1)   Find all the frequent itemsets by C4.5 execution (execution logs)
    (2)   Find all the possible ARs by C4.5 execution
    (3)   Save the obtained ARs
    End
    (4)   Save the rules set containing only correct rules
    End

The input corresponds to the set of historical data generated by the previous SLA violation of CSBA. The AR contains QoS property of SLA in the antecedent and the violated SLA as consequent of the rule. The proposed formula of the AR is given in Eq. 4.1:

$$IF\ event\ AND\ condition\ THEN\ (perform)action$$

The next phase is the prediction phase (described below), in that phase some historical executions of CSBA are necessary to bootstrap the prediction. The concrete amount of instances that are necessary, depend both on the expected quality of prediction and on the size and complexity of the service composition.

Second Phase: SLA Prediction:
The objective of this phase is to (i) predict potential violation, and to (ii) construct/build the best configuration as the recommendation from what have been detected and

learned based on the previous learning phase. The set of such entries is presented in the Decision Tree of the output is the frequent set of dependencies. The Prediction algorithm gives precise predictions and avoids unnecessary adaptations. Generally, the approach that predicts SLA violations is based on the idea of predicting concrete SLO values based on whatever monitoring information is already available. In order to identify which data should be used to train which model, some domain knowledge is necessary. However, dependency analysis can be used to identify the factors that have the biggest influence on the respective SLOs.

The association rules prediction is explained as follows:

Given: ARs extracted.
Find: Predictive ARs.

The process of this phase is shown in Algorithm 2:

Algorithm 2: Prediction of predictive AR
Inputs: ARs
Outputs: Predicted ARs
  Reject-rule: = Boolean
Begin
  Reject-rule: =False
Do
 (5) Compare all the obtained ARs in different time intervals
 (6) Suggest predicted Rules
End
 (7) Save the predicted rules set containing only correct rules
End

6.2   Exemplifying on the Running Scenario

In this section, the prediction of violations is applied on the Travel agency CSBA scenario (described in Sect. 3). The rules below are an outcome of the Decision Tree mining the data sets sent as inputs from an Excel file of the Travel-Agency scenarios is described in (Sect. 3).

Example of Rule:
IF sum RT2 + RT3 $\leq$ 10 = no **And** RT3 $\leq$ 6**And**RT2 $\leq$ 11 **And** RT1 $\geq$ 5 **Then** Violation = Viol(P1, P6), the configuration will be: {Response time $\leq$ **20, 1 CPU, 2 GB RAM**}.

As presented in Fig. 2, IaaS is an infrastructure as a service that is a rented service from amazon service provider with 1 CPU and 2 GB Ram and it promises PaaS is a platform as a service that presents a rented IIS (internet information services), with 1 CPU and 2 GB Ram and it promises to satisfy response time < 20 s. The Global SLA service promises to satisfy response time < 20 s, giving 10 s as a response time for each of 'Reserve Hotel' and 5 s for 'Reserve Flight', and 5 s for 'Payment Application' service. The response time for example could be violated when any of these application services has an internal violation in response time at the level of SaaS, PaaS or IaaS. In order to avoid such situation, the SLA manager acts proactively based on history of completed activities. At the same time, another monitoring component detects that there is an I/O failure at the SaaS layer as S1 has produced a wrong output.

A specific rule is triggered that derives the best strategy which consists of executing another instance of *the web service* on a more powerful server with a better memory and CPU allocation. (Amazon (3 CPU, 3 GB RAM)). Assume that a Monitoring Component, running at the server where *the web service* is executed, detects that the available main memory is not sufficient (IaaS layer) for *the web service*. At this stage, proactive actions are suggestions to be taken based on some predefined suitable actions for each type of violation the system may encounter. In Table 2, we can see that each violation has a violation type that could be availability or security depending on the type of violation. The actions taken are of two types, namely surgery and elastic actions.

Table 2. Proactive actions.

Violation ID	Violation type	Action type	Action
Viol 1	Violation Qos (Response Time, KPI)	Reparation	Raise the RAM capacity to 2 GB
Viol 2	Availability	Substitution	Change violated service
Viol 3	Security	Reparation	Adapt the service to security police

6.3   Cross Layer Rules

The DM technique uses event rules to reason about the events that occur at a specific service layer. Based on the events received *RemoteFault* caused by (*AuthorizationDenied* fault) at a higher layer, a specific rule is fired which derives that the best strategy is to execute another instance of the failed at a more powerful server and with a better memory and CPU allocation. But if the new service requires different protocol/resources, this may lead to adaptation in the PaaS and IaaS layer as well.

Indeed, monitoring events at each layer in an isolated manner is not enough as events may depend on each other. For this purpose, aligning events and defining relationships between them allows a correct diagnosis and avoids the execution of a conflicting adaptation. Events are derived based on rules in the form: *Antecedent → Consequent*.

Example: a composition fault caused by a missing or an extra role may trigger the *ReplayScope* action. Since reexecuting the process activity from its beginning with a missing *partnerLink* may cause the same problem, a SWRL rule may recommend adaptation at SaaS layer by selecting a substituting service for the corresponding role, to achieve reexecuting all of the activities inside the faulty scope.

hasActivity (ws, act) ∧ causesFault (act, ExtraRole) ∨ causesFault(act, Missing Role) → isManagedBy(ws, Substitute) ∧ isManagedBy(act, ReplayScope) → causesFault (ws, ProcessFault).

Figure 5, simulated or real data are presented as a form of Excel or text file (delaminated file) to the Decision Tree. These data contains entries for process instances with set of violation at different services, below is an example of an entry:

Set of such entries are presented to the Decision Tree, where the output is a **frequent set** of dependencies (in other word the most violating services that happens together). But the output needs to be filtered since the time constraint of the violating services is important.

For example, the rules below are an outcome of the Decision Tree mining the data sets inputs from the Excel file:

Example of Rule:
IF somme RT2 + RT3 ≤ 10 = no **And** RT3 ≤ 6 **And** RT2 ≤ 11 **And** RT1 ≥ 5 **Then** Violation = Viol (P1, P6), the configuration will be: {Response time ≤ **20, 1 CPU, 2 GB RAM}.**

We extracted the best configuration available in the database to proactive the global violation before it come.

The configuration manager is responsible for configuring the CSBA. The proactive actions suggested by the proactive engine are mapped into the configuration manager to take the action. The action taken is recorded in the knowledge.

Figure 7 shows the form when "suggest action" is pressed. Proactive actions are suggested based on an excel file containing the suitable action corresponding to the predicted violation.

7   Implementation

To demonstrate the applicability and feasibility of our approach, we developed a prototype[2] using JAVA. We trigger the execution of 100 process instances using a test client. For each of these instances we select the concrete supplier service and shipper service randomly in order to ensure that history data used for learning contains metrics data on each of these services. During process instance execution, the previously specified metrics are measured and saved in the knowledge database. Then, for each checkpoint a decision tree is learned using the J48 algorithm.

For the implementation of the Predictor, we rely on the WeKa J48 implementation of the C4.5 algorithm, which takes as input a '.arff' file and builds a decision tree as shown in Fig. 6: Text files of real data. The '.arff' file contains a list of typed variables (including the target variable) and, for each trace prefix (e.g., for each data snapshot), the corresponding values are also maintained. The resulting Decision Tree is then analyzed to generate predictions and recommendations as shown in Fig. 6.

---

[2] A video demonstration is available at: https://www.youtube.com/watch?v=oDEFYGBPdH0.

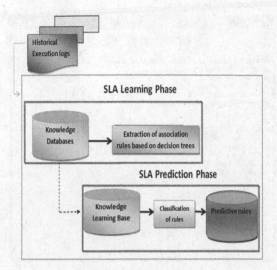

Fig. 6. Learning phase.

The configuration manager is responsible for configuring the CSBA. The proactive actions suggested by the proactive engine are mapped into the configuration manager to take the action. The action taken is then stored in the knowledgebase. The algorithm searches in the database (as shown in (Fig. 7) for suitable actions that can be used. Below in Fig. 8 is a part of the Excel file that we used for our decision. For example, as shown in the Table 2 since the violation is Response Time, then the suitable action is to add 1 CPU to the violating service.

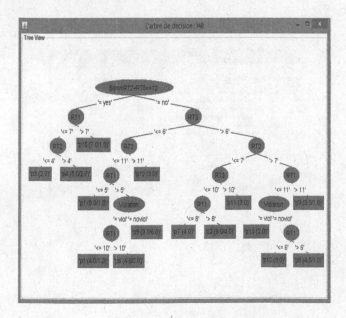

Fig. 7. Decision tree.

Fig. 8. Text files of real data.

We evaluated experimentally the model's performance and accuracy. The experiments were performed on a machine with quad-core CPU 2.6 GHz, 8 GB RAM and Mac OS X operating system. This experiment evaluates the algorithm's raw relevant and absolute accuracy. The static metrics precision and recall is measured while fluctuating the interval size from 4 to 20 events. Figure 8 shows that relevant precision is 1 for small intervals and falls while increasing the interval size, while absolute precision fluctuates similarly at lower levels (as more irrelevant sub-patterns are discovered) (Fig. 9).

Fig. 9. A screenshot of ARs extraction.

8 Conclusion and Future Work

In this paper we studied adaptation technologies particularly the detection and proactive technologies for CSBA. We discussed the outcomes of our analysis. In particular, we discussed the limitations of different approaches of cross-layer service adaptations.

The major limitation we found is the lack of coordination between adaptation activities that may lead to conflicts or incompatibilities. According to our study, the current solutions do not consider the fact that adaptation in a layer may affect adversely the other layers of service based systems. According to our study, current cross-layer adaptation approaches lack efficient coordination which leads to conflict and incompatibilities. We believe that these problem must be addressed for an efficient cross-layer service adaptation. We presented the results of a brief study on adaptive agent based systems. We found in our study that the agent based adaptive systems have some advanced, features such as context-awareness, self-adaptation, etc. The adaptive SBAs can be benefited by these features especially, the service based adaptive systems can be more intelligent and autonomous.

Additionally, based on our understanding we presented some research directions in the area of cross layer service adaptations. We strongly believe that the research in this area should focus on context awareness, self-adaptation, and performance etc. to develop highly high-performance solutions. We also presented a proposal of a solution which are currently working on.

There are a few limitations of our study. Firstly, this is merely a literature review. However, the state of the art could be better reviewed or understood by benchmarking the existing solutions. A comparison of adaption technologies in different contexts can be done by following a set of rigorous protocols. This paper is missing such an comparison. In our future work, we plan to conduct an empirical study with the current cross-layer adaptation technologies. Also, we plan to conduct a study by covering more contexts (Fig. 10).

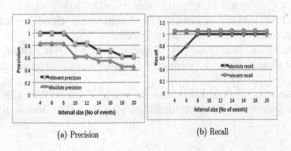

(a) Precision   (b) Recall

Fig. 10. Evaluation results.

References

1. Boniface, M., Phillips, S.C., Sanchez-Macian, A., Surridge, M.: Dynamic service provisioning using GRIA SLAs. In: Nitto, E., Ripeanu, M. (eds.) ICSOC 2007. LNCS, vol. 4907, pp. 56–67. Springer, Heidelberg (2009). doi:10.1007/978-3-540-93851-4_7
2. Brandic, I.: Towards self-manageable cloud services. In: 33rd Annual IEEE COMPSAC 2009 (2009)
3. Bodenstaff, L., Wombacher, A., Reichert, M., Jaeger, C.: Analyzing impact factors on composite services. In: SCC 2009, pp. 218–226. IEEE SCC (2009)
4. Breiman, L., Friedman, J.H., Olshen, R.A., Stone, C.J.: Classification and Regression Trees. Wadsworth & Brooks, Montery (1984). ISBN 0-412-04841-8, 358 pages
5. Chelghoum, N.: Fouille de données spatiales, Un problème de fouille de donnéesmulti-tables. Thèse de doctorat présentée et soutenue publiquement à l'université deVersailles Saint-Quentin-en-Yvelines U.F.R de sciences Par Nadjim CHELGHOUM le16 décembre 2004 (2004)
6. Fugini, M., Siadat, H.: SLA Contract for Cross-Layer Monitoring and Adaptation. In: Rinderle-Ma, S., Sadiq, S., Leymann, F. (eds.) BPM 2009. LNBIP, vol. 43, pp. 412–423. Springer, Heidelberg (2010). doi:10.1007/978-3-642-12186-9_39
7. Han, J., Kamber, M., Pei, J.: Data Mining: Concepts and Techniques, 3rd edn. Morgan Kaufmann, USA (2011)
8. Leitner, P., Michlmayr, A., Rosenberg, F., Dustdar, S.: Monitoring, prediction and prevention of SLA violations in composite services. In: ICWS 2010, pp. 369–376. IEEE SCC (2010)
9. Peter, M., Timoth, G.: The NIST definition of cloud computing (2011)
10. Schmieders, E., Micsik, A., Oriol, M., Mahbub, K., Kazhamiakin, R.: Combining SLA prediction and cross layer adaptation for preventing SLA violations. In: Proceedings of the 2nd Workshop on Software Services: Cloud Computing and Applications based on Software Services, Timisoara, Romania, June 2011
11. Tao, C., Rami, B., Xin, Y., Online QoS modeling in the cloud: a hybrid and adaptive multi-learners approach. In: The 7th IEEE/ACM UCC, London, UK (2014)
12. Vaitheki, K., Urmela, S.: A SLA violation reduction technique in Cloud by resource rescheduling algorithm (RRA). Int. J. Comput. Appl. Eng. Technol. 3(3), 217–224 (2014)
13. Emeakaroha, V.C., Netto, M.A.S., Brandic, I., De Rose, C.A.F.: Application-level monitoring and SLA violation detection for multi-tenant cloud services. In: Emerging Research in Cloud Distributed Computing Systems (2015)
14. Yehia, T., Rafiqul, H., Dinh Khoa, N., Béatrice, F.: PAEAN4CLOUD: a framework for monitoring and managing the sla violation of cloud service-based applications. In: CLOSER 2014, pp. 361–371 (2014)
15. Zaki, M.: Generating non-redundant association rules. In: ACM SIGKDD International Conference on Knowledge Discovery and Data Mining, KDD, Boston, pp. 34–43 (2000)

Map Reduce Autoscaling over the Cloud with Process Mining Monitoring

Federico Chesani, Anna Ciampolini, Daniela Loreti$^{(\boxtimes)}$, and Paola Mello

DISI - Department of Computer Science and Engineering, Università di Bologna,
Viale del Risorgimento 2, Bologna, Italy
`{federico.chesani,anna.ciampolini,daniela.loreti,paola.mello}@unibo.it`

Abstract. Over the last years, the traditional pressing need for fast and reliable processing solutions has been further exacerbated by the increase of data volumes – produced by mobile devices, sensors and almost ubiquitous internet availability. These big data must be analyzed to extract further knowledge.

Distributed programming models, such as Map Reduce, are providing a technical answer to this challenge. Furthermore, when relaying on cloud infrastructures, Map Reduce platforms can easily be runtime provided with additional computing nodes (e.g., the system administrator can scale the infrastructure to face temporal deadlines). Nevertheless, the execution of distributed programming models on the cloud still lacks automated mechanisms to guarantee the Quality of Service (i.e., autonomous scale-up/-down behavior).

In this paper, we focus on the steps of monitoring Map Reduce applications (to detect situations where the temporal deadline will be exceeded) and performing recovery actions on the cluster (by automatically providing additional resources to boost the computation). To this end, we exploit some techniques and tools developed in the research field of Business Process Management: in particular, we focus on declarative languages and tools for monitoring the execution of business process. We introduce a distributed architecture where a logic-based monitor is able to detect possible delays, and trigger recovery actions such as the dynamic provisioning of a congruent number of resources.

Keywords: Business Process Management · Map Reduce · Cloud computing · Autonomic system

1 Introduction

The exponential increase in the use of mobile devices, the wide-spread employment of sensors across various domains and, in general, the trending evolution towards an Internet of Things, is constantly creating ever growing volumes of data that must be processed to extract knowledge. This pressing need for fast analysis of large amounts of data calls the attention of the research community and fosters new challenges in the big data research area [7]. Since data-intensive

© Springer International Publishing AG 2017
M. Helfert et al. (Eds.): CLOSER 2016, CCIS 740, pp. 109–130, 2017.
DOI: 10.1007/978-3-319-62594-2_6

applications are usually costly in terms of CPU and memory utilization, a lot of work has been done to simplify the distribution of computational load among several (physical or virtual) nodes and take advantage of parallelism.

Map Reduce programming model [9] has gained significant attraction for this purpose. The programs implemented according to this model can be automatically split into smaller tasks, parallelized and easily executed on a distributed infrastructure. Furthermore, data-intensive applications requires a high degree of elasticity in resource provisioning, especially if we deal with deadline constrained applications. For this reason, most of the current platforms for Map Reduce and distributed computation in general [2,3,10] allow to scale the infrastructure at execution time.

If we assume that the performance of the overall computing architecture is stable and a minimum Quality of Service (QoS) is guaranteed, Map Reduce parallelization model makes relatively simple to estimate a job execution time by on-line checking the execution time of each task in which the application has been split – as suggested in [16]. This estimation can be compared to the deadline and used to predict the need for scaling the architecture.

Nevertheless, the initial assumptions are not always satisfied and the execution time can differ from what is expected (a) architectural factors (e.g., performance variability of the physical/virtual computation nodes or the fluctuation of the network bandwidth between the nodes), or (b) domain-specific factors (e.g., a task is slowed down due to the input data content or location). This unpredictable behaviour could be run-time corrected if the execution relays on an elastic set of computational resources as that provided by cloud computing systems.

By offering "the illusion of infinite computing resources available on demand" [4], cloud computing is the ideal enabler for tasks characterized by a large and variable need for computational power. Cloud computing is indeed knowing a wide success in a plethora of different applicative domains, thanks to the maturity of current standards and implementations.

Usually, the cloud is the preferred choice for applications that must comply to a set of contract terms and functional and non-functional requirements specified by a Service Level Agreement (SLA). However, the complexity of the resulting overall system, as well as the dynamism and flexibility of the involved processes, often require an on-line operational support checking compliance. Such monitor should detect when the overall system deviates from the expected behaviour, and raise an alert notification immediately, possibly suggesting/executing specific recovery actions. This run-time monitoring/verification aspect – i.e., the capability of determining during the execution if the system exhibits some particular behaviour, possibly compliant with the process model we have in mind – is still matter of an intense research effort in the emergent Process Mining area. As pointed out in [25], applying Process Mining techniques in such an online setting creates additional challenges in terms of computing power and data quality.

Starting point for Process Mining is an event log. We assume that in the architecture going to be monitored it is possible to sequentially record events.

Each event captures information about the execution of the activities/steps in the process/task, usually distinguishing between the initiation and the termination of the activity execution. Typical data captured by event logs comprises the identifier of the activity, time stamps, and process instance identifiers. In case of distributed computation, also identifiers of the executing resources/nodes, and other network-related data are of interest.

While in an cloud architecture several tools exist for performing a generic, low-level monitoring task [1,19], we also advocate the use of an application-/process-oriented monitoring tool in the context of Process Mining in order to run-time check the conformance of the overall system w.r.t. the expected behaviour. Essentially, the goal of this work is to apply well-known Process Mining techniques to the monitoring of complex distributed applications, such as Map Reduce, in a cloud environment. In turn, monitoring outcomes and alarms will open the possibility of run-time recovery from undesired situations.

Map Reduce applications typically operate in dynamic, complex and interconnected environments demanding high flexibility: thus, providing a detailed and complete description of any possible behaviour of the overall system at design time can be quite difficult. A different approach focuses instead on the elicitation of the (possibly minimal) set of the desired behavioural properties that the overall system should exhibit during its execution. In the Process Mining research field, the graphic notation Declare [21] has been proposed with the aim of defining these properties in a declarative manner, in terms of constraints. MOBUCON EC (MOnitoring BUsiness CONstraints with Event Calculus [18]) is a verification framework able to dynamically monitor streams of events characterizing the process executions (i.e., running cases). The observed events are compared to the system's expected behaviour, which is specified in Declare notation. With respect to the original Declare notation, MOBUCON EC supports also time-aware and "data-aware" constraints [17]: e.g., it is possible to specify deadlines and pose constraints on the data that characterize the execution of an activity.

This works investigates the adoption a monitoring system (specifically, MOBUCON EC), for dynamically monitor each node of a distributed infrastructure for data processing running on a cloud environment. The resulting information is used for taking scaling decisions and dynamically recovering from critical situations with a best effort approach (by means of an underlying previously implemented infrastructure layer). In particular, the adoption of MOBUCON EC allows us to specify the system properties including time and data constraints: in a Map Reduce and cloud-based scenario, data and time perspectives (concerning with the timing and frequency of events) becomes fundamental to check deadlines, discover bottlenecks, measure service levels and monitor the utilization of resources.

The paper is organized as follows. In Sect. 2, after introducing the applicative scenario based on the Map Reduce model, we present the overall architecture, describing the main components and their relationships. A special emphasis is given to the monitoring block, based on declarative constraints. Section 3 presents the use case scenario, based on the execution of a well-known benchmark over the popular Map Reduce platform Hadoop. This section also includes the experimental results demonstrating the potential of our approach. Related work and Conclusions follow in Sects. 4 and 5, respectively.

2   System Context and Specifications

In this section, we propose a framework architecture to online detect user-defined critical situations in a Map Reduce environment (accordingly to high-level rules definable in a declarative language) and autonomously react by providing or removing resources.

2.1   Applicative Scenario

Map Reduce is a programming model able to simplify the complexity of parallelization. Following this approach, the input data-set is partitioned into an arbitrary number of parts, each exclusively processed by a different computing task, the *mapper*. Each *mapper* produces intermediate results (in the form of *key/value* pairs) that are collected and processed by other tasks, called *reducers*, in charge of calculating the final results by merging the *values* associated to the same *key*. The most important feature of Map Reduce is the possibility to automatically parallelize the programs implemented according to this model.

In this scenario, the estimation of the execution time can be crucial to check deadlines, detect bottlenecks, and to guarantee SLA. However, the prediction of the execution time for Map Reduce applications is not a trivial task. The duration of each *mapper* or *reducer* task can vary depending on different factors – e.g., the content of the analysed data block, the performance of the machine on which the task is executed, the location of the input data (local to the task or on another machine), the bandwidth between the physical nodes of the distributed infrastructure.

Since elasticity is so crucial in the data-intensive scenario, all the main platforms that implement the Map Reduce model [2,3,10] offer application scale-up/-down as a feature, making relatively simple to add (or remove) computational nodes to the distributed infrastructure while performing a data-intensive analysis. Currently, this operation must be requested by a human system administrator when the need for further resources is somehow identified. On the other hand, providing the infrastructure with the ability to autonomously detect (1) when a scaling-up/-down operation is necessary, and (2) how many resources are needed, is still an open problem.

The monitoring capability is a mandatory function for the achievement of such autonomy, since it can be viewed as the functional block that triggers the scaling-up/-down operations. However, properties like, e.g., mappers' execution times being under a threshold, can be highly dependent on the specific Map Reduce application, and can greatly vary from case to case. Thus, to be effective, the monitored property should be customizable on the basis of the specific needs of the Map Reduce application domains. Moreover, the property to be monitored might be user-dependent, i.e., it might be affected also by other aspects such as, e.g., user's needs, SLAs, and legal/business terms.

2.2 Framework Architecture

The main component of the proposed architecture is the Map Reduce Auto-scaling Engine. This application-level software consists of three main subcomponents (grey blocks in Fig. 1): the Monitoring block, the Recovery block, and the Platform Interface block. These elements interacts with the Map Reduce platform to detect and react to anomalous sequences of events in the execution flow.

The Monitoring component takes as input a high-level specification of the desired system properties: i.e., they describe the desired behaviour of a Map Reduce workflow in terms of the expected on-line sequence of events from the Map Reduce platform's log. Given these input data, the Monitoring component is able to rise alerts whenever the execution flow violates user-defined constraints.

The alarms raised by the Monitoring block are then evaluated by the Recovery component, which estimates how many computational nodes must be provisioned (or de-provisioned) to face the critical condition.

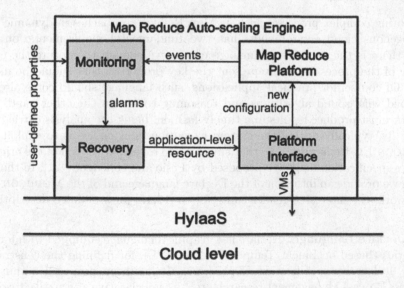

Fig. 1. Framework architecture.

Finally, the Platform Interface is in charge of translating the requests for new Map Reduce nodes into virtual machine (VM) provisioning requests to the infrastructure manager. The Platform Interface is also responsible for the installation of Map Reduce-specific software on the newly provided virtual machines (VMs). The output of this subcomponent is a new configuration of the distributed computing cluster with a different number of working nodes/resources.

As shown in Fig. 1, Map Reduce Auto-scaling Engine relays on a lower level component called Hybrid Infrastructure as a Service (HyIaaS) for the provisioning of VMs [15]. This layer encapsulates the cloud functionality and interacts with different infrastructures to realize a hybrid cloud: if the resources of a private (company-owned) on-premise cloud are no longer enough, HyIaaS redirects the scale-up request to an off-premise public cloud. Therefore, thanks to HyIaaS, the resulting cluster of VMs for Map Reduce computation can be composed by VMs physically deployed on different clouds. Further details about HyIaaS can be found in [15].

The hybrid nature of the resulting cluster can be very useful, especially if the on-premise cloud has limited capacity. However, it can also further exacerbate the problem of Map Reduce performance prediction. If part of the computing nodes is available through a higher latency, the execution time can be substantially afflicted by the allocation of the tasks and the amount of information they trade with each other.

In the following section, we provide an introduction to the chosen declarative approach to monitoring.

2.3   Monitoring the System Execution w.r.t. Declarative Constraints

Monitoring complex processes such as Map Reduce approaches in dynamic and heterogenous cloud environments has two fundamental requirements: on one hand, there is the need for a language expressive enough to capture the complexity of the process and to represent the key properties that should be monitored. Of course, for practical applications, such language should come already equipped with sound algorithms and reasoning tools. On the other hand, any monitor must produce results in a timely fashion, being the analysis carried out on the fly, typically during the system execution. Our choice is to exploit the MOBUCON EC framework [18], that supports an extended version of the original Declare specification language proposed by Pesic and colleagues [21]. In the following we provide an intuition of the Declare language and of the MOBUCON EC framework: the interested reader can refer to [17,18] for a detailed description.

The Declare Language. Declare is a graphical language equipped with formal semantics (based on Linear Temporal Logic, LTL), for defining the constraint-based, declarative specification of processes. It has been proposed within the Business Process Management research area, as a declarative alternative to existing procedural approaches that usually focus on the many aspects of control flows.

Declare focuses on the (minimal) set of rules/constraints that must be satisfied in order to correctly execute the process, thus accommodating *flexibility by design*, and providing a set of modelling abstractions that suitably mediate between control and flexibility. The main elements of the language are the *activities* (representing atomic units of work envisaged by the process), and the *constraints* that each activity (its execution) has w.r.t. other activities.

The constraints range from classical sequence patterns between activity executions, to loose relations, prohibitions and cardinality constraints. They are grouped into four families:

i. *existence/absence* constraints, used to constrain the number of times an activity must/can be executed;
ii. *choice* constraints, requiring the execution of some activities selecting them among a set of available alternatives;
iii. *relation* constraints, expecting the execution of some activity when some other activity has been executed;
iv. *negation* constraints, forbidding the execution of some activity when some other activity has been executed.

In Table 1 few Declare constraints are shown. Activities are depicted as boxes labelled with the activity name, while relations/constraints between activities are depicted as a solid line between activities. For example, the *Responded Existence* constraint specifies that upon the execution of activity a, it must be verified that b has been executed before a, or it will be executed afterwards. The constraint is "triggered" (i.e., must be verified) if and only if activity a is executed: this is indicated in the graphical notation by the full dot placed beside the activity a. In other words, if activity a is never executed, the constraint should not enforced/verified.

Table 1. Some Declare constraints.

Constraint	Description
0 1..* [a] [b]	**Absence** The target activity a cannot be executed **Existence** Activity b must be executed at least once
[a]•—[b]	**Responded Existence** When a is executed, b must be executed either before or afterwards
[a]•—▶[b]	**Response** Every time the source activity a is executed, the target activity b must be executed after a
[a]—▶•[b]	**Precedence** Every time the source activity b is executed, a must have been executed before
[a]•—‖▶[b]	**Negation response** Every time the source activity a is executed, b cannot be executed afterwards
[a]•—$(X..Y)$—▶[b]	**Response with metric temporal deadline** Every time the source activity a is executed, the target activity b must be executed after a, at least after X time units after a execution time stamp, and no later than Y time units from the execution time of a

Another interesting example shown in Table 1 is the *Response* constraint: graphically, it possible to notice the addition of a arrow head, meaning that the execution of activity b is expected *after* the execution of activity a. Table 1 shows also a *Response with metric temporal deadline*, whose semantics is the same of the *Response* constraint, but the execution of activity b is bound to happen within the time interval $[X..Y]$, where X and Y indicates a finite number of time units from a execution time.

Finally, we might notice that Declare specifications are "open": any activity can be freely executed, unless it is explicitly stated otherwise by means of a constraint. If openness guarantees flexibility, it calls also for explicit support for *prohibitions*, so as to allow the definition of desired behaviours, as well as of unwanted ones. As an example of prohibition, Table 1 shows the *Negation Response* constraint, that forbids the execution of an activity b after the execution of an activity a.

The MOBUCON EC Framework. MOBUCON EC[1] is a framework for the runtime monitoring of event-based systems with respect to Declare constraints. It tracks a running execution trace showing the instantiations and evolutions of the modelled constraints, reporting for each instance whether it is currently *satisfied* (i.e., the execution trace is compliant with it), *pending* (i.e., further events are needed to comply with it) or *violated* (i.e., the execution trace is non-compliant with it).

MOBUCON EC exploits an axiomatization of the Declare constraints based on the Event Calculus [14], a well-known logic framework for describing the state of a system, and the effects that happening events have on the state. Moreover, constraints in MOBUCON EC have been extended with metric temporal aspects and data-aware conditions. In MOBUCON EC the constraints are being monitored, and their state changes depending on the observed events.

MOBUCON EC comes equipped with a Graphical User Interface that allows human users to inspect the state of each constraint, as well as an overall "system health index" that reports a user-defined measure of the health of the system under monitoring. Beside outputting a system status through the GUI, MOBUCON EC can be easily customized with user-defined rules for defining actions/reaction upon the occurrence of specific states/situations. Such possibility, in conjunction with the fact that the whole framework is Java-based, allows to easily define and implement any envisaged reaction.

3   Use Case Scenario

The architecture shown in Fig. 1 has been implemented and analysed using a test bed framework. In particular, a simulation approach has been adopted to create specific situations, and to verify the run-time behaviour of the whole architecture. To this end, synthetic data has been generated, with the aim of stressing the Map Reduce implementation.

---

[1] Available   for   download   at   https://www.inf.unibz.it/~montali/tools.html#MOBUCON.

3.1   Testbed Architecture and Data

The Map Reduce model is implemented and supported by several platforms. In this work we opted for Apache Hadoop [2], one of the most used and popular frameworks for distributed computing. Hadoop is an open source implementation consisting of two components: Hadoop Distributed File System (HDFS) and Map Reduce Runtime. The input files for Map Reduce jobs are split into fixed size blocks (default is 64 MB) and stored in HDFS. Map Reduce runtime follows a master-worker architecture. The master (Job-Tracker) assigns tasks to the worker nodes. Each worker node runs a Task-Tracker daemon that manages the currently assigned task. Each worker node can have up to a predefined number of *mappers* and *reducers* simultaneously running. This concurrent execution is controlled through the concept of *slot*: a virtual container that can host one running task at a time. The user can specify the number s_w of *slots* for each worker w. This number should reflect the maximum number of processes that the worker can concurrently run (e.g., on a dual core with hyperthreading s_w is suggested to be 4). The Job-Tracker will assign to each worker a number n_w of tasks to be concurrently executed, such that the relation $n_w \leq s_w$ is always guaranteed.

We define S as the total number of *slots* in the Map Reduce cluster:

$$S = \sum_w s_w \tag{1}$$

The value in Eq. 1 also addresses the degree of parallelism of the cluster (i.e., total number of tasks that the platform can concurrently execute).

For the sake of simplicity, we start focusing only on map phase deadlines because all the map tasks usually operates on similar volumes of data and we can assume that in a normal execution they will require similar amount of time – as also suggested by [16]. Conversely, the amount of data processed by the reduce phase is unknown until all the *mappers* have completed, thus complicating the estimation of a deadline for each *reducer*.

Under these hypothesis, the deadline d_M for each *mapper* can be evaluated as:

$$d_M = \frac{D_M \cdot S}{M} \tag{2}$$

where D_M is the deadline for the execution of the whole map phase and M is the total number of mappers to be launched (equal to the number of data blocks to be processed).

Our Hadoop test bed is composed of 4 VMs: 1 master and 3 worker nodes. Each VM has 2 VCPUs, 4 GB RAM and 20 GB disk. At the cloud level we use 5 physical machines, each one with a Intel Core Duo CPU (3.06 GHz), 4 GB RAM and 225 GB HDD. Since a dual core machine (without hyperthreading) can concurrently execute at most two tasks, we assigned two *slots* to each worker. Our Map Reduce platform can therefore execute up to six concurrent tasks ($S = 6$).

As for the task type, we opted for a word count job, often used as a benchmark for assessing performances of a Map Reduce implementation. In our scenario we prepared a collection of 20 input files of 5 MB each. Consequently, Hadoop Map Reduce Runtime launches $M = 20$ *mappers* to analyse the input data. In this test bed, we would like to complete the map phase in $D_M = 200$ s, so (before the computation begins) we state that every map task should not exceed one minute execution ($d_M = 60$ s). The d_M deadline is later updated with a feedback mechanism from the recovery system (see Sect. 3.4).

According to the default Hadoop configuration, the output of all the *mappers* is analysed by a single *reducer*. To emulate the critical condition of some tasks showing an anomalous behaviour, we artificially modified 8 input files, so as to simulate a dramatic increase of the time required to complete the corresponding tasks. The *mappers* analysing these blocks resulted to be 6 times slower than the normal ones.

Note that, as other Map Reduce platforms, Hadoop has a fault tolerance mechanism to detect the slow tasks and relaunch them from the beginning on other – possibly more performing – workers. This solution is useful in case the problem is caused by architectural factors (poor performance or bandwidth saturation on the original worker), but is likely to be counter-productive when the execution slow down is related to the content of the data blocks involved. In that case indeed, the problem will occur again on the new worker. The only way to speed up the computation is by assigning to the newly provided workers other pending tasks in the queue, thus to runtime increase the value of S for the Map Reduce platform.

3.2   Properties to Be Monitored

In this work, we mainly focus on time-constrained data insight: the aim is to identify as soon as possible the critical situation of the Map Reduce execution going to complete after a predefined deadline. Practically speaking, this corresponds to situations where the total execution time of the Map Reduce is expected to stay within some (business-related) deadline: e.g., banks and financial bodies require to perform analyses of financial transactions during night hours, and to provide outcomes at the next work shift.

In MOBUCON EC framework, a number of properties to be monitored are already directly supported. In particular, a support for non-atomic activities execution is proposed within the MOBUCON EC framework, where for each start of execution of a specific ID, a subsequent end of execution (with same ID) is expected. This feature has been particularly useful during the verification of our test bed, to identify exceptions and worker faults due to problems and issues not directly related to the Map Reduce approach. For example, during our experiments we ignored fault events generated by power shortages of some of the PC composing the cloud. The out-of-the-box support offered by MOBUCON EC was exploited to identify these situations and rule them out.

To detect problematic mappers, we decided to monitor a very simple Declare property between the start and the end of the elaboration of each mapper, as shown in Fig. 2: after an event Map start, a corresponding event Map end should be observed, within zero and d_M seconds (we opted here for considering the minimum time unit as one second, although depending on the application domain minutes or milliseconds might be better choices). In the following we will refer to this constraint as the *Waiting Task* constraint. Notice that MOBUCON EC correlates different events on the basis of the *case*: i.e., it requires that every observed event belongs to a specific case, identified by a single case ID. To fulfil such requirement, we fed the MOBUCON EC monitor with the events logged by the Hadoop stack, and exploited the Map identifier (assigned by Hadoop to each mapper) as a *case ID*. This automatically ensures that each Map start event is indeed matched with the corresponding Map end event.

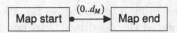

Fig. 2. *Waiting Task*, a Declare Response constraint with metric temporal deadline.

The constraint shown in Fig. 2 allows us to detect mappers that are taking too much time to compute their task. The deadline d_M is set to 60 s at the beginning of our test bed. This value has been chosen on the basis of the total completion time we want to respect while analysing the simulation data (see Eq. 2). Obviously, when dealing with real application cases, some knowledge about the application domain is required to properly set such deadline. Mappers that violate the deadline are those that, unfortunately, were assigned a long task. This indeed would not be a problem for a single mapper. However, it could become a problem if a considerable number of mappers gets stuck on long tasks, as this might undermine the completion of the whole bunch of data within a certain deadline.

Another interesting situation arises when the user doesn't have any knowledge of the volume of data to be processed and, consequently, the number of map tasks to be launched is not known *a priori*. The methodology presented in this paper allows the user to still detect anomalies in the Map Reduce process that would require additional resources to speed up the computation. For example, the deadline for each map task can be estimated by considering the execution history (e.g., taking into account the average completion time of each completed mapper). The same approach can be used for the runtime estimation of the reduce phase deadline compliance.

MOBUCON EC monitors and reasons upon each constraint, such as for example the *Waiting Task* shown in Fig. 2. To this end, it explicitly represents the states of each constraint (more precisely, of each *instance* of each constraint): the happening of events trigger the state transitions of the constraint instance related to the happened event. Figure 3 shows the state diagram for constraints of the type "Response with deadline", and specifically for the *Waiting Task* constraint: upon the detection of the Map start event, an instance of the constraint enters the *pending* state, meaning that the constraint is waiting to be verified. Upon the happening of the Map end, the constraint switches into the *satisfied* state. Otherwise, if the deadline expires, the *violated* state is reached.

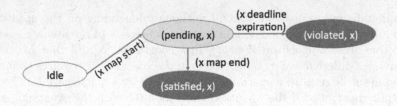

Fig. 3. State diagram of the property (*waiting_task*, *x*). The diagram gives a representation of the evolution in time of the map task *x*.

The *Waiting Task* constraint introduced so far allows to answer the question "How many maps violated the deadline?": by simply counting the *violated* constraints, it is possible to know how many mappers have exceeded the deadline *from the beginning of the Map-Reduce process*. However, in order to reason upon possible unwanted situations, we are more interested on knowing how many mappers passed over the deadline *and are still executing*. Indeed, the aim of the monitor is to detect and react to possible delays and bottlenecks at run-time. Roughly speaking, we would say that a "possibly dangerous" situation happens when a certain number of slots (out of the total available slots) passed over the deadline and are still busy in the computation.

MOBUCON EC is based on the Event Calculus logic framework [14], and can be easily expanded by simply adding further EC axioms. We exploited such possibility, and defined a further constraint, named *Long Execution Map*, which simply captures those maps that are computing and that are over the deadline. Figure 4 shows the state diagram followed by the *Long Execution Map* constraint: differently from *Waiting Task* (where the violated property remains so until the end of the execution), the state of *Long Execution Map* constraint moves from *too_long* to *freed* once a mapper *x* (that has previously violated the *response* constraint) ends.

3.3 Study of the System Health

We might notice that the *Long Execution Map* property captures the mappers that got stuck on some task, but it is still focused on each single mapper. To establish if a problem occurs to the overall system, we must aggregate this information, and consider for each time instant *how many* mappers are stuck w.r.t. the total number of available mappers. To this end, MOBUCON EC also supports the definition of a *health function* to combine the previously specified properties.

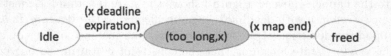

Fig. 4. State diagram of the property (*long_execution_maps*, *x*).

The *health function* allows the user to aggregate and monitor different aspects of the distributed execution. For example, in our scenario, it can be interesting to consider the number of mappers that have violated the deadline w.r.t. the total number of mappers to be launched:

$$V = \frac{\#viol}{M} \tag{3}$$

where *#viol* is the number of *Waiting Task* constraints in *violated* state. This quantity could be used to calculate the *health function* as follows:

$$H = 1 - V \tag{4}$$

Equation 4 takes into account the portion of mappers that didn't got stuck from the beginning of the execution until the current tick of the clock. H takes values in $[0, 1]$. The lower the value, the higher the risk that the overall Map Reduce framework violates some business deadline.

Comparing the number of healthy mappers (i.e., with regular execution time) with the total M, makes the solution not very reactive to the event of a long running task. Furthermore, at every clock tick, Eq. 4 estimates the health by intrinsically assuming that the following mappers will not expire their deadlines.

For these reasons, we modify the health function as follows:

$$H = 1 - \frac{\#too\_long}{S} \tag{5}$$

where *#too_long* represents the number of *Long Execution Map* properties in *too_long* state.

Equation 5 considers the portion of non-stuck mappers over a window of observed mappers. The window is chosen to be equal to the number of slots S (i.e., maximum number of concurrently running mappers). In other words, the system health is expressed as the fraction of mappers that are not busy with a long task, over the degree of parallelism S. In this way, at every clock tick, the health of the system gives an indication of how many cores of the Map Reduce cluster are executing a long task.

Note that, thanks to MOBUCON EC and the properties we defined, it is also possible to understand in every moment how many cores are not directly involved in the computation (i.e., not executing a map task). This is possible by simply monitoring the following function:

$$W = \frac{S - \#too\_long - \#pend}{S} \tag{6}$$

where *#pend* is the number of *Waiting Task* constraints in *pending* state.

The portion of infrastructure cores identified by W can be either waiting for the data to be exchanged between nodes, performing a context switch from a mapper to another or being used for other operations.

In our use case scenario, the alarms raised by the Monitoring component are simply the effect of the comparison of the system health value (as calculated through Eq. 5) with a threshold fixed to 0.3. This signifies that the Recovery system is requested to intervene when at least 30% of the launched mappers exceed the d_M deadline.

3.4   The Recovery Component Logic

Given the trend of the user-defined health function calculated by the Monitoring component, the Recovery block in Fig. 1 is in charge of determining how many new computing nodes must be provided to the Map Reduce cluster (i.e., how to update the degree of parallelism S) in order to satisfy the deadline constraint.

At the beginning of the execution ($t = t_0$), the computing scenario is identified by the status vector $[t_0, M_0, D_{M0}, S_0]$, where M_0 is the number of mappers to be launched, D_{M0} is the deadline for the whole map phase and S_0 is the total degree of parallelism of the cluster (i.e., the number of execution slots).

If the Recovery component receives an alarm from the Monitoring at time $t = t_i$, we can redefine the status of the computing scenario as $[t_i, M_i, D_{Mi}, S_i]$. The deadline D_{Mi} can be easily calculated as

$$D_{Mi} = D_{M0} - (t_i - t_0) \tag{7}$$

while the number M_i of remaining mappers can be calculated as

$$M_i = M_0 - M_{Ci} \tag{8}$$

where M_{Ci} is the number of mappers completed up to t_i.

Given the status at time t_i, the degree of parallelism S_i (necessary in order to accomplish the execution within the new deadline D_{Mi}) can be recomputed as follows:

$$S_i = \left\lceil \frac{M_i \cdot \bar{t}_{Ci}}{D_{Mi}} \right\rceil \tag{9}$$

where $\bar{t}_{Ci}$ is the average completion time of the M_{Ci} mappers. Note that $\bar{t}_{Ci}$ can be either runtime updated every time the Monitoring system detects a mapper completion or computed at time $t = t_i$ as follows:

$$\bar{t}_{Ci} = \frac{(t_i - t_0) \cdot S_0}{M_{Ci}} \tag{10}$$

Note that Eq. 9 assumes that the remaining mappers (going to be executed after t_i) will have the same average completion time $\bar{t}_{Ci}$ as those before.

The difference $S_i - S_0$ suggests the number of computation slots to be provisioned at time t_i.

After the provisioning is completed, the Recovery component updates the value of the deadline d_M checked by the Monitoring for each mapper. The new value of d_M is the average completion time $\bar{t}_{Ci}$. This feedback mechanism allows to runtime adapt the computation of the system health and can be further refined by considering the distribution function of the completion times.

The recovery policy can be either implemented with a traditional procedural approach or expressed through Declare language. The definition of declarative properties for recovery is matter of future work.

3.5 The Output from Mobucon EC Monitor

In Fig. 5, we show what happens when we analyse a word count execution on the Hadoop architecture described in Sect. 3.1, with respect to the properties discussed in Sect. 3.2. Note that, as we focus on the performance of the system when data-specific factors slow down the computation, the declarative semantic employed and the results of the following evaluation are independent from the specific Map Reduce-encoded problem (e.g., word count, terasort, inverted index etc.).

Figure 5 is composed of four strips, representing the evolution of different properties during the execution. From top to bottom of the figure we have: the health function, graphical representation of the *Waiting Task* constraint, *Long Execution Maps* constraint and description of the events occurred in each time interval. In the latter in particular (bottom part of Fig. 5), the observed events have starting labels `ts` or `tc` to represent the start and the completion of a task, respectively. There are also a number of events starting with the label `time`: these events represent the ticking of a reference clock, used by MOBUCON EC to establish when deadlines are expired.

The health function on top of Fig. 5 is the one defined in Eq. 5: indeed, the system healthiness dramatically decreased when six over seven of the first launched mappers in our testbed got stuck in a long execution task. The *Long Execution Map* strip (third strip from the top in Fig. 5) further clarifies the intervals during which the long map tasks exceed their time deadline.

Finally, the *Waiting Task* constraint strip (second strip from the top in Fig. 5) shows the status of each mapper: when the mapper is executing, the status is named *pending* and it is indicated with a yellow bar. As soon as there is information about the violation of a deadline (because of a tick event from the reference clock), the horizontal bar representing the status switched from *pending* to *violated*, and the color is changed from bright yellow to red. Notice that once violated (red color), the response constraint remains as such: indeed, this is a consequence of the Declare semantics where no compensation mechanisms are considered.

For reasons of space, we provide in Fig. 6 the evolution of our test (subsequent to what shown in Fig. 5). As expected, the total number of mapper violating the deadline constraint is 8, as we provided 8 modified files in the input dataset. MOBUCON EC is therefore able to suddenly and efficiently identify any anomaly in the Hadoop execution (according to simple user-defined constraints).

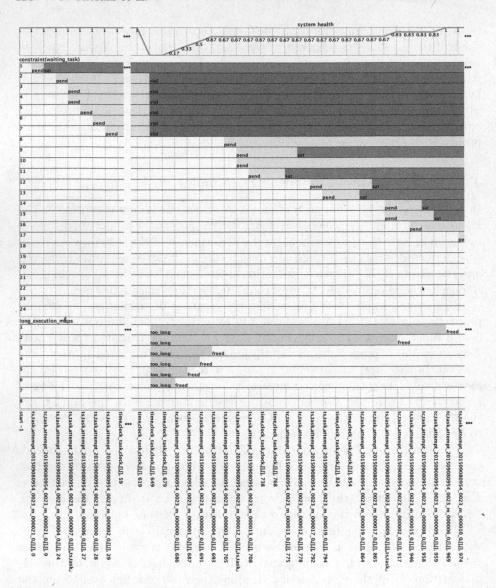

Fig. 5. The output of the MOBUCON EC monitor for the execution of word count job on the given testbed. (Color figure online)

The health function values in the output of MOBUCON EC monitor can be used to determine when a recovery action is needed. The intervention can be dynamically triggered by a simple threshold mechanism over the health function or by a more complex user-defined policy (e.g., implementing an hysteresis cycle), possibly specified with a declarative approach.

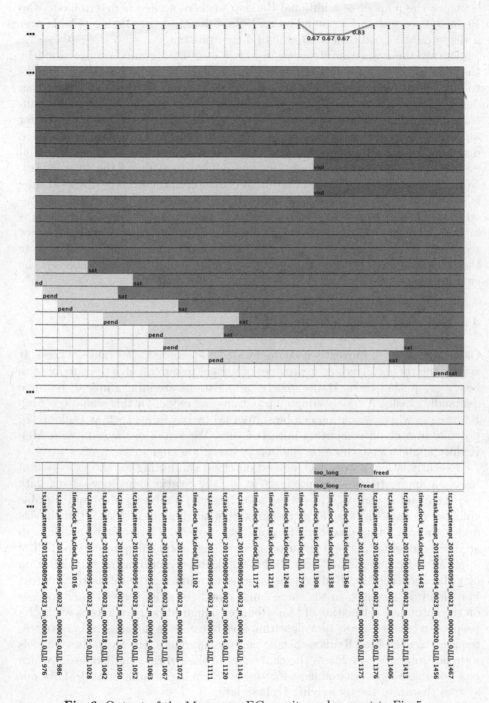

Fig. 6. Output of the MOBUCON EC monitor subsequent to Fig. 5.

Once the number of additional Hadoop workers needed is determined, Map Reduce Auto-scaling Engine relays on HyIaaS for the provisioning of VMs over a single public cloud or federated hybrid environment.

During the evaluation depicted in Figs. 5 and 6, 85 events are processed by the MOBUCON EC monitor in 285 ms (worst case over 10 evaluations). Thanks to the high expressiveness of the adopted declarative language, the user can define complex constraints, thus increasing the computational cost of the run-time monitoring. We are aware that, under this condition, the system can suffer a penalty in the execution time and the described method can show limits when dealing with fast monitored tasks (i.e., the time between task start and task end events is too short for the Monitoring component to evaluate the compliance). Nevertheless, in the envisioned Map Reduce scenario, the average duration time is in general higher than the time required by MOBUCON EC to check the constraints. Furthermore, since the recovery action to provide additional workers is intrinsically time consuming (tens of minutes), the Monitoring component is not requested to be responsive in the order of sub-seconds. Therefore, we can state that the time to detect anomalies shown in Figs. 5 and 6 is acceptable for the envisioned scenario.

4   Related Work

Cloud computing is currently used for a wide and heterogeneous range of tasks. It is particularly useful as elastic provider of virtual resources, able to contribute to heavy computing tasks. Data-intensive applications are an example of resource demanding tasks. A widely adopted programming model for this scenario is Map Reduce [9], whose execution can be supported by platforms such as Hadoop [2], possibly in a cloud computing infrastructure. We tested our system with Map Reduce applications, choosing Hadoop as execution engine.

Recently, a lot of work has focused on cloud computing for the execution of big data applications: as pointed out in [8], the relationship between big data and the cloud is very tight, because collecting and analysing huge and variable volumes of data require infrastructures able to dynamically adapt their size and their computing power to the application needs. The work [6] presents an accurate model for optimal resource provisioning useful to operate Map Reduce applications in public clouds. Similarly, [20] deals with optimizing the allocation of VMs executing Map Reduce jobs in order to minimize the infrastructure cost in a cloud data center. In the same single-cloud scenario, the work [22] focuses on the automatic estimation of Map Reduce configuration parameters, while [26] proposes a resource allocation algorithm able to estimate the amount of resources required to meet Map Reduce-specific performance goals. However, these models were not intended to address the challenges of the hybrid cloud scenario, which is a possible target environment for the provisioning of additional VMs in our system thanks to the underlying HyIaaS layer.

More similarly to our approach, cloud bursting techniques has been adopted for scaling Map Reduce applications in the work [16], which presents an online provisioning policy to meet a deadline for the Map phase. Differently from our approach, [16] focuses on the prediction of the execution time for the Map phase with a traditional approach to monitoring, which introduces complexity in the implementation and tuning, whereas our solution can benefit from a simple enunciation of the system properties relaying on Declare language.

Also the work presented in [13] deals with cloud monitoring/management for big data applications. It proposes an extension of the Map Reduce model to avoid the shortcomings of high latencies in inter-cloud data transfer: the computation inside the on-premise cloud follows the batch Map Reduce model, while in the public cloud a stream processing platform called Storm is used. The resulting system shows significant benefits. Differently from [13], we chose to keep complete transparency and uniformity with respect to the allocation of the working nodes and their configuration.

As regards the use of declarative and logic-based frameworks (such as MOBUCON EC) for verification and monitoring, several examples can be found in literature in different application domains but we are not aware of any work applying it to the monitoring of Map Reduce jobs in a cloud environment. Event Calculus-based approaches have been used in various fields to verify the compliance of a system to user-defined behavioural properties. For example, [11,23] exploit ad-hoc event processing algorithms to manipulate events and fluents, written in JAVA. Differently from MOBUCON EC they do not have an underlying formal basis, and they cannot take advantage of the expressiveness and computational power of logic programming.

Several authors [5,12] have investigated the use of temporal logics – Linear Temporal Logic (LTL) in particular – as a declarative language for specifying properties to be verified at runtime. Nevertheless, these approaches lack the support of quantitative time constraints, non-atomic activities with identifier-based correlation, and data-aware conditions. These characteristics – supported by MOBUCON EC – are instead very important in our application domain.

5 Conclusions and Future Work

Distributed programming models – such as Map Reduce – implemented over cloud infrastructures are providing a technical answer to the challenge of big data processing but still lack mechanisms to autonomously scale the clusters when further computational power is required.

In this paper, we present a framework architecture that encapsulates an application level platform for Map Reduce and lends the infrastructure the ability to autonomously check the execution, detecting and reacting to bottlenecks and constraint violations through Business Process Management techniques. The system operates in a *best effort* fashion.

Focusing on *activities* and *constraints*, the use of the Declare language and the MOBUCON EC framework has shown significant advantages in the monitoring system implementation and customization.

Although this work represents just a first step towards an auto-scaling engine for Map Reduce, its declarative approach to the issue shows promising results, both regarding the reactivity to critical conditions and simplification in constraint definition. Indeed, by allowing the user to specify the system properties/constraints in a graphical fashion, MOBUCON EC framework highly simplifies the implementation process.

For the future, we plan to employ the defined architecture to test various diagnosis and recovery policies and verify the efficacy of the overall auto-scaling engine in a wider scenario (i.e., with a higher number of Map Reduce workers involved). Furthermore, we are going to adapt the system to different – possibly more performing – platforms for distributed data processing (e.g., Apache Spark [3]) and to translate the presented recovery model into declarative constraints that can be easily implemented in a graphical fashion through the MOBUCON EC interface.

Particular attention will be given to the hybrid cloud scenario, where the HyIaaS component is employed to transparently perform VM provisioning either on an on-premise internal or an off-premise public cloud. In case of a hybrid deploy, several additional constraints will need to be taken into account (e.g., the limited inter-cloud bandwidth), thus further complicating the implemented monitoring and recovery policies. Nevertheless, we believe that a declarative approach to the problem can contribute to significantly simplify the implementation of the solution.

Finally, in the area of Process Mining, the growing number of available event data (that have to be processed for monitoring and conformance checking) is considered a computational challenge. Therefore, in order to reduce the time for event processing, different decomposition approaches can be considered (see [24]). How to map a Process Model, suitably decomposed, in a distributed architecture - such as Map Reduce implemented over cloud-infrastructures - is a subject for future work.

References

1. Amazon Cloud Watch (2016). https://aws.amazon.com/it/cloudwatch/. Accessed July 2016
2. Apache Hadoop (2016). https://hadoop.apache.org/. Accessed July 2016
3. Apache Spark (2016). http://spark.apache.org. Accessed July 2016
4. Armbrust, M., Fox, O., R., G.: Above the clouds: a Berkeley view of cloud computing. Technical rep., Electrical Engineering and Computer Sciences, University of California at Berkeley (2009)
5. Bauer, A., Leucker, M., Schallhart, C.: Runtime verification for LTL and TLTL. ACM Trans. Softw. Eng. Methodol. **20**(4), 14:1–14:64 (2011). http://doi.acm.org/10.1145/2000799.2000800
6. Chen, K., Powers, J., Guo, S., Tian, F.: CRESP: towards optimal resource provisioning for MapReduce computing in public clouds. IEEE Trans. Parallel Distrib. Syst. **25**(6), 1403–1412 (2014)
7. Chen, M., Mao, S., Liu, Y.: Big Data: a survey. Mob. Netw. Appl. **19**(2), 171–209 (2014)
8. Collins, E.: Intersection of the Cloud and Big Data. IEEE Cloud Comput. **1**(1), 84–85 (2014)

9. Dean, J., Ghemawat, S.: MapReduce: simplified data processing on large clusters. Commun. ACM **51**(1), 107–113 (2008). http://doi.acm.org/10.1145/1327452.1327492
10. Ekanayake, J., Li, H., Zhang, B.: Twister: a runtime for iterative Map Reduce. In: Proceedings of the First International Workshop on Map Reduce and its Application of ACM HPDC Conference (2010)
11. Farrel, A., Sergot, M., Sallè, M., Bartolini, C.: Using the event calculus for tracking the normative state of contracts. Int. J. Coop. Inf. Syst. **14**(02n03), 99–129 (2005). http://www.worldscientific.com/doi/abs/10.1142/S0218843005001110
12. Giannakopoulou, D., Havelund, K.: Automata-based verification of temporal properties on running programs. In: Proceedings of 16th Annual International Conference on Automated Software Engineering (ASE 2001), pp. 412–416, November 2001
13. Kailasam, S., Dhawalia, P., Balaji, S., Iyer, G., Dharanipragada, J.: Extending MapReduce across clouds with BStream. IEEE Trans. Cloud Comput. **2**(3), 362–376 (2014)
14. Kowalski, R.A., Sergot, M.J.: A logic-based calculus of events. New Gener. Comput. **4**, 67–95 (1986)
15. Loreti, D., Ciampolini, A.: A hybrid cloud infrastructure of Big Data applications. In: Proceedings of IEEE International Conferences on High Performance Computing and Communications (2015)
16. Mattess, M., Calheiros, R., Buyya, R.: Scaling MapReduce applications across hybrid clouds to meet soft deadlines. In: 2013 IEEE 27th International Conference on Advanced Information Networking and Applications (AINA), pp. 629–636, March 2013
17. Montali, M., Chesani, F., Mello, P., Maggi, F.M.: Towards data-aware constraints in declare. In: Shin, S.Y., Maldonado, J.C. (eds.) Proceedings of the 28th Annual ACM Symposium on Applied Computing, SAC 2013, Coimbra, Portugal, 18–22 March 2013, pp. 1391–1396. ACM (2013). http://doi.acm.org/10.1145/2480362.2480624
18. Montali, M., Maggi, F.M., Chesani, F., Mello, P., van der Aalst, W.M.P.: Monitoring business constraints with the event calculus. ACM TIST **5**(1), 17 (2013). http://doi.acm.org/10.1145/2542182.2542199
19. OpenStack Ceilometer (2016). https://wiki.openstack.org/wiki/Ceilometer. Accessed July 2016
20. Palanisamy, B., Singh, A., Liu, L.: Cost-effective resource provisioning for MapReduce in a cloud. IEEE Trans. Parallel Distrib. Syst. **26**(5), 1265–1279 (2015)
21. Pesic, M., Aalst, W.M.P.: A declarative approach for flexible business processes management. In: Eder, J., Dustdar, S. (eds.) BPM 2006. LNCS, vol. 4103, pp. 169–180. Springer, Heidelberg (2006). doi:10.1007/11837862_18
22. Rizvandi, N.B., Taheri, J., Moraveji, R., Zomaya, A.Y.: A study on using uncertain time series matching algorithms for MapReduce applications. Concurrency Comput. Pract. Experience **25**(12), 1699–1718 (2013). http://dx.doi.org/10.1002/cpe.2895
23. Spanoudakis, G., Mahbub, K.: Non-intrusive monitoring of service-based systems. Int. J. Coop. Inf. Syst. **15**(03), 325–358 (2006). http://www.worldscientific.com/doi/abs/10.1142/S0218843006001384
24. Van Der Aalst, W.M.P.: Distributed process discovery and conformance checking. In: Lara, J., Zisman, A. (eds.) FASE 2012. LNCS, vol. 7212, pp. 1–25. Springer, Heidelberg (2012). doi:10.1007/978-3-642-28872-2_1

25. Van Der Aalst, W., et al.: Process mining manifesto. In: Daniel, F., Barkaoui, K., Dustdar, S. (eds.) BPM 2011. LNBIP, vol. 99, pp. 169–194. Springer, Heidelberg (2012). doi:10.1007/978-3-642-28108-2_19
26. Verma, A., Cherkasova, L., Campbell, R.H.: Resource provisioning framework for MapReduce jobs with performance goals. In: Kon, F., Kermarrec, A.-M. (eds.) Middleware 2011. LNCS, vol. 7049, pp. 165–186. Springer, Heidelberg (2011). doi:10.1007/978-3-642-25821-3_9

Detecting Anomaly in Cloud Platforms Using a Wavelet-Based Framework

David O'Shea[1], Vincent C. Emeakaroha[1(⊠)], Neil Cafferkey[1],
John P. Morrison[1], and Theo Lynn[2]

[1] Irish Centre for Cloud Computing and Commerce,
University College Cork, Cork, Ireland
{d.oshea,vc.emeakaroha,n.cafferkey,j.morrison}@cs.ucc.ie
[2] Irish Centre for Cloud Computing and Commerce,
Dublin City University, Dublin 9, Ireland
theo.lynn@dcu.ie

Abstract. Cloud computing enables the delivery of compute resources as services in an on-demand fashion. The reliability of these services is of significant importance to their consumers. The presence of anomaly in Cloud platforms can put their reliability into question, since an anomaly indicates deviation from normal behaviour. Monitoring enables efficient Cloud service provisioning management; however, most of the management efforts are focused on the performance of the services and little attention is paid to detecting anomalous behaviour from the gathered monitoring data. In addition, the existing solutions for detecting anomaly in Clouds lacks a multi-dimensional approach. In this chapter, we present a wavelet-based anomaly detection framework that is capable of analysing multiple monitored metrics simultaneously to detect anomalous behaviour. It operates in both frequency and time domains in analysing monitoring data that represents system behaviour. The framework is first trained using over seven days worth of historical monitoring data to identify healthy behaviour. Based on this training, anomalous behaviour can be detected as deviations from the healthy system. The effectiveness of the proposed framework was evaluated based on a Cloud service deployment use-case scenario that produced both healthy and anomalous behaviour.

Keywords: Multi-dimensional anomaly detection · Wavelet transformation · Cloud monitoring · Data analysis · Cloud computing

1 Introduction

Cloud computing has transformed the delivery of IT resources into services that are accessible through the Internet. The large-scale and abstract nature of Cloud platforms is intimidating to both consumers and administrators. This is reflected in the difficulty of managing such systems to provision consistent and reliable Cloud services in order to gain consumer trust and high adoption of Clouds. Anomalous

© Springer International Publishing AG 2017
M. Helfert et al. (Eds.): CLOSER 2016, CCIS 740, pp. 131–150, 2017.
DOI: 10.1007/978-3-319-62594-2_7

behaviour endangers the consistent performance of virtual machines in providing resources for the execution of consumer applications. It is therefore important to have a strategy for detecting anomalous events in Cloud environments to prevent such inconsistency and improve the reliability of Cloud services. To detect anomaly in Clouds, a number of challenges such as the differentiation between normal and anomalous behaviours need to be addressed. In addition, Cloud environments are dynamic. This means that normal behaviour can be continuously unfolding and a current model of normal behaviour may be different in the future.

Moreover, Cloud service provisioning management, based on monitoring, has focused on detecting performance issues and has largely ignored anomaly detection. Most of the existing anomaly detection solutions tend to address a particular fixed formulation of the problem [8,19]. A recent survey on anomaly detection in Clouds [20] has shown the lack of multi-level detection techniques to adequately address reliability issues in Clouds.

In this chapter, we propose a novel anomaly detection framework for detecting anomalies in the behaviour of services hosted on Cloud platforms. The framework consists of a monitoring tool to supervise service execution on Cloud infrastructures, and a wavelet-inspired anomaly detection technique for analysing the monitoring data across Cloud layers and reporting anomalous behaviour. Based on a service-deployment use-case scenario, the detection technique is evaluated to demonstrate its efficiency. The achieved results are compared against existing algorithms to show the technique's significance.

The rest of the chapter is organised as follows: Sect. 2 presents some background knowledge on anomaly detection and discusses categories of anomaly. In Sect. 3, we analyse the related work and differentiate our contributions to it. Section 4 presents the proposed framework, focusing on the monitoring and anomaly detection components, while Sect. 5 describes its implementation details. In Sect. 6, we present the evaluation of the framework and Sect. 7 concludes the chapter.

2   Background

Anomaly detection (or outlier detection) is the identification of items, events or observations that do not conform to an expected pattern or to other items in a data set. In a regular and repeatable time series, a profile of expected behaviour should be easily obtainable. In medicine, one such example is an electrocardiogram (ECG). This is used to classify a patient's heart activity. A medical doctor has been trained to quickly identify anomalous ECGs, or indeed anomalies in an ECG, by studying a large number of healthy ECGs. An analogous health monitor would therefore be desirable in other areas of science, including the health of shared network resources and Cloud-based services [9,14] that are subject to demands that vary greatly and experience periodic growth, seasonal behaviour and random variations. Anomalous behaviour can be the result of unprecedented user requirements, malicious (hacking) activities, or can be symptomatic of issues with the system itself. Before identifying the cause of anomalous behaviour, one must first identify anomalous behaviour by detecting a measurable deviation from the expected behaviour.

2.1   Anomaly Categories

Based on existing research [8], anomalies can be grouped into the following categories:

1. Point Anomaly: This is a situation where an individual data instance can be considered as anomalous with respect to the rest of the data. It is seen as the simplest form of anomaly and most of the existing research on anomaly detection is focused on this category [8].

2. Contextual Anomaly: This represents a data instance that is anomalous in a particular context. It is also known as a conditional anomaly [27]. The context is mostly derived from the structure in the data set and should be included in the problem formulation. Mostly, the choice to use a contextual anomaly detection technique depends on its meaningfulness in the target application domain. For example, where an ambient temperature measurement would be at the lowest during the winter (e.g., $-16\,°C$) and peak during the summer (e.g., $38\,°C$), a temperature of $80\,°C$ would be anomalous. However, $80\,°C$ is an acceptable value in a temperature profile of boiling water. In this category, the availability of contextual attributes is a key factor. In some cases, it is easy to define context and therefore the use of a contextual anomaly detection technique would be appropriate. In some other cases, context definition and the application of such techniques are challenging.

3. Collective Anomaly: This represents a situation where a collection of related data instances is anomalous with respect to the entire data set. The single data instances in this collection may not be anomalous individually but when they occur together, they are considered anomalous. The following sequence of events in a computer network provides an example:

... http-web, **buffer-overflow**, http-web, smtp-mail, **ssh**, smtp-mail, http-web, **ftp**, smtp-mail, http-web ...

The occurrence of the above highlighted events together could signify an attack where the attacker caused a *buffer-overflow* to corrupt the network, and then remotely accessed the machines using SSH and copied data through FTP. In this example, note that each of these events could be normal but their clustering is anomalous.

3   Related Work

Previously, extensive research has been done for anomaly detection in large-scale distributed systems such as Clouds [15,16,20,23,25]. Ibidunmoye *et al.* [20] present a review of the work done in performance anomaly detection and bottle-neck identification. They describe the challenges in this area and the extent of the contributed solutions. In addition, they pointed out the lack of multi-level anomaly detection in Clouds. Mi *et al.* [23] present a hierarchical software-orientated approach to anomaly detection in Cloud systems, tracing user requests through VMs (virtual machines), components, modules and finally functions. The authors attempt to identify those software modules that are responsible for system degradation by identifying those that are active during abnormal and normal behaviour of the system. For example, this approach considers a module to be responsible for abnormal system behaviour if its response latency exceeds the required threshold. However, it does not consider a module that finished quickly because of a software crash as contributing to system degradation.

Some current research in anomaly detection relies on fixed thresholds [7,30]. In [30], particular mention is made of the assumption of Gaussian distributed data with defined thresholds. Unfortunately, this places assumptions (and therefore limitations) on the data being analysed if it must fit (or is assumed to fit) a particular distribution. Typically, these thresholds must be calculated completely *a priori* and therefore require anomaly-free time-series data of the system. MASF [7] is one of the more popular threshold-based techniques of use in industry, where thresholds are defined over precise time-intervals (hour-by-hour, day-by-day, etc.). Lin *et al.* [21] firstly use a global locality-preserving projection algorithm for feature extraction, which combines the advantages of PCA (Principal Component Analysis) with LPP (Locality Preserving Projection). They then use an LOF (Local Outlier Factor) based anomaly detection algorithm on the feature data. LOF assigns a factor to each data point that measures how anomalous it is, and considers points whose factor exceeds a certain threshold to be anomalies.

Statistical approaches have also been developed in prior academic work [1,2,5] to extend to multi-dimensional data as well as reducing false positives. However, these methods often require knowledge of the time-series distribution or may not adapt well to an evolving distribution. On the other hand, probabilistic approaches, such as Markov chains [6,26], can produce excellent predictions of a system's behaviour, particularly if the system is periodic with random, memory-less transition between states. The size of the probability matrix will grow with the number of defined Markov states, and this may present an issue when extending to multi-metric analysis. However, all of these methods only rely on time-domain information while more information exists in the frequency domain. Considering the distributed nature of Clouds, it is a prime target for sophisticated intrusion attacks [18] and therefore merits the consideration of all information available.

Recent works [17,22,23,29] have begun to use wavelet transforms (which utilise time and frequency domain information) as part of their pre-processing techniques to identify and characterise anomalies in Cloud-based network systems. Wang *et al.* [29] describe EbAT - an anomaly detection framework that performs real-time wavelet-based analysis to detect and predict anomalies in the behaviour of a utility Cloud. Their system does not require prior knowledge of normal behaviour characteristics, and is scalable to exascale infrastructure. Using the RUBiS benchmark to simulate a typical website, it achieves 57.4% better accuracy than threshold-based methods in detecting uniformly distributed injected anomalies. It aggregates metric data before analysis in order to achieve better scalability. However, it does not consider multi-level anomaly detection. Guan *et al.* [17] present a wavelet-based anomaly detection mechanism that exhibits 93.3% detection sensitivity and a 6.1% false positive rate. The algorithm requires normal runtime Cloud performance training data. However, it is not indicated how transient anomalies in the training data are identified to prevent false negatives. A subset of metrics that optimally characterises anomalies is chosen. Metric space combination is then applied to further reduce the metric space. It is unclear from this approach what would occur if a metric that was initially stable – and therefore excluded from the metrics under consideration – began to exhibit anomalous behaviour. Once an anomaly is detected using this method the metric responsible cannot be identified.

To the best of our knowledge, none of the existing solutions present a wavelet-based multi-level anomaly detection technique that can detect and diagnose the root causes of anomalies across Cloud resource and application layers.

4   Anomaly Detection Framework

This section describes the architecture of our proposed anomaly detection framework. It is designed to address the previously identified challenges. The architecture is capable of handling the service provisioning lifecycle in a Cloud environment, which includes service scheduling, application monitoring, anomaly detection and user notification.

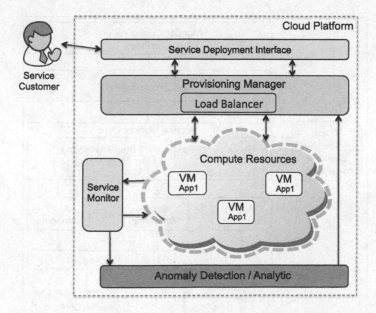

Fig. 1. System architecture.

Figure 1 presents an abstract view of our architecture and its operations. Customers place their service requests through a defined interface (*Service Deployment Interface*), which acts as the front-end in the Cloud environment. The received requests are validated for format correctness before being forwarded to the *Provisioning Manager* for further processing. The provisioning manager includes a *Load Balancer* that is responsible for equally distributing the service/application deployment for optimal performance. The applications are deployed on the *Compute Resources* for execution. The *Service Monitor* supervises the execution of the applications on the compute resources. The resulting monitoring data are forwarded to the *Detection/Analytic* component for analysis. Any anomaly detection from the analysis is communicated to the provisioning manager to take appropriate action.

The proposed architecture is generic to support a wide range of applications, varying from traditional web services to parameter sweep and bag-of-task applications. In this chapter, we detail the monitoring and anomaly detection components.

4.1 Service Monitor Design

The service monitor comprises individual configurable monitoring tools in a decentralised fashion. It is capable of monitoring Cloud resources and applications, which gives it an advantage over resource-monitoring only tools such as *LoM2HiS framework* [10, 11]. At the application level, the service monitor supports event-based monitoring of activities. Figure 2 presents an overview of the service monitor.

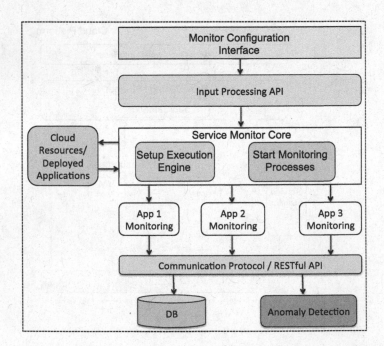

Fig. 2. Service monitor.

As shown in Fig. 2, the service monitor has a modular design. The configuration of the tools is done through the *Monitor Configuration Interface*. It allows the parameterisation of the individual monitoring tools, for example to specify different monitoring intervals.

The *Input Processing API* is responsible for gathering the configurations from the previous component and parsing them into a suitable format for the back-end service monitor core engine to understand. The *Service Monitor Core* instantiates the necessary monitoring tools with the proper configuration parameters and supervises them while monitoring the deployed Cloud services. The monitoring tools are executed in parallel and each sends its monitored data using the *Communication Protocol* into a database as well as to the anomaly detection module.

In designing the service monitor, we strived to make it non-intrusive, scalable, interoperable and extensible. These qualities have been associated with efficient monitoring tools as described in a recent monitoring survey [12]. The separation of the service monitor components into modules makes it easily extensible with new functionalities. To achieve non-intrusiveness, we host the monitoring software on separate Cloud nodes to the ones used to execute the customer services. However, we deploy light weight monitoring agents on the compute node for gathering the monitoring data and sending it back to the server. This helps to avoid resource contention between the monitoring server and the deployed Cloud services that might degrade customer service performance. In addition, this separation increases the scalability of the monitoring tool since it facilitates the creation of clusters of monitoring agents with decentralised control servers. The communication protocol uses a platform-neutral data interchange format for formatting and serialising data to achieve interoperability.

4.2 Anomaly Detection Algorithm

Our anomaly detection algorithm is highly configurable for different detection stages. This extended version can tolerate data of an arbitrary rank. In our use case, we consider multiple one-dimensional metrics simultaneously. The wavelet algorithm could also consider matrix and other forms of data streams (e.g. matrix/tensor) if available.

In the algorithm's operation as shown in Fig. 3, we first configure the rank to one (vector data) for N data streams. It currently uses the Morlet wavelet form. Other waveforms can also be used and may be more efficient. An optimal configuration is part of our ongoing work.

The second stage details the training by taking the pre-processed data and performing the wavelet-transform on each metric. The returned spectrograms are then passed to a machine learning technique that has a knowledge of the history of the Cloud system. The newest spectrograms are used to update the running estimate of the mean and standard deviation of an ideal performance. In this way, a profile of behaviour can be extracted, and a deviation from this profile can be identified as an anomaly.

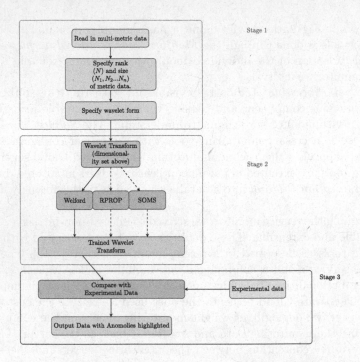

Fig. 3. Flow chart of anomaly detection technique.

Thirdly, the trained multi-metric spectrogram is then compared to the multi-metric spectrogram of data under investigation. The steps are outlined in the pseudo code included in Algorithm 1. One benefit of this multi-layer approach is that after having inspected a given data sample for anomalies, the new data can be easily used to extend the usefulness of the trained model and tolerances for normal behaviour can be updated.

The threshold scaling parameter m allows for a specific tolerance to be set for each metric. The wavelet transform is computed using Eq. 1.

$$CWT_x^\psi(\tau, s) = \Psi_x^\psi(\tau, s) = \frac{1}{\sqrt{|s|}} \int x(t)\psi^*(\tfrac{t-\tau}{s})dt \qquad (1)$$

The mother wavelet (ψ) is a windowing function that scales (s is the scaling parameter and a scalar quantity) and translates (τ is the translation parameter) the time trace ($x(t)$, a vector of metrics recorded simultaneously). A two-dimensional spectrogram (of the complex coefficients) is generated from varying s and τ, for each metric. As s is increased, the time window becomes smaller. This in turn fafects the resolution of frequencies detected in the time traces. The wavelet transform offers superior temporal resolution of the high frequency components and scale (frequency) resolution of the low frequency components. The values of s and τ range from 0 to the length of the time trace undergoing transformation. The exact configuration of the anomaly detection algorithm is introduced in a broad sense so that it can be further optimised without major restructuring.

Algorithm 1. Pseudo code for Multi-metric Wavelet Transform and Welford training.

1: **function** $Train\_data(timetraces, metric)$
2:    **for** day **in** $timetraces$ **do**
3:        $SPEC = wavelet\_transform(rank, N_1, N_2, ...N_n, metric, \Delta t, \Delta \omega, ...)$
4:        $M_T, S_T = Welford\_N\_rank(M_T, S_T, SPEC, day)$
    return $M_T, \sqrt{\frac{S_T}{days}}$

1: **function** $Check\_for\_anomaly(metrics_A, M_T, S_T, m)$
2:    $M_A = wavelet\_transform(metrics_A, \Delta t, \Delta \omega, ...)$
3:    **if** $M_A > (M_T + m \cdot |S_T|)$ **then**
4:        anomaly found
5:        Record location to $locs$
6:    **else if** $M_A < (M_T - m \cdot |S_T|)$ **then**
7:        anomaly found
8:        Record location to $locs$
9:    **else**
10:        no anomalies found
        $Ratio = \frac{M_A}{M_T}$
    return $Ratio, locs$

1: **function** $Welford\_N\_rank(M, S, SPEC, day))$
2:    $M_{Temp} = M$
3:    $M \mathrel{+}= \frac{(SPEC - M)}{day}$
4:    $S \mathrel{+}= (SPEC - M_{Temp})(SPEC - M)$
    return M, S

5 Implementation Details

This section describes the implementation of the proposed anomaly detection framework. Our focus is on the monitoring and anomaly detection components.

5.1 Service Monitor Implementation

The monitor configuration interface was realised using Ruby on Rails technology, which enabled rapid development and facilitates its compatibility with other components. A key feature of Ruby on Rails is its support for modularity. We used this feature to make it easily extendible with new functionality. Ruby on Rails also has a rich collection of open source libraries. Based on this, we used the JSON library to aggregate the input configuration data before transferring them down to the next component.

The input processing API component is implemented as a RESTful service in Java. Since Ruby on Rails supports RESTful design, it integrates seamlessly with this component in passing down the input data. The input processing API extracts these data and makes them available to the service monitor core component.

The service monitor core component sets up and manages the execution of user selected and configured monitoring tools. We use multi-threading to achieve parallel execution of the monitoring tools since they are developed as individual applications.

Each monitoring tool incorporates communication protocols for transferring the monitoring data to head components. The communication protocols comprise a messaging bus based on RabbitMQ [28], HTTP and RESTful services. This combination achieves interoperability between platforms. We use a MySQL database to store the monitoring data. Hibernate is used to realise the interaction between the Java classes and the database. With Hibernate, it is easy to exchange database technologies. Thus, the MySQL database could be easily exchanged for another database platform.

5.2   Anomaly Detection Algorithm Implementation

The wavelet transform is implemented in a similar fashion to the wavelet module in MLPY [3]; however, several modifications have been made to accommodate dynamic Cloud metric data. Firstly, the Fast Fourier Transform(FFT) backend is specified using libraries such as scipy, numpy, GSL or FFTW [13]. Secondly, our implementation overcomes dimensionality restriction of data. This means that we are capable of considering a matrix form of data as opposed to the single dimension required by the previous system. We perform a one-dimension, column-wise, multi-scale wavelet transform. That is, the data of an arbitrary N metrics recorded in time are transformed from the time domain to the frequency-time domain across all metrics simultaneously. There exists also an opportunity to use this extended routine further to bring the multi-scale wavelet transform to near real-time applications by executing each individual scale calculation in parallel.

The wavelet transform is usually implemented as part of a larger routine that includes some pre-processing [6,24] and is often trained using an advanced neural network (ANN) such as RPROP (Resilient BackPRoPagation) or SOMS [31]. The routine employed here for the machine learning based on the wavelet transform is outlined in Fig. 3.

In Fig. 3, the solid black arrows indicate the elements of the routine currently available. The dashed black arrows indicate features still in development.

The use of additional ANNs after the Welford Algorithm could allow the identification or correlation of anomalies between metrics.

6   Evaluation

The goal of our evaluation is to demonstrate the efficacy of the proposed framework to monitor Cloud service execution, analyse the monitoring data and detect anomalous behaviour. It is based on a use-case scenario that describes the service interactions. First, we present the evaluation environment and the use-case descriptions.

6.1   Experimental Environment

To set up the experimental environment, an OpenStack Cloud platform installation running Ubuntu Linux was used. The basic hardware and virtual
machine configurations of our OpenStack platform are shown in Table 1. We
use the Kernel-based Virtual Machine (KVM) hypervisor for hosting the virtual
machines.

Table 1. Cloud environment hardware.

Machine Type = Physical Machine				
OS	CPU	Cores	Memory	Storage
OpenStack	Intel Xeon 2.4 GHz	8	12 GB	1 TB
Machine Type = Virtual Machine				
OS	CPU	Cores	Memory	Storage
Linux/Ubuntu	Intel Xeon 2.4 GHz	1	2048 MB	50 GB

As shown in Table 1, the physical machine resources are capable of supporting on-demand starting of multiple virtual machines for hosting different Cloud
services.

6.2   Use Case Scenario

This use case scenario describes a Cloud service deployment, the monitoring of
the service and the analysis of the monitoring data to detect anomalous behaviour. To realise this, we set up Apache web servers with back-end MySQL databases on our OpenStack platform as the demonstrator Cloud service. On the web
servers, we deploy a transactional video-serving web application that responds
to requests and makes queries to back-end databases. Video data were uploaded
to the web servers that could be rendered on request. The service is designed to
receive and process different queries and workloads generated by users.

In this scenario, we simulate user behaviour in terms of generating queries
and placing them to the Cloud service using Apache JMeter [4]. The workload
consists of three HTTP queries and two video rendering requests. With these
queries, we generate approximately 15 requests per second, representing light to
moderate load on a real-world service. The video requests invoke playback of
music video data on the web servers. We generate five requests per second for
two videos in a mixed sequence.

The execution of this service on the web servers was monitored using the service monitor described in Sect. 4.1. The application-level monitor is event-based.
Therefore, it can continuously monitor the performance of each request/query
placed to the web application. We monitor 74 metrics (such as *BytesReceived,
ByteSent, ResponseTime, CPUUserLevel, CPUIdle, FreeDisk, FreeMemory etc.*)
from this service deployment.

For this evaluation, we gathered 10 days' worth of data from this service execution monitoring. Since the workload is simulated, the load distribution on each particular machine was repeated each day, therefore the recorded metrics should vary in similar ways each day. No seasonal or periodic effect of the environment on the machines should have occurred; therefore the metric distribution should be normal apart from the presence of small amount of random noise.

Due to the velocity, volume, and real-time nature of Cloud data, it is difficult to obtain time-series data with true labelled anomalies. Moreover, to aptly test our N-metric anomaly detector, a multi-metric anomaly is needed to be present in the monitoring data. To address this, we chose to simulate a Distributed Denial of Service (DDoS) attack by injecting appropriate chosen values into the time trace data from one of the monitored virtual machines (post training). To simulate this attack, a simple anomaly injector was written to inject an anomaly into several metrics including CPUIDLE, CPUUSER, CPUSYSTEM and INBYTES. The DDoS attack was ramped up in 40 s (one time step between measurements), had a duration of around 100 min and then returned to a typical behaviour for the system. Figures 4 and 5 demonstrate this visually.

Given that there is a redundancy between CPUUSER, CPUIDLE and CPUSYSTEM, we will leave out CPUIDLE in this evaluation. We compare the trained spectrograms against the data injected with anomalies in the following sections.

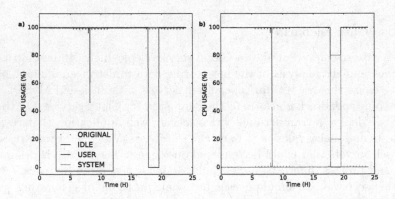

Fig. 4. Sample anomaly injection: **(a)** compares a sample trace (blue) of CPUIDLE against an average time trace with an anomaly injected (black). Here the CPUIDLE drops to zero in under 40 s (one time step between measurements). **(b)** Demonstrates the conservation of total CPU usage between CPUIDLE, CPUUSER and CPUSYSTEM (their sum equals 100%). Here a ratio of 0.8 is used to divide the CPUUSER and CPUSYSTEM levels. (Color figure online)

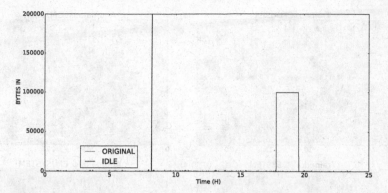

Fig. 5. Sample anomaly injection of INBYTES comparing a sample trace (blue) of INBYTES against an average time trace with an anomaly injected (black). Here the INBYTES increases rapidly over 40 s and maintains a large height for 100 min before dropping back to typical levels. (Color figure online)

6.3   Data Analysis and Results

This section presents the achieved results of our multi-metric anomaly detection framework.

Based on the wavelet algorithm, we generate a separate spectrogram for each day of data. Given that the system load is approximately similar from one day to the next, a typical presence (or absence) of frequency-time events can be detected through the comparison of the individual spectrograms.

To determine if a point is anomalous, we perform a simple test including the comparison of the ratio of $\frac{M_{New}}{M_{Trained}}$ with the relative magnitudes of the spectrograms, where M_{New} is the spectrogram from data under investigation and $M_{Trained}$ is the expected form of the spectrogram based on past behaviour. First, we present the trained and the anomalous spectrograms for visual inspection.

Figure 6 presents the trained and injected data spectrograms. From a visual inspection, it is easy to identify regions of similarity and difference. According to these results, we were able to detect the DDoS attack, which affected the CPU usage and the number of incoming bytes. As shown in Fig. 6, we can see that the anomalous behaviours in the three metrics occurred at the same time point. To deeply analyse the location of the anomaly, Fig. 7 presents the ratio of the trained and injected data spectrograms. This further confirms the consistency of the anomalous behaviour across the multiple metrics.

The injected anomaly has successfully been detected in each of the presented metrics. However, given the symmetry of the anomaly shape further explanation is required.

Wide side-bands are seen at lower frequencies (frequencies ≤ 0.004); this is because the wavelet transform is based on the FFT and therefore expects periodic data. Cloud-based data can contain periodic behaviour, as the use of cloud-based services is user-need driven. In the data monitoring and collection,

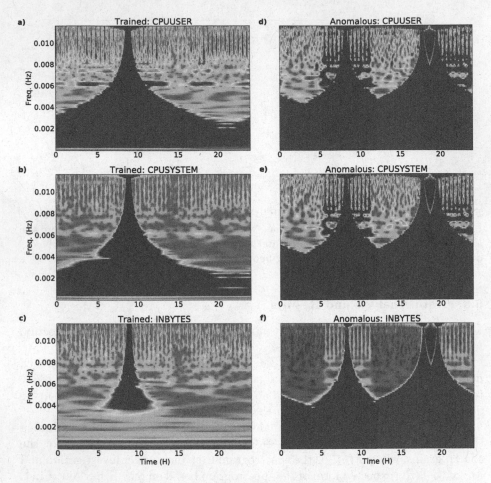

Fig. 6. Visual comparison of trained spectrograms (left) and spectrograms that contain a simulated DDoS attack (right).

similar loads were observed on the monitored VMs from one day to the next. This is appropriate as it reflects a real-world scenario with periodic user behaviour.

Additionally, the wavelet method employed here is multi-scale. This means that we can reduce time-resolution to increase frequency resolution, and vice-versa. One effect of this means that the side bands of the highlighted anomaly will increase due to reduced time resolution.

Finally, the size of the DDoS attack is also a point for discussion. It is possible that the duration of the simulated attack was unnecessarily large. The simulated DDoS lasted for 100 min and occupied 7% of the total modelled day.

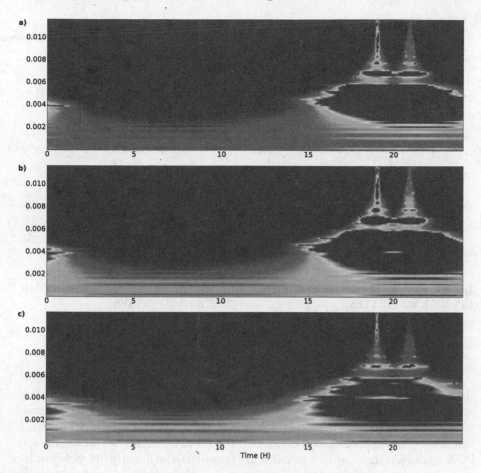

Fig. 7. Plot of the ratio of the anomalous spectrograms **(a)** CPUUSER, **(b)** CPUSYS-TEM and **(c)** INBYTES to the trained spectrograms shown in Fig. 6. The high ratios (red) indicate the presence of anomalous behaviour. (Color figure online)

6.4   Principal Component Analysis

In this section, we present comparisons between the results of the wavelet inspired method and a pure statistical approach to show the former's significance. In this case, we calculate the mean of the data column and determine the distance of each element from the centroid. Outliers are determined to be any points that lie outside a confidence interval of 99%.

Figure 8(a) displays two of the injected metrics versus REPORTTIME. The detected anomalies are highlighted in red. This verifies that the injected anomalies satisfy the definition of an anomaly sought by this method (they lie outside the strict confidence interval).

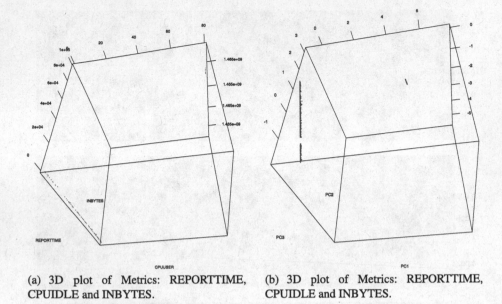

(a) 3D plot of Metrics: REPORTTIME, CPUIDLE and INBYTES.

(b) 3D plot of Metrics: REPORTTIME, CPUIDLE and INBYTES.

Fig. 8. 3D plots of sample injected metrics and PCA analysis.

Figure 8(b) contains the results of the Principal Component Analysis (PCA). Here it is shown that for the much reduced system, the data is not normally distributed and two of the PCA axes returned would be sufficient to describe the data.

A limitation of this statistical method is that for a given confidence interval, it will always identify points as anomalous (false positives). Another limitation with PCA analysis is that the data supplied are required to be linearly independent (no redundant data) and therefore this requires some knowledge of the data to know which metrics are co-linear. For a low number of metrics, this can addressed quite quickly; however, this can become tedious for higher numbers of metrics recorded.

Figure 9(a) presents a 3D plot of the three analysed example metrics from one of the training data-sets used to train the wavelet-based anomaly detector. The three metrics FREEDISK, OUTPACKETS and CPUIDLE are plotted. From the plot it is clear that anomalous points are identified.

Figure 9(b) shows the PCA reduction performed on 10 metrics. The axes (PC1, PC2 and PC3) are the three most significant vectors demonstrating the extent of the non-normality of the data. Taking a confidence interval of 99% yielded many anomalies. This is expected, as a purely statistical approach will, by construction, always discover anomalous points regardless of whether the points are in fact anomalous or not. Furthermore, the number of outliers will be determined by the confidence interval selected. The interesting thing is that this approach failed to detect some of the injected anomalies in the gathered data. This demonstrates an advantage of the wavelet method over a pure statistical approach.

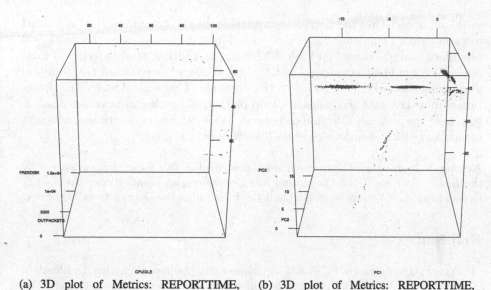

(a) 3D plot of Metrics: REPORTTIME, CPUIDLE and INBYTES.

(b) 3D plot of Metrics: REPORTTIME, CPUIDLE and INBYTES.

Fig. 9. 3D plots of sample injected metrics and PCA analysis.

7   Conclusion

This paper presented an anomaly detection framework for detecting anomalous behaviour of services hosted on Cloud platforms. It contains a monitoring tool to monitor service executions in Clouds and gather monitoring data for analysis. A wavelet-based detection algorithm was implemented to provide a multi-level analysis of the monitoring data for anomaly detection. It uses frequency domain and time domain information to estimate an anomaly-free spectrogram. The healthy spectrogram is trained (removing seasonality and noise/randomness) by using an extended two-dimensional Welford algorithm to create two-dimensional mean and standard deviations. These quantities are then used to check for the presence of anomalies by comparing the trained mean and standard deviation with those of the new data.

The framework was evaluated based on a Cloud service deployment use-case scenario in an OpenStack evaluation testbed. We used 10 days of gathered monitoring data from the service execution from which a day's data were systematically injected with anomalies for the evaluation. The wavelet-inspired method successfully detected the injected anomalies, and a brief comparison was made with a pure statistical approach, reinforcing the success of our technique.

In the future, we aim to progress this work to a near real-time implementation where the anomaly detection will be carried out on the monitoring data at runtime. The effect of moving to real time will mean the introduction of a time-window, which will be continuously updated as the monitoring platform reports updated metric values. Moving to real time will also distribute the computational workload as each time the metrics are updated, the spectrogram will be appended to rather than being entirely recalculated. Further extensions to this work will permit the comparison of multiple ANNs across multiple (and individual) metrics, which would allow for cross-metric comparison while retaining the ability of identifying the metric(s) containing the anomaly. This will allow for the detection of more complex anomalies in Cloud platforms. Further work is also possible to extend the anomaly injection techniques so we can better determine the limits of this and other anomaly detection frameworks.

Acknowledgements. The research work described in this paper was supported by the Irish Centre for Cloud Computing and Commerce, an Irish national Technology Centre funded by Enterprise Ireland and the Irish Industrial Development Authority.

References

1. Agarwal, S., Mozafari, B., Panda, A., Milner, H., Madden, S., Stoica, I.: BlinkDB: queries with bounded errors and bounded response times on very large data. In: Proceedings of the 8th ACM European Conference on Computer Systems, pp. 29–42. ACM (2013)
2. Agarwala, S., Alegre, F., Schwan, K., Mehalingham, J.: E2EProf: automated end-to-end performance management for enterprise systems. In: 37th Annual IEEE/IFIP International Conference on Dependable Systems and Networks, DSN 2007, pp. 749–758, June 2007
3. Albanese, D., Visintainer, R., Merler, S., Riccadonna, S., Jurman, G., Furlanello, C.: mlpy: machine learning Python (2012). http://mlpy.sourceforge.net/. Accessed 22 Feb 2016
4. Apache Software Foundation. Apache JMeter (2016). http://jmeter.apache.org/. Accessed 06 Jan 2016
5. Bahl, P., Chandra, R., Greenberg, A., Kandula, S., Maltz, D., Zhang, M.: Towards highly reliable enterprise network services via inference of multi-level dependencies. In: SIGCOMM. Association for Computing Machinery Inc., August 2007
6. Bakhtazad, A., Palazoglu, A., Romagnoli, J.A.: Detection and classification of abnormal process situations using multidimensional wavelet domain hidden Markov trees. Comput. Chem. Eng. **24**(2), 769–775 (2000)
7. Buzen, J.P., Shum, A.W.: MASF - multivariate adaptive statistical filtering. In: International CMG Conference, pp. 1–10 (1995)
8. Chandola, V., Banerjee, A., Kumar, V.: Anomaly detection: a survey. ACM Comput. Surv. **41**(3), 15:1–15:58 (2009)
9. Doelitzscher, F., Knahl, M., Reich, C., Clarke, N.: Anomaly detection in IaaS clouds. In: 2013 IEEE 5th International Conference on Cloud Computing Technology and Science (CloudCom), pp. 387–394, December 2013
10. Emeakaroha, V.C., Brandic, I., Maurer, M., Dustdar, S.: Low level metrics to high level SLAs - LoM2HiS framework: bridging the gap between monitored metrics and SLA parameters in cloud environments. In: 2010 International Conference on High Performance Computing and Simulation (HPCS), pp. 48–54, July 2010
11. Emeakaroha, V.C., Netto, M.A.S., Calheiros, R.N., Brandic, I., Buyya, R., De Rose, C.A.F.: Towards autonomic detection of SLA violations in cloud infrastructures. Future Gener. Comput. Syst. **28**(7), 1017–1029 (2012)

12. Fatema, K., Emeakaroha, V.C., Healy, P.D., Morrison, J.P., Lynn, T.: A survey of cloud monitoring tools: taxanomy, capabilities and objectives. J. Parallel Distrib. Comput. **74**, 2918–2933 (2014)

13. Frigo, M.: A fast Fourier transform compiler. ACM Sigplan Not. **34**, 169–180 (1999). ACM

14. Gander, M., Felderer, M., Katt, B., Tolbaru, A., Breu, R., Moschitti, A.: Anomaly detection in the cloud: detecting security incidents via machine learning. In: Moschitti, A., Plank, B. (eds.) Trustworthy Eternal Systems via Evolving Software, Data and Knowledge, pp. 103–116. Springer, Heidelberg (2013)

15. Guan, Q., Fu, S.: Adaptive anomaly identification by exploring metric subspace in cloud computing infrastructures. In: 2013 IEEE 32nd International Symposium on Reliable Distributed Systems (SRDS), pp. 205–214, September 2013

16. Guan, Q., Fu, S.: Wavelet-based multi-scale anomaly identification in cloud computing systems. In: 2013 IEEE Global Communications Conference (GLOBECOM), pp. 1379–1384, December 2013

17. Guan, Q., Fu, S., DeBardeleben, N., Blanchard, S.: Exploring time and frequency domains for accurate and automated anomaly detection in cloud computing systems. In: 2013 IEEE 19th Pacific Rim International Symposium on Dependable Computing (PRDC), pp. 196–205. IEEE (2013)

18. Gul, I., Hussain, M.: Distributed cloud intrusion detection model. Int. J. Adv. Sci. Technol. **34**, 71–82 (2011)

19. Hodge, V.J., Austin, J.: A survey of outlier detection methodologies. Artif. Intell. Rev. **22**(2), 85–126 (2004)

20. Ibidunmoye, O., Hernández-Rodriguez, F., Elmroth, E.: Performance anomaly detection and bottleneck identification. ACM Comput. Surv. **48**(1), 1–35 (2015)

21. Lin, M., Yao, Z., Gao, F., Li, Y.: Toward anomaly detection in IaaS cloud computing platforms. Int. J. Secur. Appl. **9**(12), 175–188 (2015)

22. Liu, A., Chen, J.X., Wechsler, H.: Real-time timing channel detection in an software-defined networking virtual environment. Intell. Inf. Manag. **7**(06), 283 (2015)

23. Mi, H., Wang, H., Yin, G., Cai, H., Zhou, Q., Sun, T., Zhou, Y.: Magnifier: online detection of performance problems in large-scale cloud computing systems. In: 2011 IEEE International Conference on Services Computing (SCC), pp. 418–425, July 2011

24. Penn, B.S.: Using self-organizing maps to visualize high-dimensional data. Comput. Geosci. **31**(5), 531–544 (2005)

25. Reynolds, P., Killian, C., Wiener, J.L., Mogul, J.C., Shah, M.A., Vahdat, A.: PIP: detecting the unexpected in distributed systems. In: Proceedings of the 3rd Conference on Networked Systems Design and Implementation, NSDI 2006, Berkeley, CA, USA, vol. 3. USENIX Association (2006)

26. Sha, W., Zhu, Y., Chen, M., Huang, T.: Statistical learning for anomaly detection in cloud server systems: a multi-order Markov chain framework. IEEE Trans. Cloud Comput. (2015). https://doi.org/10.1109/TCC.2015.2415813

27. Song, X., Wu, M., Jermaine, C., Ranka, S.: Conditional anomaly detection. IEEE Trans. Knowl. Data Eng. **19**(5), 631–645 (2007)

28. Videla, A., Williams, J.J.W.: RabbitMQ in Action: Distributed Messaging for Everyone. Manning Publications Company, Grand Forks (2012)

29. Wang, C., Talwar, V., Schwan, K., Ranganathan, P.: Online detection of utility cloud anomalies using metric distributions. In: 2010 IEEE Network Operations and Management Symposium (NOMS), pp. 96–103, April 2010

30. Wang, C., Viswanathan, K., Choudur, L., Talwar, V., Satterfield, W., Schwan, K.: Statistical techniques for online anomaly detection in data centers. In: 2011 IFIP/IEEE International Symposium on Integrated Network Management (IM), pp. 385–392, May 2011
31. Zhang, Z., Wang, Y., Wang, K.: Fault diagnosis and prognosis using wavelet packet decomposition, Fourier transform and artificial neural network. J. Intell. Manuf. 24(6), 1213–1227 (2013)

Security SLA in Next Generation Data Centers, the SPECS Approach

Massimiliano Rak[1], Valentina Casola[2(✉)], Silvio La Porta[3],
and Andrew Byrne[3]

[1] Dipartimento di Ingegneria dell'Informazione,
Seconda Università di Napoli, Aversa, Italy
massimiliano.rak@unina2.it
[2] Dipartimento di Ingegneria Elettrica e Tecnologie dell'Informazione,
Università di Napoli Federico II, Napoli, Italy
casolav@unina.it
[3] EMC Ireland COE Innovation, Cork, Ireland
{silvio.laporta,andrew.byrne}@emc.com

Abstract. Next generation Data Centers (ngDC) provide a significant evolution how storage resources can be provisioned. They are cloud-based architectures offering flexible IT infrastructure and services through the virtualization of resources: managing in an integrated way compute, network and storage resources. Despite the multitude of benefits available when leveraging a Cloud infrastructure, wide scale Cloud adoption for sensitive or critical business applications still faces resistance. One of the key limiting factors holding back larger adoption of Cloud services is trust. To cope with this, datacenter customers need more guarantees about the security levels provided, creating the need for tools to dynamically negotiate and monitor the security requirements. The SPECS project proposes a platform that offers security features with an *as-a-service* approach, furthermore it uses Security Service Level Agreements (Security SLA) as a means for establishing a clear statement between customers and providers to define a mutual agreement. This paper presents an industrial use case from EMC that integrates the SPECS Platform with their innovative solutions for the ngDC. In particular, the paper illustrates how it is possible to negotiate, enforce and monitor a Security SLA in a cloud infrastructure offering.

Keywords: Cloud · ngDC · Cloud security · Security SLA

1 Introduction

Storage services, as many IT services, are increasingly moving toward the virtualized, distributed cloud model. Indeed, recent concepts like ngDC or Software-Defined Data Centers (SDDC) [6] are grounded in the ideas of virtualization, offering the capability to run multiple independent virtual servers using a set of shared, physical resources. Resource Pooling is a significant advantage of the

© Springer International Publishing AG 2017
M. Helfert et al. (Eds.): CLOSER 2016, CCIS 740, pp. 151–169, 2017.
DOI: 10.1007/978-3-319-62594-2_8

ngDC and SDDC solutions, enabling the automatic allocation of storage, network and compute resources to meet the demand of incoming requests. This is one of the foundational concepts cloud computing is built on. In fact, through ngDC provisioning models, a Cloud Service Provider (CSP) may offer on demand, scalable, secure and cost effective cloud infrastructures or services upon which Cloud Service Customers (CSC) can develop their own services.

This solution is very attractive from an organisation's perspective offering economic benefits as well as providing increased resilience and accessibility of services. However, one of the main challenges hindering broad-scale adoption is the perception of loss of security and control over resources that are dynamically acquired in the cloud and that reside on remote providers.

This is primarily due to concerns over the privacy and security of the data, workloads, and applications, outsourced to the Cloud provider's data centers. In cloud computing the tangible assets of the CSC become intangible, virtual resources, that are dynamically acquired via a 3rd Party provider. With these provisioning models, the loss of control over their own services and assets (data), CSCs are naturally hesitant to place critical business applications or sensitive data in the cloud. Cloud providers typically offer a set of security measures, advertised to potential customers; however, without the ability to provide assurance of those security measures, or to maintain visibility over the service, the Cloud customer has no way to verify that the service is being provided as described. A possible solution to this challenge could be the adoption of SLAs, clearly stating what services are provided by the CSP and the related responsibilities in case of violation.

For example, Amazon and Google offer, on their Cloud storage services, an SLA that details the remediation process (partial reimbursement of payment, credit for extended service period, etc.) available to the customer, in the event that the monthly uptime of their services does not meet the 99.9% offering. However, customers are unable to request more specific requirements and are reliant on the CSP itself to report the actual measurement of the service uptime.

Despite the intense research efforts into developing standards and frameworks for SLAs [5,9,11,14,23], at the state of the art few solutions allow CSPs to offer practical, implementable Security SLAs. Moreover, there are very few services able to concretely enforce and monitor the security features which would enable CSCs to verify the status of guaranteed SLAs.

In such a context, the SPECS project[1] proposes a framework which aims to facilitate the automated negotiation, monitoring and enforcement of Security SLAs. In this paper, we aim to address both the needs of the CSP (offering secure services on an ngDC to customers according to an agreed SLA), and the CSC (negotiating the level of security granted by providers). In particular, we integrate SPECS with vSphere[2] and ViPR storage controller[3], two commercial products offered as a service by EMC to fulfil the ngDC components by offering

---

[1] http://www.specs-project.eu/.
[2] https://www.vmware.com/it/products/vsphere.
[3] http://www.emc.com/vipr.

virtual machines running services on top of storage resources. The acquisition and configuration of these services are managed through SPECS and are formally defined in Security SLAs which guarantee the performance and security requirements of the services.

The implementation of security capabilities, enforced and monitored through SPECS demonstrates the effectiveness of Security SLAs as a concrete solution that can be adopted by commercial products and services. The effectiveness and adaptability of this approach is further strengthened by use of standardized security metrics to identify capabilities delivered by SPECS to support the storage service. The outcome of this process is to improve trust in the capabilities of the CSP to protect and guarantee their assets services in the cloud.

The remainder of this paper is organized as follows: Sect. 2 introduces the SPECS framework for the provisioning of cloud services guaranteed by Security SLAs. Section 3 describes the main features and limitations of ngDC that motivate the need for more flexible and secure Data Centers. Section 4 introduces our proposal of enhancing the ngDC with Security SLAs based on the adoption of SPECS. In particular, this section focuses on EMC storage solutions. Section 5 describes the architecture of the ngDC storage testbed and the enhanced security features. Finally, Sect. 6 gives an overview of related work on frameworks and guidelines for SLAs, while Sect. 7 summarizes the conclusions and provides direction for future work.

2   The SPECS Framework

The SPECS framework provides services and tools to build applications offering services with security features defined in, and granted by, a Security SLA [1,20].

The framework addresses both CSPs' and users' needs by providing tools for (1) enabling user-centric negotiation of security parameters in a Security SLA; (2) providing a trade-off evaluation process among CSPs; (3) real time monitoring of the fulfilment of SLAs agreed with CSPs; (4) notifying both End-users and CSPs in the event that an SLA is violated; (5) enforcing agreed SLAs in order to maintain the agreed security levels. The SPECS framework is also able to "react and adapt" in real-time to fluctuations in the security level by applying the required countermeasures.

In order to provide security capabilities granted by Security SLAs, a SPECS Application orchestrates the so called *SPECS Core Services* dedicated to the *Negotiation*, *Enforcement* and *Monitoring* of an SLA. Through these core services, the cloud service is enhanced with security capabilities guaranteed by the signed SLA.

In SPECS, four primary actors have been defined:

- **End-user:** The CSC of a Cloud service;
- **SPECS Owner:** The Cloud service provider;
- **External CSP:** An independent (typically public) CSP, unaware of the SLA, providing only basic resources without security guarantees;
- **Developer:** Supports the SPECS Owner in the development of SPECS applications.

As illustrated in Fig. 1, the interactions among the parties are very simple: the End-user uses the cloud services offered by the SPECS Owner, which acquires resources from External CSPs, enriched with capabilities to meet the End-user's security requirements. SPECS then monitors and enforces the End-user's security requirements to ensure the agreed security levels and alert the End-user of any breaches in the terms of the Security SLA.

3   Next Generation Data Center Storage

The ngDC is a highly efficient and optimized data center that allows organisations to achieve more within the confines of the available resources (physical servers, power, cooling, facilities, etc.). The key advantage of the ngDC is its

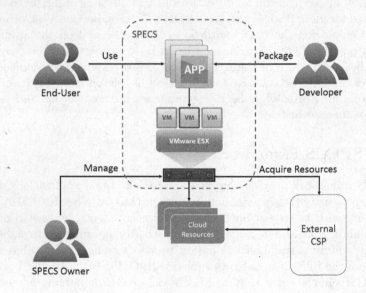

Fig. 1. SPECS entity relationships.

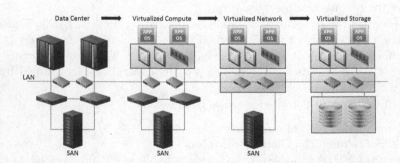

Fig. 2. Evolution of the data center towards a fully virtualized environment.

agility and ability to adapt rapidly to changes in an organizations business and workload requirements.

This efficiency is achieved in a ngDC by consolidating the physical resources, in other words virtualizing it. As illustrated in Fig. 2, we view the classical data center model as one in which there are dedicated physical resources for each application. The first step in evolving towards the ngDC is to move away from dedicated resources to consolidated resources virtualizing the physical compute resources though the use of Hypervisor technologies.

The second evolutionary step consists of virtualizing the network resources, dividing them into discrete segments to isolate and segregate the traffic and/or the service by creating virtual networking components (e.g. virtual LANs, virtual SANs, virtual switches, etc.) that are part of a Hypervisor, logical links, and even converged networks.

The final phase before achieving a completely virtualized data center is to abstract out the physical storage resources. In the classic storage model, an Intelligent Storage System is used to group disks together and then partition those physical disks into discrete logical disks. These logical disks are assigned a Logical Unit Number (LUN), and are presented to a host, or hosts, as a physical device. Redundancy of the data stored on the disks is provided by RAID (Redundant Array of Independent Disks) technology, which is applied at either the physical disk layer or the logical disk layer.

This classic model has several limitations however. For example, there is an upper limit to the maximum number of physical disks that can be combined to form a logical disk. Another issue is that often the amount of storage provisioned for each application is greater than what is actually needed in order to prevent application downtime. Both of these situations results in inefficient usage of the physical storage resources that needlessly remain idle.

These kinds of inefficiencies in the management of storage can be resolved by introducing Software Defined Storage (SDS) applications such as EMC's ViPR in order to virtualize the tiered storage resources. The outcome is a more efficient usage of resources which result in reduced power and space costs in the data center as well as reduced workloads for storage and server administrators. The power of these SDS solutions is the abstraction of the physical resources by creating resource pools designed to support more generalized workloads across applications. This enables the capacity usage and requirements to be more closely monitored and aligned to the available resources - further reducing the operational costs.

But what about software-defined security? Most (or all) of the security controls can be automated and managed through software, depending on how virtualized the infrastructure is. Such an approach requires any service to be *controlled* under some security policy. The innovative idea proposed in SPECS to enhance the ngDC, is to provide Security-as-a-Service (SecaaS) according to agreed Security SLA. To achieve this, in following sections, the integration of the ngDC with the SPECS framework is proposed such that the full Security SLA life cycle can be managed.

4   Integrating SPECS with ngDC

In an ngDC, the infrastructure is virtualised, delivered as a service and controlled by management applications. This shift in the architectural approach for the data center offers a more agile, flexible and scalable model. The core idea is the decoupling of the hardware from the software layer. Indeed, services (OSs, applications and workloads) view these abstracted, virtual resources as though they were physical compute, storage and network resources. Figure 3 illustrates the relationships between the infrastructure layer, including different physical resources, and the service and application layers.

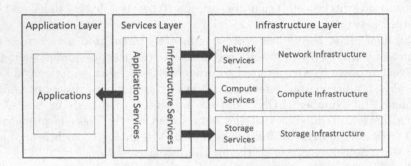

Fig. 3. Next generation data center architecture.

Despite the multitude of benefits available when leveraging a Cloud infrastructure, wide scale Cloud adoption for sensitive or critical business applications still faces resistance due to concerns over the privacy and security of the data, workloads and applications outsourced to the Cloud provider's data centers. Cloud providers typically offer a set of security measures advertised to potential customers, however without the ability to provide assurance of those security measures or to maintain visibility over the service, the Cloud customer has no way to verify the service is being provided as described.

To provide validation of the SPECS platform as a solution to address these concerns, through the use of Security SLAs to guarantee services, the use case illustrated in Fig. 4 was established. This use case represents a typical scenario in which an End-user wishes to deploy a service on a Virtual Machine on storage configured to their performance and security requirements. The candidate technologies used here are vSphere and ViPR storage controller. While the technologies are effective in achieving their objectives, they require advanced technical expertise in order to deploy services correctly. Typically, this requires dedicated management from IT personal to handle End-user requests for resources and services.

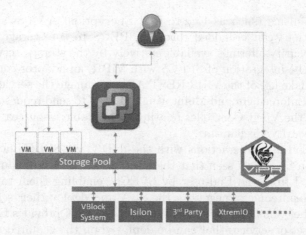

Fig. 4. EMC ngDC use case.

Integrating SPECS with the ngDC offers a SecaaS solution, not only by establishing a process to negotiate and monitor services running in the data center, but also building on the native security features present by providing additional security features delivered through virtual machines dynamically allocated and instantiated on the data center to meet the security requirements. Combined with the guarantees provided through a Security SLA, these security features improve the confidence with which end users can migrate their applications and data to the Cloud.

Figure 5 illustrates where the SPECS platform integrates with a typical ngDC architecture. In EMC's use case, ViPR is used to offer software defined storage management for the ngDC storage resources while vSphere provides central management of the compute resources.

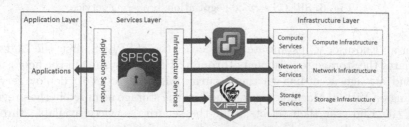

Fig. 5. SPECS enhanced data center architecture.

The SPECS framework provides the core functionality (negotiation, enforcement and monitoring) and the interface between the SPECS application presented to the End-user and the services they are requesting (e.g., storage service through ViPR). Operating at this level in the overall architecture, additional security mechanisms (such as End-to-End Encryption, AAA-as-a-Service and Data Geolocation) can be added through SPECS to the storage service that enhance the security offerings available natively by the storage service.

To enable the integration of SPECS with ViPR, an adaptor component was developed to make use of the ViPR REST API to manage the service. The adaptor allows the Enforcement and Monitoring modules to send requests and receive response from the ViPR Controller relating to available resources, performance options and security mechanisms.

The new End-user interactions with the ngDC, via SPECS, are illustrated in Fig. 6. It can be clearly seen that the interactions with individual tools have been abstracted from the End-user by SPECS, enabling them to quickly and easily define requirements, negotiate the SLA and deploy their service. A new component, Chef[4], has been added to the use case. Chef provides End-user with an extensible set of services that can be deployed on the acquired resources.

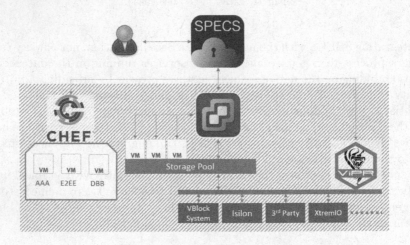

Fig. 6. SPECS service negotiation/brokering service.

For an End-user to instantiate a service on top of SPECS, first, the End-user negotiates the storage requirements through SPECS, resulting in the acquisition of cloud storage according to the Security SLA. In order to deploy services (running on Virtual Machines provisioned on the storage acquired in the previous step), the End-user then selects from the security services available through SPECS. The following subsections provide more detail on these processes.

4.1   Providing Storage Service Using SPECS Core Services

Figure 7 outlines the End-user interactions with SPECS, and subsequent interactions between SPECS and ViPR, to provide cloud storage resources. The initial step involves the negotiation of the SLA for storage resources through the

---

[4] https://www.chef.io/chef/.

SPECS platform. Here, we consider only ViPR to provide storage resources though SPECS is capable of support other providers such as AWS S3. Once the SLA has been negotiated and submitted, SPECS queries the storage services, namely ViPR, to determine if the End-user requirements can be satisfied. SPECS returns the negotiated offer to the End-user, who can sign the SLA, triggering the Enforcement phase of SPECS. This phase brokers the resources and makes the newly acquired resources available to the End-user. In parallel to this, the Monitoring module is configured so that the resources are continuously monitored. The three phases are described in detail in the following subsections.

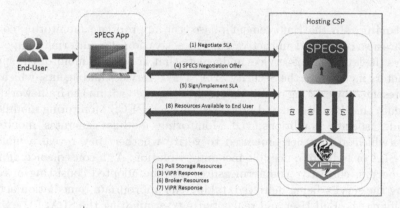

Fig. 7. SPECS service negotiation/brokering service.

Negotiation. Formatted according to the SPECS Security SLA format [7], security features are represented using few simple concepts: *Security capabilities*, the set of security controls [15] that a security mechanism is able to enforce on the target service; *Security metrics*, the standard of measurement adopted to evaluate security levels of the services offered; *Security Level Objectives (SLOs)*, the conditions, expressed over security metrics, representing the security levels that must be respected according to the SLA.

Security-related SLOs are negotiated based on the SPECS Customer's requirements. A set of compliant and feasible offers, each representing a different supply chain to implement, is identified and validated. The agreed terms are included in a Security SLA that is signed by the SPECS Customer and the SPECS Owner.

Enforcement. Once the SLA has been successfully negotiated, it is implemented through the Enforcement services, which acquire resources from External CSPs and activate the appropriate components (that implement security capabilities). This approach provides security capabilities *as-a-Service* to fulfil the SLOs included in the signed Security SLA.

Each identified security capability is implemented by an appropriate security mechanism able to cover a set of pre-defined security controls. In SPECS, a security mechanism is a piece of software dedicated to implementing security features on the target service. The information associated with a security mechanism is included in the mechanism's metadata, prepared by the mechanism's developer and includes all information needed to automate the security mechanism's deployment, configuration and monitoring.

Cloud-automation tools, such as Chef, can be used to automatically implemented the security capabilities required in the SLA through the Enforcement services.

Monitoring. In the Enforcement phase, the appropriate monitoring components are also configured and activated. The activation of monitoring components includes the launching of services and agents that are able to monitor the specific parameters included in the Security SLA. These services and agents, which may be represented by existing monitoring tools integrated within the framework, generate data that is collected and processed by the SPECS Monitoring module [2].

Under specific conditions, the Monitoring module generates monitoring events, which are further processed to verify whether they reveal a violation of the SLA or indicate a possible incoming violation. As a consequence, if a violation occurs, corrective countermeasures may be adopted consisting of reconfiguring the service being delivered, taking the appropriate remediation actions, or notifying the End-User and renegotiating/terminating the SLA.

4.2   Providing SPECS Applications on Top of ngDC

In order to illustrate the process of offering services through SPECS on top of ngDC we focus on a simple example related to the acquisition by a web developer (the End-user) of a web container enhanced with a set of security features (refer to [19] for a deeper description of the SPECS application development).

It is reasonable to suppose that the End-user is not a security expert, in that he/she is aware of the technologies that may be involved (SSL, authentication and authorization protocols, etc.), but has no detailed knowledge of the best practices and of how to protect their application from malicious attacks. For this reason, the acquisition of VMs hosting the web container and the enforcement of security features are accomplished through SPECS.

It should be noted that, at the state of the art, even if the web developer acquires VMs from a public CSP, he/she is the only person responsible for setting up any security configuration. Existing appliances offer predefined services (for example, a pre-configured web server), but checking and comparing the security features offered by different CSPs is not an easy task. The web developer has to (i) manually find the security features provided by each CSP, (ii) evaluate and compare existing offers, (iii) apply a suitable configuration, if not natively supported, and (iv) implement a monitoring solution to verify at runtime the respect of the security features.

The SPECS ecosystem provides a turnkey solution to the above issues, as it (i) offers a single interface to choose among multiple offerings on multiple providers, (ii) enables the web developer to specify explicitly the needed security capabilities on the target web container, (iii) automatically configures the VMs in order to enforce the security controls requested, (iv) offers a set of security metrics to monitor the respect of the security features requested, (v) enables continuous monitoring of the security metrics negotiated, and (vi) can automatically remediate to (some of the) alerts and violations that may occur to the SLA associated to the web container.

Moreover, running on top of the ngDC it is possible to negotiate the storage pool on top of which the Web Container will run, granting that the web server files will reside on a storage that have a set of security features agreed through the SLA.

In order to deliver web servers, pre-configured according to security best practices, we developed a mechanism named *WebContainerPool* devoted to automatically deploying and configuring a pool of web servers over a set of Virtual Machines acquired on the hosting ngDC. The developed mechanism not only provides the web server cloud service, but it also offers some reliability features, measured in terms of the two metrics (i) `LevelofRedundancy` and (ii) `LevelofDiversity`. The former ensures resiliency to failures through replication of the web container instances (used transparently by the End-user), while the latter ensures resiliency to vulnerability-based attacks by employing different software and/or hardware instances of the same web container.

In practice, the *WebContainerPool* mechanism has been developed as a security mechanism, and includes a set of pre-configured web servers (at the state of the art, Apache and Ngnix) and a load balancer (based on HAProxy). Web servers are synchronized through Memchached, so that the accesses to the web application are synchronized. The Chef cookbook associated to the *WebContainerPool* can be used also independently of the SPECS framework. It is worth pointing out that, if the aim is to apply the same process to a different cloud service (e.g., a Secure CMS), it is first necessary to develop a Cloud service cookbook dedicated to offer the CMS, and later on to select the security mechanisms that can be offered for it, possibly developing custom ones.

The proposed service (web container), as outlined above, relies on (a pool of) virtual machines, hosting synchronized web servers. The service offers some integrated security features (redundancy and diversity), but a lot of additional security capabilities can be provided. In SPECS three main security mechanisms are already available:

- **TLS:** it is a preconfigured TLS server, configured according to security best practices.
- **SVA (Software Vulnerability Assessment):** it regularly performs vulnerability assessment over the virtual machines, through software version checking and penetration tests.
- **DoSprotection:** it consists in a solution for denial of service attacks detection and mitigation based on the OSSEC tool.

5   The SPECS Enhanced ngDC Testbed

A core objective of the SPECS framework is to deploy cloud services with user defined security capabilities guaranteed by a Security SLA. SPECS allows the End-user to express the desired capabilities in the Security SLA using a reference standard for guidance and clarity (e.g. Cloud Control Matrix (CCM) 3.0 [4] and NIST SP 800-53 [15]).

This section describes the testbed used to integrate SPECS and ViPR, the ViPR usage model and the additional security features delivered *as-a-Service* and guaranteed by a Security SLA.

5.1   Physical and Software Testbed

The core components of the architecture, illustrated in Fig. 8, are: (i) ESXi server; (ii) Cisco Switch; (iii) VMAX array; (iv) Management Server.

The management server, running Windows Server 2008 R2 (W2K08), is used to set up and manage the network to which the VMAX array is connected. Additionally, EMCs SMI-S provider[5] must be installed on this system in order to enable the management of VMAX via ViPR. The VMAX array itself consists of the VMAX controller and two VMAX bays equipped with 800 disks for a total available storage space of 200 TB. ESXi[6] is VMware's bare-metal hypervisor that virtualizes servers and is installed directly on top of the physical server, partitioning resources into multiple virtual machines. These core components are connected via high speed fiber channel managed by a Cisco switch.

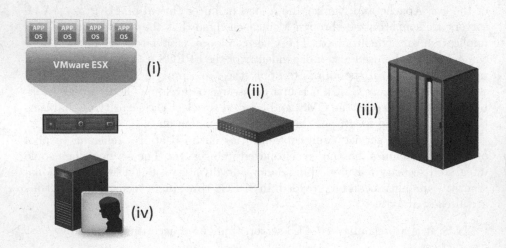

Fig. 8. Physical architecture.

---

[5] https://community.emc.com/docs/DOC-19629.
[6] https://www.vmware.com/products/vsphere-hypervisor.

In addition to these core components, the following technologies are also used to support the test environment:

– VMWare vSphere ESXi: Delivers industry-leading performance and scalability with benefits including improved reliability and security, streamlined deployment and configuration, higher management efficiency, simplified hypervisor patching and updating. It is the platform in which the entire system runs.
– VCenter Server: Provides a centralized and extensible platform for managing virtual infrastructure. vCenter Server manages VMware vSphere environments, giving IT administrators simple and automated control over the virtual environment to deliver infrastructure with confidence.
– EMC ViPR Controller: Storage automation software based on the open source development project CoprHD[7]. It centralizes and transforms multi-vendor storage into a simple and extensible platform. It abstracts and pools resources to deliver automated, policy-driven storage services on demand via a self-service catalogue. With vendor neutral, centralized storage management, customers can reduce storage provisioning costs up to 73%[8], provide greater choice and deliver a path to the cloud through storage-as-a-service. It is used to abstract the physical layer and to manage the storage resource of the data center.
– Chef: Powerful automation platform that is able to automate the configuration, deployment, and management of VMs across different networks.

Figure 9 illustrates the high level configuration of the Chef Server alongside the SPECS core provider. In this configuration, Chef is used to install and configure applications, and to deploy VMs. Services selected from the list of available Chef cookbooks via SPECS are deployed from the Chef Server as a Chef recipe, launching a VM preconfigured to execute the security service.

As can also be seen in Fig. 9, the ViPR controller is run as a virtual appliance (vApp) on the ESXi server with the VMAX storage in the back-end supplying the physical resources. The ViPR vApp is deployed using the $3+2$ configuration file for redundancy purposes (available from EMC support[9]). In this deployment configuration, five virtual machines are used to run the vApp, with two VMs able to fail without affecting the availability of the vApp.

5.2   ViPR Storage Service

The ViPR controller, accessible as a web application running as a vApp on an ESXi server, offers End-users virtual pools of storage resources using the *as-a-Service* model. Using the ViPR REST API to execute commands, the SPECS

---

[7] https://github.com/CoprHD.
[8] https://www.emc.com/collateral/data-sheet/h11750-emc-vipr-software-defined-storage-ds.pdf.
[9] https://support.emc.com/downloads/32034_ViPR.

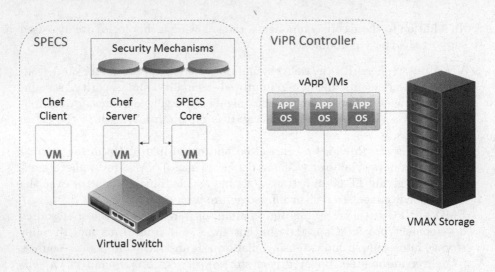

Fig. 9. Software architecture.

framework can access all the functionality available through the controller. Furthermore, this enables SPECS to allocate storage services with additional security controls that can be negotiated, for example Business Continuity Management and Operational Resilience (BCR-01, BCR-09 and BCR-11 in the CCM control framework).

While it is possible for the End-user to set up security features for the storage directly via ViPR, this requires significant technical and security expertise and should be restricted to IT or Security administrators. In contrast to this, the SPECS negotiation interface is intuitive to personnel without specific administration expertise, enabling them to select security requirements. Furthermore, the End-user can monitor the enforcement of the security metrics over which the SLOs have been defined.

For the evaluation and testing of the SPECS framework, the security control baselines defined in NIST SP 800-53 were selected to provide a standardised mapping to the security mechanisms offered by SPECS. This paper presents different security controls across categories such as *Access Control, Identification and Authentication, Physical and Environmental Protection*, and *System and Information Integrity*.

Table 1 shows some of the possible mappings used between the NIST security controls and the security features for Cloud storage services, expressed through security metrics and their description.

On selecting the *EMC ngDC* SPECS application from the SPECS portal, the End-user can choose from different service configuration parameters categorised into the following *Security Capabilities*:

- Secure Storage Capabilities: Security capabilities added to support services offered by storage providers.
- Availability Capabilities: Capabilities providing redundancy and business continuity in the event of security incidents involving the storage service.

Table 1. Sample mappings between NIST security controls and SPECS security features.

Metric	Description	NIST mapping
RAID level(s)	Defines the RAID level the volumes in the virtual pool will consist of	SA-2, SC-6, CP-9, CP-10, SI-17
SAN multi-path	The number of paths that can be used between a host and a storage volume	SC-6, SI-17
Data geolocation	Defines in which data center the virtual storage and its copies are located	PE-17, PE-18, PE-20, SI-12
Max mirrors	Defines the Maximum number of data storage mirrors	SC-5, SC-6, SI-13

Each type of capability is responsible for a specific security aspect and is associated with a group of security controls. For example, the *Availability Capabilities* are associated with SC-6 (Resource Availability) and SI-17 (Fail-Safe Procedures) as defined in NIST SP 800-53 for Security and Privacy Controls [10].

On selecting the type of capabilities required for the storage service, the End-user is then presented with the available capabilities under the selected categories. For example, the *Availability Capabilities* category includes capabilities such as RAID level, High Availability, Maximum Snapshots, etc.

Once the End-user has specified their requirements, the SPECS portal will display an overview of the capabilities requested for the storage service, as shown in Fig. 10. This form displays the metric, the associated value and the importance weight associated with each. The *importance* is an additional specifier, selected by the user, that enables SPECS to make more informed decisions about the service a provider is offering.

The final step in the negotiation process is for the End-user, on review of the negotiated SLA, to sign and, then, implement the SLA. Once the Agreement has been submitted and signed, the Implementation function (of the Enforcement module) implements the signed SLA by making a series of requests via the ViPR REST API to set up the storage service according to the requested capabilities and security mechanisms. Furthermore, during the implementation phase, the ViPR monitoring agents are configured according to the metrics specified in the SLA. Once the resources have been acquired through SPECS and the SLA has been signed and implemented, End-users can observe the implemented SLAs. The SPECS Monitoring module can then make requests via ViPR's REST API to continuously check the status of the allocated storage and verify if any SLA violation occurs.

Metric Name	Operation	Value	Importance
Raid Level (s)	eq	RAID5	HIGH
Multi-volume Consistency	eq	TRUE	MEDIUM
High Availability (Type)	eq	Array	MEDIUM
Maximum Snapshots	eq	1	MEDIUM
Max Native Continuous copy	eq	1	MEDIUM
HA Max Mirrors	eq	1	MEDIUM
Provisioning Type	eq	Thick	MEDIUM
Protocols	eq	iSCSI	MEDIUM
Drive Type	eq	SATA	MEDIUM
System Type	eq	vmax	MEDIUM
Min SAN Multi Path	eq	1	MEDIUM
Max SAN Multi Path	eq	2	MEDIUM
Data Geolocation	eq	GEOLOC-EU-IRE	HIGH

SUBMIT AGREEMENT   DOWNLOAD AGREEMENT

Fig. 10. SLA negotiated through SPECS.

Once the enforcement process is successfully completed, the resources are available through the ViPR administration interface. The capabilities requested through SPECS are reflected in the provisioned storage.

6   Related Work

The drive towards the adoption of SLAs by CSPs is an important initiative in strengthening the trust in services. SLA management frameworks like SLA@SOI [23] associate services with an SLA, detect SLA violations and are even able to recover from them.

Many research projects, like Contrail [14], Optimis[10] and mOSAIC include SLAs in their framework.

ENISA ([3,8,13]) outlined the need for a Security SLA that offers clear guarantees with respect to the security provided by CSPs to services. Projects like CUMULUS [17], A4Cloud [18], SPECS [20], SLAReady[11], SLALOM[12], MUSA [21] are actively working on this topic, attempting to clearly model and represent security into an SLA.

Security and compliance issues in the ngDC are a primary concern due to the reliance on traditional security models that have not adapted to virtualization. In the ngDC, and cloud computing as a whole, new, significant risks have been

---

[10] http://www.optimis-project.eu/.
[11] http://www.sla-ready.eu/.
[12] http://slalom-project.eu/.

introduced to IT services. Organisations who traditionally hosted workloads and data in internal data centres running on their own infrastructure, now face a loss of visibility and control. The trust boundaries that were clearly established in physical infrastructures are now blurred as virtualized resources are increasingly used.

These new security issues have spawned several research activities into novel solutions to address the security of data in the cloud. For example, Nithiavathy proposes a framework to check the data integrity on CSP using homomorphic token and distributed erasure-coded data [16]. Alternatively, a secure multi-owner data sharing scheme for cloud users using group signature and broadcast encryption was proposed in [12].

Other works focus on the security of the communication between distributed Data Centers such as [22], that proposed a security framework to manage a large number of secure connections implemented using Kinetic[13] and Pyretic[14] as a centralized middleware tool. Each of these solutions address a part of the problem, but do not offer a platform on which the CSC can combine security requirements that can be addressed in different application layers, or give formal assurance to End-user through SLAs.

7   Conclusions

Next generation Data Centers provide a significant evolution how resources, such as storage, network and compute, can be dynamically provisioned. The ngDC offers the possibility to virtualize resources and dynamically pool them according to customer needs in an *Infrastructure-as-a-Service* provisioning model.

To achieve broader adoption, datacenter customers need more guarantees about the security levels provided, creating the need for tools to negotiate security requirements and to be able to monitor their enforcement. This paper has investigated the potential of integrating the SPECS platform with the commercially available ViPR storage solution from EMC. This integration was shown to offer security *as-a-service*, enhancing the security provided by the ViPR, and guaranteed by a Security SLA.

By integrating the ViPR Controller API with SPECS, End-users may negotiate the performance and security capabilities of a Cloud storage service with resources managed transparently through ViPR. The Storage Adaptor created for this integration is platform agnostic and also supports Amazon S3 and the open source CoprHD project. In addition to negotiating capabilities that are native to the target service (e.g. ViPR), the SPECS application offers new security capabilities that enhance the overall security of the target service.

The application described in this paper fully implements the ngDC paradigm by offering storage services protected by Security SLAs through the negotiation, enforcement and monitoring phases.

---

[13] http://resonance.noise.gatech.edu/.
[14] http://frenetic-lang.org/pyretic/.

Potential future directions from this work should focus on the definition of new security metrics that can be easily measured and monitored in order to provide new security capabilities to a storage infrastructure and enable a CSP to enrich its security service offerings.

Acknowledgements. This research is partially supported by the EC FP7 project SPECS (Grant Agreement no. 610795).

References

1. Casola, V., De Benedictis, A., Rak, M., Villano, U.: Preliminary design of a platform-as-a-service to provide security in cloud. In: Proceedings of the 4th International Conference on Cloud Computing and Services Science, CLOSER 2014, Barcelona, Spain, 3–5 April 2014, pp. 752–757 (2014)
2. Casola, V., De Benedictis, A., Rak, M.: Security monitoring in the cloud: an SLA-based approach. In: 10th International Conference on Availability, Reliability and Security, ARES 2015, Toulouse, France, 24–27 August 2015, pp. 749–755 (2015)
3. Catteddu, D.: Security and resilience in governmental clouds. Technical report CSA (2011)
4. CSA: Cloud controls matrix v3.0 (2015). https://cloudsecurityalliance.org/download/cloud-controls-matrix-v3/
5. CSCC: The CSCC practical guide to cloud service level agreements. Technical report, CSCC (2012)
6. Davidson, E.A.: The Software-Defined-Data-Center (SDDC): concept or reality? [VMware] (2013). http://blogs.softchoice.com/advisor/ssn/the-software-defined-data-center-sddc-concept-or-reality-vmware/
7. De Benedictis, A., Rak, M., Turtur, M., Villano, U.: Rest-based SLA management for cloud applications. In: 2015 IEEE 24th International Conference on Enabling Technologies: Infrastructure for Collaborative Enterprises (WETICE), pp. 93–98, June 2015
8. Dekker, M.: Critical cloud computing a CIIP perspective on cloud computing services. Technical report, ENISA (2012)
9. EC: Unleashing the potential of cloud computing in Europe. Technical report, EC (2011)
10. Force, J.T., Initiative, T.: Security and privacy controls for federal information systems and organizations. NIST Spec. Publ. **800**, 53 (2013)
11. ISO: ISO/IEC NP 19086–1, Information Technology-Cloud computing-Service level agreement (SLA) framework and technology-Part 1: Overview and concepts (2014)
12. Marimuthu, K., Gopal, D.G., Kanth, K.S., Setty, S., Tainwala, K.: Scalable and secure data sharing for dynamic groups in cloud. In: 2014 International Conference on. Advanced Communication Control and Computing Technologies (ICACCCT), pp. 1697–1701. IEEE (2014)
13. Dekker, G.H.M.: Survey and analysis of security parameters in cloud slas across the European public sector (2011). http://www.enisa.europa.eu
14. Morin, C.: Open computing infrastructures for elastic services: contrail approach. In: Proceedings of the 5th International Workshop on Virtualization Technologies in Distributed Computing, pp. 1–2. ACM (2011)
15. NIST: SP 800–53 Rev 4: Recommended Security and Privacy Controls for Federal Information Systems and Organizations. Technical report, NIST (2013)

16. Nithiavathy, R.: Data integrity and data dynamics with secure storage service in cloud. In: 2013 International Conference on Pattern Recognition, Informatics and Mobile Engineering (PRIME), pp. 125–130. IEEE (2013)
17. Pannetrat, A., Hogben, G., Katopodis, S., Spanoudakis, G., Cazorla, C.: D2.1: security-aware SLA specification language and cloud security dependency model. Technical report, certification infrastructure for multi-layer cloud services (cumulus) (2013)
18. Pearson, S.: Toward accountability in the cloud. IEEE Internet Comput. **15**(4), 64–69 (2011)
19. Rak, M., Ficco, M., Battista, E., Casola, V., Mazzocca, N.: Developing secure cloud applications. Scalable Comput. Pract. Exp. **15**(1), 49–62 (2014)
20. Rak, M., Suri, N., Luna, J., Petcu, D., Casola, V., Villano, U.: Security as a service using an SLA-based approach via specs. In: IEEE Proceedings of IEEE CloudCom Conference 2013 (2013)
21. Rios, E., Iturbe, E., Orue-Echevarria, L., Rak, M., Casola, V.: Towards self-protective multi-cloud applications - MUSA - a holistic framework to support the security-intelligent lifecycle management of multi-cloud applications. In: CLOSER 2015 - Proceedings of the 5th International Conference on Cloud Computing and Services Science, Lisbon, Portugal, 20–22 May 2015, pp. 551–558 (2015)
22. Talpur, S.R., Abdalla, S., Kechadi, T.: Towards middleware security framework for next generation data centers connectivity. In: Science and Information Conference (SAI), pp. 1277–1283. IEEE (2015)
23. Theilmann, W., Yahyapour, R., Butler, J.: Multi-level SLA management for service-oriented infrastructures. In: Mähönen, P., Pohl, K., Priol, T. (eds.) ServiceWave 2008. LNCS, vol. 5377, pp. 324–335. Springer, Heidelberg (2008). doi:10.1007/978-3-540-89897-9_28

Dynamically Loading Mobile/Cloud Assemblies

Robert Pettersen(✉), Håvard D. Johansen, Steffen Viken Valvåg,
and Dag Johansen

University of Tromsø—The Arctic University of Norway, Tromsø, Norway
{robert,haavardj,steffenv,dag}@cs.uit.no

Abstract. Distributed applications that span mobile devices, computing clusters, and cloud services, require robust and flexible mechanisms
for dynamically loading code. This paper describes LADY: a system that
augments the .NET platform with a highly reliable mechanism for resolving and loading assemblies, and arranges for safe execution of partially
trusted code. Key benefits of LADY are the low latency and high availability achieved through its novel integration with DNS.

Keywords: Mobile · Cloud · Latency · Extensible distributed systems

1 Introduction

Distributed applications require their executable code to be available in local
memory of each participating CPU. Traditionally, application code has been
distributed and installed in advance with the underlying Operating System (OS),
and fully loaded into local memory when a member process starts executing.
Code can be distributed and installed manually by the machine administrators,
or on a shared file system trusted by all member processes. More recently, code
is commonly distributed over the Internet using some OS specific application
distribution service like Ubuntu's apt-get system [8], the Google Play Store, the
Apple iTunes Store, or the Windows Store.

Unlike application distribution services that require processes or even the
entire OS to be restarted, run-time code injection enables long-lived distributed
systems to be updated on-the-fly, minimizing service interruption. This allows
systems to evolve over time as requirements change and bugs are fixed. Run-
time code injection is also a foundation for extensibility in component systems
like Sapphire [23] and Kevoree [3], and enable popular end-user applications like
Firefox and Eclipse to be extended with third-party plugins on demand. In big-
data systems like MapReduce [4], Pig [16], DryadLINQ [22], and Cogset [21],
the underlying code for user-defined functions, which play a central role in distributed data processing, are typically distributed and loaded dynamically at
run-time. Transferring code between processes is also a recurring requirement for
actor-based distributed computing and mobile agent computing [10,11]. Whenever an object is serialized and transferred over the wire, the code required for

M. Helfert et al. (Eds.): CLOSER 2016, CCIS 740, pp. 170–186, 2017.
DOI: 10.1007/978-3-319-62594-2_9

de-serialization must be somehow available at the recipient. If object schemas are allowed to evolve over time, so must the serialization code.

Dynamic run-time code injection in distributed applications demands a robust system for distributing code, managing dependencies, and resolving version conflicts. If security related updates are to be distributed, the mechanism must also be resilient to attacks [13].

This paper presents LADY , a .NET library for loading assemblies dynamically and an associated cloud service for maintaining and looking up meta-information about assemblies.[1] LADY provides a robust and generic infrastructure for code distribution, allowing applications to locate, obtain, and dynamically load code—in the form of .NET assemblies—from remote repositories. LADY aims to do just a few things, but do them well:

- Provide a reliable and highly available service for finding up-to-date information about assemblies, including available versions and ways of obtaining them.
- Implement the required mechanisms for obtaining assemblies, for example through direct downloads or through a package installation system.
- Discover dependencies between assemblies, resolve the versioning conflicts that may result, and hide latency by caching and prefetching assembly data.
- Arrange for safe execution of partially trusted code, while retaining the ability to load new assemblies into sandboxed environments.

LADY offers a simple and unobtrusive interface focused on the central task of loading assemblies. LADY does not impose any particular architecture on the application, and can be combined and coexist comfortably with dependency injection frameworks like Microsoft's Managed Extensions Framework or Ninject, for systems that focus on extensibility. It can also be used as an auto-updater to check for bugfixed versions of libraries, a deserializer that automatically resolves references to missing assemblies, or as a utility to load and safely execute user-defined functions.

2   System Overview

LADY targets the .NET platform, and therefore revolves around assemblies: containers for compiled code, and the unit of deployment in .NET. A .NET application is compiled into one or more assemblies. Each application has exactly one *main assembly*, which contains that application's entry point, and is stored as an executable .EXE file. Non executable assemblies are stored as .DLL files.[2]

Assemblies can include functionality from other assemblies by referencing them. Assemblies can be cryptographically signed by their creators, which lets the .NET runtime verify that they are authentic before loading them. When an assembly is signed, the public key of the signer is combined with the assembly's

---

[1] LADY is an acronym for Loading Assemblies Dynamically.
[2] Both .EXE and .DLL files have the same Portable Executable file format.

short name, its localization culture, and its version number to produce a so-called *strong name*. Strong names are globally unique and therefore allow assemblies to reference each other by name without ambiguity.

2.1   The Assembly Lookup Service

A central feature of LADY is its globally available and resilient Assembly Lookup Service (ALS), which resolves assembly names to URIs. LADY can also determine if an assembly has been superseded by a more recent version, and provides the functionality for obtaining assemblies in a number of ways, for example by downloading them via HTTP. Additional features include caching of assemblies, automatic resolution of assembly references (for example during deserialization), prefetching of assemblies based on dependencies, and creation of sandboxed environments to execute partially trusted code.

LADY stores meta-information about assemblies in a cloud database. To load an assembly and proceed with execution, an application may have to wait for a database lookup to complete. We therefore value predictable and low-latency lookup performance. Currently, we use Amazon's DynamoDB [5] as our database backend as it has an official API for C#, boasts scalability, and offers predictable performance. LADY does not rely on other advanced database features and can therefore easily be adapted for other database systems.

LADY manages a database D containing information on all known assemblies. Each assembly m has a name, a culture, a public-key token, and a version number. The *Strong Name (SN)* of m uniquely identifies m in the universe Ψ of all assemblies, and is defined by the tuple:

$$SN(m) = (m.name, m.culture, m.publicKeyToken, m.version)$$

We define the *Packet Name (PN)* of m similarly to the strong name tuple, but without the version number:

$$PN(m) = (m.name, m.culture, m.publicKeyToken)$$

Any two assemblies $m, n \in \Psi, SN(m) \neq SN(n)$ are members of an implicitly related sequence of assemblies if they have the same packet name $PN(m) = PN(n)$. Because version numbers are assumed to be monotonically increasing, packet names define ordered set of assemblies that relate in name, culture, and public key tokens; typically used for different version of the same code-base as it evolves over time.

Let $D \subseteq \Psi$ be the set of all assemblies known to LADY. For each assembly $m \in D$, LADY maintains two types of information: base records and assembly records. *Base records* maps the packet name $PN(m)$ to the strong names of all assemblies $m' \in D$ with $PN(m) = PN(m')$. This enables LADY to support wildcard queries for specific version on an assembly, like for the most recent one in D. *Assembly records* maps $SN(m)$ to the network location and protocol for downloading the assembly code for m. For instance, the record may contain a download URL, or a NuGet[3] package identifier and version.

2.2 Security

The *public-key token* of assembly m is defined by the .NET framework as an 8-byte hash of the public key matching the private key used to sign m; commonly displayed as a 16-digit hexadecimal number. Since the public-key token is determined by the signer of m, and also incorporated into $SN(m)$, it is not possible to modify m without knowing the private key (or breaking cryptography building blocks). With access to the source code for m, the code can be recompiled and signed it with a different key. However, the resulting assembly m' would have a different packet name $PN(m') \neq PN(m)$, and thus also a different strong name. As such, an attacker cannot successfully trick correct processes to execute m' instead of m.

While the assembly naming scheme protects against malicious tampering with the code, we would also like to guarantee that LADY can provide genuine and valid download locations. Even if an attacker cannot manufacture fake assemblies, he could potentially register an assembly with invalid meta-information, rendering it unobtainable through LADY. To guard against such attacks, we require registrations of assemblies to be in the form of signed messages, where the message signer's public key must correspond to the public key token of the assembly in question. This ensures that the person or program registering the assembly is the same as the signer of the assembly, and third parties cannot register invalid information about assemblies.

2.3 Assembly Registration

To add new assemblies to LADY, we provide the `register` command-line utility. The `register` utility does not directly update LADY's cloud database D, but instead constructs a registration message $\{REG, m\}_k$, signed with key k, provided by the software vendor. The signed message can then be sent to LADY. Upon receiving $\{REG, m\}_k$, LADY verifies that k matches $m.publicKeyToken$, before adding m to D. For instance, to register the MyLib assembly, the vendor will run:

```
$ lady register -a MyLib.dll -p MyLib -v 1.2 -k mykey.pfx
```

Here, the assembly file is specified with the -a option. The utility uses reflection to extract the strong name $SN(\texttt{MyLib.dll})$. The -p and -v options specify a NuGet package identifier and version, respectively, so the assembly will be registered as obtainable by using NuGet to install version 1.2 of the MyLibrary package. Finally, the `mykey.pfx` file, specified with the -k option, contains the

---

[3] NuGet is a package management system, closely integrated with Microsoft Visual Studio.

key pair for signing the registration message. If the public key does not match the public key token of the assembly, registration will fail. The `register` utility is implemented to be suitable for scripting and integration into existing build systems.

We have settled initially on this model where assemblies must be explicitly added to LADY by their vendors. It would be possible to create automated tools for registering assemblies that have been created by others. For example, we could integrate with existing package management systems like NuGet and scan all newly uploaded packages for strong-named assemblies, automatically registering them with LADY. This could improve the coverage of our lookup service, and possibly be more convenient for developers, but we have deferred that investigation to future work.

2.4   Loading Assemblies Explicitly

LADY provides the `LoadAssembly` method to applications for explicit loading of assemblies at runtime. For example usage, consider the configuration parser in Code Listing 1. Here, the code calls `LoadAssembly` at an early point in the program execution[4] to load the YamlDotNet library, identified by its assembly name and public key token. The culture is left unspecified and defaults to "neutral". The latest release with major version 3 is requested by specifying "3.*" as the version number.

YamlDotNet provides functionality for parsing of YAML—a human-friendly serialization language commonly used in configuration files. Bugs in configuration parsing can be unpleasant and can potentially render the application exploitable. By loading the code dynamically with LADY, the application can ensure that it always has the latest available version of YamlDotNet library, thereby picking up any bugfix releases promptly and automatically. It will not be necessary to deploy a new version of the application just because a bug has been discovered and fixed in one of the libraries that it depends on.

Once an assembly is loaded, its functionality can be accessed programmatically in two ways. The first approach is to use reflection to instantiate objects and invoke methods. This is exemplified in the method `AccessUsingReflection`, which parses a YAML string into a `Config` object. The implementation instantiates a `Serializer` object using reflection, before invoking its `Deserialize` method. This approach certainly works, but there are some factors that make it cumbersome:

1. Types must be specified as strings with fully-qualified type names: a verbose and error-prone task. The verbosity stacks up when multiple types are involved; in the example, the `Serializer` constructor requires a `CamelCaseNamingConvention` object, which must be instantiated first.

---

[4] In this example, the call happens in the static constructor of the `ConfigParser` class.

Code Listing 1. Example to illustrate dynamic loading of an assembly, and how to access its functionality.

```
using YamlDotNet.Serialization;
using YamlDotNet.Serialization.NamingConventions;

class ConfigParser
{
    static readonly ILady lady = LadyFactory.Init();
    static readonly Assembly yaml = lady.LoadAssembly(
        name: "YamlDotNet", publicKeyToken: "ec19458f3c15af5e", version: "3.*");

    public class Config
    {
        public string ConferenceName { get; set; }
        public DateTime Deadline { get; set; }
    }

    public static Config AccessUsingReflection(string data)
    {
        var namingConvention = yaml.NewInstance(
            "YamlDotNet.Serialization.NamingConventions.CamelCaseNamingConvention");
        dynamic d = yaml.NewInstance("YamlDotNet.Serialization.Deserializer",
                            null, namingConvention, false);
        return d.Deserialize<Config>(new StringReader(data));
    }

    public static Config StaticallyTypedAccess(string data)
    {
        var d = new Deserializer(namingConvention: new CamelCaseNamingConvention());
        return d.Deserialize<Config>(new StringReader(data));
    }
}
```

2. Constructor arguments are specified as `object` instances, without static type checking. Method calls have similar constraints. The invocation of `Deserialize` looks superficially as if it might be type-checked by the compiler, but in fact the code relies on `dynamic` variables, which are assumed to support any and all operations, and defer actual type checking until run-time.
3. Named and default arguments cannot be used. Combined with the lack of static type checking, this often leads to long lists of `null` arguments where any non-default arguments must be positioned with great care.
4. Finally, reflective invocations add overhead, which may be an issue if they end up sitting on the critical path.

Also note that the `NewInstance` method used in this example is itself an extension method that we have implemented as a convenience in a utility library. `NewInstance` fills in default values for various optional hooks and packs the constructor arguments into an array. Without relying on such helpers, the object instantiation code would have to be even more verbose.

The drawbacks of reflection might call into question the practical utility of loading assemblies dynamically. Fortunately, there is a way to get the best of both worlds, and benefits from static type checking and related IDE features like code completion while still using LADY under the hood. This second approach is to compile the application with the most recent assembly versions that are available at build time, and override the assembly resolution mechanism at runtime so that LADY gets a chance to load any newer versions that may have been released since then.

The `StaticallyTypedAccess` in Code Listing 1 demonstrates this approach. The method is similar to `AccessUsingReflection`, but with the clarity and safety of normal syntax, with type checking, and without the overhead of reflective calls. This works because the compiler has access to the YamlDotNet assembly at compile time, but also means that a reference to that specific version of YamlDotNet is included in the application's assembly. However, assembly references are not resolved immediately when an application starts. The .NET runtime resolves assemblies on demand, when a method that references the assembly is first entered. On startup, LADY hooks into the assembly loading mechanism by overriding certain event handlers, and therefore gets to decide how exactly to resolve an assembly reference. By the time `StaticallyTypedAccess` is invoked, LADY has already been instructed to load the *latest* version of the YamlDotNet assembly, so that is the version that will be used.

2.5 Loading Referenced Assemblies

In additional to explicitly loaded assemblies, as described in Sect. 2.4, LADY supports loading assemblies by references in the code. For example, an application might load a plug-in assembly through LADY, and the plug-in might contain references to other assemblies that have not been loaded, or even installed. Another scenario that may trigger assembly resolution is during object deserialization when data contain references to types defined in unresolved assemblies. A prime advantage of LADY is that any blob of serialized data can be deserialized at any node and at any time, so long as all of the referenced assemblies have been registered with LADY.

Resolving assemblies on demand raises the question of how to deal with conflicting versions. If plug-ins A and B both reference assembly C, but demand different versions of C, or if two blobs of data were serialized with different versions of an assembly, a potential conflict will result. One technical possibility is to load multiple versions of the same assembly. However, this is not a recommended practice, due to the confusion that may arise when types have the same name but different identities [15].

In some cases, the right thing to do is simply to load the most recent assembly version that exists. Of course, this only works for versions that are backwards-compatible. There is a standard called *semantic versioning* [1] that would resolve this issue if it was adopted universally. With semantic versioning, the major version number is bumped whenever a backwards-incompatible change is introduced. However, a 2014 survey indicates that this standard remains to be widely adopted [19]. Therefore, LADY takes a more conservative approach and does not attempt to infer automatically if two versions of an assembly are compatible. Instead, we rely on hints from the client, in the form of a compatibility policy, which is a simple boolean-valued function that may be specified programmatically. Whenever LADY must determine if a given pair of assembly versions should be considered compatible, it consults the compatibility policy by invoking this function. The default compatibility policy is a slightly stricter version of semantic versioning: if both the major and the minor version numbers are equal, the assemblies are considered compatible. (Build and revision numbers may still differ.)

Armed with this concept of compatibility policies, LADY takes the following approach to assembly resolution: the assembly lookup service is first queried to retrieve all known versions of the assembly in question, and the most recent version that is compatible with the requested version is then selected. LADY then proceeds to obtain and load this specific version of the assembly. For example, if an assembly is registered with versions 1.0.1, 1.0.2, and 1.0.3, and three plug-ins each reference one of these versions, then the actual version that will be loaded (under the default compatibility policy) is 1.0.3, regardless of the order in which the plug-ins are loaded.

2.6   Loading Partially Trusted Code

While plug-ins might be considered trusted code by some applications, there are many cases where applications wish to load and execute partially trusted code with a limited set of permissions. For example, distributed data processing models like MapReduce rely on user-defined functions for flexibility and expressiveness. When invoking these functions, it is prudent to do so from a sandboxed environment with restricted capabilities for hazardous actions like network and file I/O. On the surface, this appears to preclude the use of LADY from user-defined functions, since network and file I/O are needed to locate and obtain an assembly, and a full, unrestricted permission set is required in order to override the assembly resolution mechanism.

LADY resolves this problem by offering a *sandbox* abstraction based on .NET application domains [14]. Application domains provide an isolation boundary for security, reliability and versioning, and for loading assemblies. They are typically created by runtime hosts—which are responsible for bootstrapping the common language runtime before an application is run—but a process can create any number of additional application domains to further separate and isolate execution of code.

LADY must be initialized (using the `LadyFactory` class) exactly once per process, and from a fully trusted application domain—typically the initial application domain that is created on startup. This singular instance of LADY thus executes with unrestricted permissions, as required. However, users may create additional sandboxes using the `MakeSandbox` method, as exemplified in Code

Code Listing 2. Example code using LADY to load partially trusted user-defined functions inside a sandbox.

```
using Microsoft.Hadoop.MapReduce;

class MapReduceSandbox : LadySandbox
{
    public MapReduceSandbox(object x) : base(x) { }

    public override void Play()
    {
        // Load an assembly with partially trusted UDFs
        Assembly myUDFs = lady.LoadAssembly(name: "MyUDFs",
            publicKeyToken: "8c11fe16618d1673", version: "*");
        var mapper = myUDFs.NewInstance("WordCountMapper") as MapperBase;
        var reducer = myUDFs.NewInstance("WordCountReducer") as ReducerCombinerBase;
        // The mapper and reducer may now be invoked in relative safety; they
        // cannot access the file system, network, environment, etc.
    }
}

class MapReduceProgram
{
    static void Main(string[] args)
    {
        LadyFactory.Init().MakeSandbox(typeof(MapReduceSandbox)).Play();
    }
}
```

Listing 2, where the user-defined functions required for a MapReduce job are loaded inside a sandbox. The sandbox is a partially trusted application domain that is initialized with a figurative umbilical cord that leads back to the fully trusted application domain. Concretely, the application must implement a subclass of **LadySandbox** with a single-argument constructor that passes on a special proxy object to its base class. The proxy object is named **x** in the example, and constitutes the umbilical cord.

Inside a sandbox, execution starts in the **Play** method. Any **LoadAssembly** calls made inside the sandbox get routed back to LADY using cross-domain remote method calls on the proxy object. LADY will determine which version to load, as described in the previous section, and retrieve the assembly data, either directly from its cache, or by first obtaining the assembly. The assembly data is then passed back to the sandbox, where it is loaded into the partially trusted application domain. Assemblies that must be loaded due to code references or during deserialization are handled similarly. This approach effectively grants sandboxes full capabilities with regards to loading of assemblies, so long as this happens through LADY. The implementation details of how to communicate across application domains are hidden. All sandboxes also share the benefit of a common cache.

3   DNS Integration

With the assembly lookup service, LADY adds a level of indirection to loading assemblies. This relieves applications of various responsibilities, and enables several useful applications. It also raises some important concerns:

Availability. If the assembly lookup service becomes unavailable, applications may also experience various forms of unavailability. For example, it might not be possible to start a scheduled MapReduce job because the assembly that contains the required user-defined map function cannot be located. (LADY does maintain a client-side cache, but it could also be missing there.)

Scalability. LADY is designed to serve numerous application instances running in many different locations all over the globe. The aggregated number of queries for assembly information is expected to be large. While cloud databases like DynamoDB generally provide great scalability, this does come with a monetary cost. As the volume of requests grows, the financial cost of operating the assembly lookup service could become prohibitive.

Latency. Applications may have a tendency to load assemblies sequentially, either as a result of logical dependencies or due to the sequential nature of their execution. Any extra latency incurred when an assembly is loaded may thus stack up and result in unwanted user-perceptible delays, for example on application startup. We should therefore strive to minimize the latency of individual assembly lookups.

To address these three concerns, we have integrated the assembly lookup service with the Domain Name System (DNS). Given that registration of new assemblies is expected to be a relatively rare event, the assembly lookup service's workload is almost read-only. This implies that caching can be an effective way to reduce both load and latency, and DNS is a globally distributed cache readily available and with extremely high availability. While the most common use of DNS is to associate globally unique host names with IP addresses, we use it to associate assemblies strong name with their meta-data, such that $\forall m \in D$, Resolve $(SN(m)) \implies m$. This approach addresses all three concerns above, since DNS is globally available, will significantly alleviate the load on the cloud database, and can generally be accessed with low latency.

For the DNS integration we rely on our previous work with the Jovaku system [17]. Jovaku mirrors database keys as labels in the DNS namespace. Database lookups can then be translated into DNS request on the client side. A relay-node is set up close to the cloud database that performs the opposite translation. Figure 1 shows how this works in *(a)* the baseline case where we have no DNS integration, *(b)* the case where we miss the DNS cache, and *(c)* the common case where we hit the DNS cache. This approach is very effective at reducing latency for read-mostly workloads like the one exhibited in LADY [17].

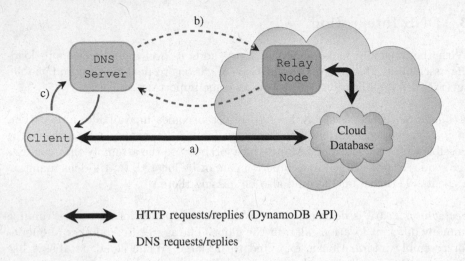

$\longleftrightarrow$ HTTP requests/replies (DynamoDB API)

$\nearrow$ DNS requests/replies

Fig. 1. Assembly lookups with and without DNS integration.

4   Evaluation

Many of the reasons to adopt LADY are anecdotal as it is hard to quantify benefits like flexibility and extensibility. However, there are concrete performance benefits, as we will demonstrate in this section. As a case study we rely on our previous work on *satellite execution* [18]: a technique to reduce latency for applications that interact repeatedly with cloud services by temporarily offloading code so it executes in closer proximity to the cloud. For this, we developed the *mobile functions* a programming abstraction, and an *execution server* capable of receiving and executing such functions.

In our original implementation of satellite execution, referred to as *baseline*, the execution server receives mobile functions as serialized objects. Before the mobile function's entry point can be invoked, the objects must be deserialized and allowed to execute. Deserialization can fail if an unresolvable assembly reference is encountered. In the baseline implementation, we handled this and other assembly resolution errors by returning an error to the client. The client would then have to upload the missing assembly to the execution server, before making a new attempt to offload the mobile function.

The baseline design for offloading mobile functions was grounded in two assumptions: (1) the client will have the code for any assemblies that are referenced by its mobile functions, and (2) the execution server has some means to resolve any assembly resolution errors it encounters. Although both assumptions are reasonable, they can cause excessive back-and-forth communication between the client and the execution server for mobile functions that depend on multiple assemblies.

Refactoring our baseline implementation to use LADY made the implementation of both the execution server and the client simpler, and the services became more robus. The modified execution server can deserialize mobile functions without having to interact with the client, trusting LADY to resolve assemblies. Similarly, a mobile function can be invoked through its entry point, and any referenced assemblies will be loaded through LADY. Additionally, we use LADY's sandboxing support, as described in Sect. 2.6, to isolate the execution of mobile functions in a separate application domain, minimizing the potential for disruption by misbehaving mobile functions.

We envision diverse applications for LADY, that may exhibit many different access and usage patterns. Performance also depends on how an application is deployed geographically, since this affects the latency to access both the execution server and DNS. Therefore, we use synthetic workloads in our experiments, and run experiments from multiple geographical locations, comparing the baseline and LADY implementation.

We set up experiments with the execution server hosted on Amazon EC2 nodes in various geographically zones. We chose Ireland, California, Singapore and Sydney to exhibit variations in the routing distance to the client located in Norway. The EC2 nodes was equipped with 4 GB memory and a dual-core 2.5 GHz 64 bit vCPU, and the client was a desktop machine equipped with 64 GB memory and a quad-core Intel Xeon E5-1620 3.7 GHz CPU. Typical observed ping latency and hop count between the client and the execution nodes is illustrated in Table 1.

Table 1. Machines involved in evaluating the effectiveness of using LADY for reducing latency in satellite execution, along with the latency and hop count to the desktop client located in Tromsø.

Location	EC2 type	Ping latency	# Hops
Ireland	t2.medium	64 ms	15
California	t2.medium	155 ms	17
Singapore	t2.medium	339 ms	20
Sydney	t2.medium	365 ms	12

We stored a set of assemblies in a separate table in the DynamoDB database in each of the locations, acting as our package management system, and registered these assemblies with LADY. We then tried executing mobile functions with a varying number of assembly dependencies that would be resolved sequentially as the execution progressed.

Figure 2 shows the difference in latency between our baseline implementation of satellite execution, and the refactored implementation using LADY. Given that the motivation for satellite execution is to reduce latency, the observed performance benefits of using LADY are highly significant. Even when a mobile function has no additional dependencies beyond its own assembly, it saves one round-trip between the execution server and the client. By comparison, LADY is only making low-latency DNS requests to resolve assemblies, coupled with

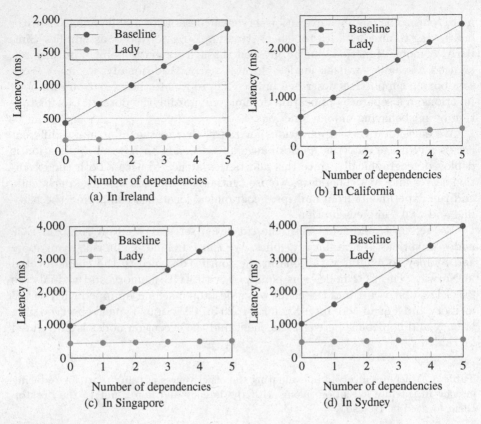

Fig. 2. Observed mean latency when executing a mobile function with a varying number of assembly dependencies with and without LADY for various geographical locations.

lookups to DynamoDB to retrieve the assembly data. The reason that we save latency in this scenario is two-fold:

1. We substitute long round-trips between Norway and Ireland with much faster DNS lookups.
2. Assembly data is stored in the cloud, instead of at the client. Since we need to load the assemblies at a node in the cloud, this data placement is more optimal.

Not every application that employs LADY will benefit from the same fortuitous circumstances. For example, if all assemblies are obtained in advance, LADY will add the overhead of one DNS lookup for each resolved assembly. In exchange, LADY guarantees that the most recent assembly version is found. Whether this is a reasonable trade-off depends on the application's requirements for flexibility and extensibility. Figure 3 shows the overhead added by LADY for the satellite execution scenario, when the experiment is set up so that all assembly data has been obtained in advance. The overhead is proportional to the number of

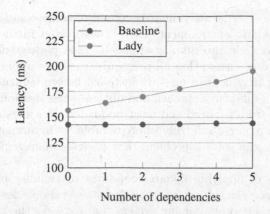

Fig. 3. Observed mean latency when executing a mobile function with all assembly data present, where LADY performs DNS lookups to ensure that the latest versions will be loaded, and the baseline does not check for updated versions.

dependencies, as each dependency adds another DNS lookup. The actual resource requirements for performing a DNS lookup are negligible.

It is worth noting that the overhead illustrated in Fig. 3 is the worst-case scenario. The mobile function used in the experiment does not do any useful work, and will be stalled while waiting for LADY to check for updated assembly versions. In a more realistic scenario the mobile function will do useful work while LADY checks for updated assemblies in the background, and might not need to stall when the next dependency boundary is crossed.

5 Related Work

Package-management systems backed by online code repositories have become common for deploying applications in many modern systems. For instance, popular operating systems like the Linux based Ubuntu and Debian systems rely on the Advanced Package Tool (APT) for software installation, upgrade, and dependency resolving [8]. To host the code online, these communities depend on donated third-party servers, known as mirrors, to distribute their software to millions of end-users. Although, these software mirroring infrastructures lack the mechanisms to deal with the wide-range of faults that can occur, solutions for resilient software mirroring has been demonstrated [12,13]. Systems distributed commercially, like Microsoft Windows, often come equipped with proprietary mechanisms for distributing software updates [7] and are generally less vulnerable to intrusions.

In the framework for code updates described by [9], semi-automatically generated software patches include both the updated code and the code for making the transition safely. By using the Typed Assembly Language, these patches can consist of verifiable native code, which is highly beneficial to system safety.

However, these systems are primarily geared towards *installing* applications into a relatively static environment. LADY goes a step further and supports dynamic loading of code into running applications. A package management system generally aims to ensure that all prerequisites for an application—e.g., the assemblies that it may depend on—are installed before launching the application. LADY takes a different approach and obtains these assemblies on demand, if and when they are referenced and must be loaded. In some cases, the assemblies that may be required are truly unpredictable, as in our satellite execution system, and LADY can solve a problem that package management systems fail to address.

The problems of applying dynamic updates of running programs is well known and has been the subject of research for several decades [20]. DYMOS [2] is perhaps the earliest programming system that explores the ideas of dynamic updates of functions, types, and data objects. DYMOS is based on the StarMod extension of the Modula language, and it is unclear to what extent the proposed mechanisms are applicable to modern application platforms like .NET. Other programming languages, like Standard ML, have also been demonstrated to support dynamic replacement of program modules during execution [6]. Our approach specifically targets the .NET platform and leverages the capabilities of the .NET application domains and their customizable assembly resolution mechanism.

The general complexity of developing and deploying modern distributed applications, which span a variety of mobile devices, personal computers, and cloud services, has been recognized as a new challenge. Users expect applications and their state to follow them across devices, and to realize this functionality, one or more cloud services must usually be involved in the background. Sapphire [23] is a recent and comprehensive system that approaches this problem by making deployment more configurable and customizable, separating the deployment logic from the application logic. The aim is to allow deployment decisions to be changed, without major associated code changes. Applications are factored into collections of location-independent objects, communicating through remote procedure calls.

We envision LADY as a particularly useful sidekick for the design and implementation of this new generation of highly flexible and extensible distributed systems. By facilitating the on-demand resolution of assemblies, system architectures can make the simplifying assumption that all participants will share a common code base, and enjoy greater freedom in their deployment decisions.

6   Conclusion

A key idea underlying LADY is to make all code live in a globally accessible namespace so that it can be referenced unambiguously by name and retrieved on demand in any context. Strong-named .NET assemblies already have globally unique names, but the ability to load code in any context is missing. LADY fills in this gap by creating a lookup service for assemblies, and by implementing the mechanisms for obtaining code on demand. This aligns with a vision where code can be deployed only once, and then instantiated anywhere, in various configurations.

The general approach of loading code on demand means that *distribution of code is decoupled from distribution of state*. In other words, code does not have to be propagated through a distributed system along the same communication paths as data. Consider, for example, a system where nodes communicate over a gossip-based protocol. A message might contain serialized data and traverse multiple edges of the gossip graph before it arrives at a node where the data must be deserialized. Any intermediate nodes will only be passing along the serialized data and may never have a need for the associated code. But the sender does not know if the target node has the requisite code installed. So to be safe, the sender will have to include the possibly redundant code as part of its outgoing message, or the design must be complicated in some other way, for example by adding additional rounds of gossip to retrieve the code.

With the separation of concerns that LADY offers, the design of such gossip-based systems could be simplified, since code would be retrieved on demand via an entirely independent mechanism whenever data was deserialized. Our satellite execution refactoring in Sect. 4 also helps to illustrate how LADY can simplify the design of other distributed systems, to improve extensibility and serve as a convenient foundation for mobile code.

References

1. Semantic Versioning. http://semver.org/
2. Cook, R.P., Lee, I.: DYMOS: A Dynamic Modification System, vol. 8, pp. 201–202. ACM, New York (1983). http://doi.acm.org/10.1145/1006140.1006188
3. Daubert, E., Fouquet, F., Barais, O., Nain, G., Sunye, G., Jezequel, J.M., Pazat, J.L., Morin, B.: A models@runtime framework for designing and managing service-based applications. In: 2012 Workshop on European Software Services and Systems Research - Results and Challenges (S-Cube), pp. 10–11, June 2012
4. Dean, J., Ghemawat, S.: MapReduce: simplified data processing on large clusters. In: Proceedings of the 6th Symposium on Operating Systems Design and Implementation, OSDI 2004, pp. 137–150. USENIX Association (2004)
5. DeCandia, G., Hastorun, D., Jampani, M., Kakulapati, G., Lakshman, A., Pilchin, A., Sivasubramanian, S., Vosshall, P., Vogels, W.: Dynamo: Amazon's highly available key-value store. In: Proceedings of the 21st ACM SIGOPS Symposium on Operating Systems Principles, SOSP 2007, pp. 205–220. ACM (2007). http://doi.acm.org/10.1145/1294261.1294281
6. Gilmore, S., Kirli, D., Walton, C.: Dynamic ML without dynamic types. Technical report, University of Edinburgh (1997)
7. Gkantsidis, C., Karagiannis, T., Rodriguez, P., Vojnović, M.: Planet scale software updates. ACM SIGCOMM Comput. Commun. Rev. **36**(4), 423–434 (2006)
8. Hertzog, R., Mas, R.: The Debian Administrator's Handbook, 1st edn. Freexian SARL (2006). https://debian-handbook.info/
9. Hicks, M., Moore, J.T., Nettles, S.: Dynamic software updating. In: Proceedings of the ACM SIGPLAN 2001 Conference on Programming Language Design and Implementation, PLDI 2001, NY, USA, pp. 13–23 (2001). http://doi.acm.org/10.1145/378795.378798

10. Johansen, D., Lauvset, K.J., van Renesse, R., Schneider, F.B., Sudmann, N.P., Jacobsen, K.: A TACOMA retrospective. Soft. Pract. Exp. **32**, 605–619 (2001)
11. Johansen, D., Marzullo, K., Lauvset, K.J.: An approach towards an agent computing environment. In: ICDCS 2099 Workshop on Middleware (1999)
12. Johansen, H., Johansen, D.: Resilient software mirroring with untrusted third parties. In: Proceedings of the 1st ACM Workshop on Hot Topics in Software Upgrades (HotSWUp), October 2008
13. Johansen, H., Johansen, D., van Renesse, R.: Firepatch: secure and time-critical dissemination of software patches. In: Proceedings of the 22nd IFIP International Information Security Conference, pp. 373–384. IFIP, May 2007
14. Microsoft: Application Domains (2015). http://msdn.microsoft.com/en-us/library/cxk374d9%28v=vs.90%29.aspx
15. Microsoft Developer Network: Best Practices for Assembly Loading. Microsoft, NET Framework 4.6 and 4.5 edn. (2016). https://msdn.microsoft.com/en-us/library/dd153782(v=vs.110).aspx
16. Olston, C., Reed, B., Srivastava, U., Kumar, R., Tomkins, A.: Pig Latin: a not-so-foreign language for data processing. In: Proceedings of the 2008 ACM SIGMOD International Conference on Management of Data, SIGMOD 2008, pp. 1099–1110. ACM (2008). http://doi.acm.org/10.1145/1376616.1376726
17. Pettersen, R., Valvåg, S.V., Kvalnes, A., Johansen, D.: Jovaku: globally distributed caching for cloud database services using DNS. In: IEEE International Conference on Mobile Cloud Computing, Services, and Engineering, pp. 127–135 (2014)
18. Pettersen, R., Valvåg, S.V., Kvalnes, A., Johansen, D.: Cloud-side execution of database queries for mobile applications. In: Proceedings of the 5th International Conference on Cloud Computing and Services Science, CLOSER 2015, pp. 586–594 (2015)
19. Raemaekers, S., van Deursen, A., Visser, J.: Semantic versioning versus breaking changes: a study of the maven repository. In: 2014 IEEE 14th International Working Conference on Source Code Analysis and Manipulation (SCAM), pp. 215–224, September 2014
20. Segal, M., Frieder, O.: On-the-fly program modification: systems for dynamic updating. Software **10**(2), 53–65 (1993). IEEE
21. Valvåg, S.V., Johansen, D., Kvalnes, A.: Cogset: a high performance MapReduce engine. Concurrency Comput. Pract. Exp. **25**(1), 2–23 (2013). doi:10.1002/cpe.2827
22. Yu, Y., Isard, M., Fetterly, D., Budiu, M., Erlingsson, Ú., Gunda, P.K., Currey, J.: DryadLINQ: a system for general-purpose distributed data-parallel computing using a high-level language. In: Proceedings of the 8th USENIX Conference on Operating Systems Design and Implementation, OSDI 2008, pp. 1–14. USENIX Association (2008). http://dl.acm.org/citation.cfm?id=1855741.1855742
23. Zhang, I., Szekeres, A., Aken, D.V., Ackerman, I., Gribble, S.D., Krishnamurthy, A., Levy, H.M.: Customizable and extensible deployment for mobile/cloud applications. In: 11th USENIX Symposium on Operating Systems Design and Implementation OSDI 2014, Broomfield, CO, pp. 97–112. USENIX Association, October 2014. https://www.usenix.org/conference/osdi14/technical-sessions/presentation/zhang

Investigation of Impacts on Network Performance in the Advance of a Microservice Design

Nane Kratzke[✉] and Peter-Christian Quint

Center of Excellence for Communication, Systems and Applications,
(CoSA) Lübeck University of Applied Sciences, 23562 Lübeck, Germany
{nane.kratzke,peter-christian.quint}@fh-luebeck.de

Abstract. Due to REST-based protocols, microservice architectures are inherently horizontally scalable. That might be why the microservice architectural style is getting more and more attention for cloud-native application engineering. Corresponding microservice architectures often rely on a complex technology stack which includes containers, elastic platforms and software defined networks. Astonishingly, there are almost no specialized tools to figure out performance impacts (coming along with this microservice architectural style) in the upfront of a microservice design. Therefore, we propose a benchmarking solution intentionally designed for this upfront design phase. Furthermore, we evaluate our benchmark and present some performance data to reflect some often heard cloud-native application performance rules (or myths).

Keywords: Cloud-native application · Microservice · Container · Cluster · Elastic platform · Network · Performance · Reference · Benchmark · REST · SDN · Software-defined network

1 Introduction

Recent popularity of cloud-native applications, container technologies, notably Docker [5], and elastic container platform solutions like Kubernetes/Borg [21], Apache Mesos [8] and Docker Swarm [6] show the increasing interest in container and container cluster technologies in cloud computing. Alongside this increasing interest the term *microservices* is often mentioned in one breath with container technologies [15].

> *"In short, the microservice architectural style is an approach to developing a single application as a suite of small services, each running in its **own process** and communicating with lightweight mechanisms, often an **HTTP resource API**. [...] Services are independently deployable and scalable, each service also provides a firm module boundary, even allowing for different services to be written in **different programming languages**."*
> (Blog Post from Martin Fowler)

© Springer International Publishing AG 2017
M. Helfert et al. (Eds.): CLOSER 2016, CCIS 740, pp. 187–208, 2017.
DOI: 10.1007/978-3-319-62594-2_10

Container technologies seem like a perfect fit for this microservice architectural approach, which has been made popular by companies like Google, Facebook, Netflix or Twitter for large scale and elastic distributed system deployments. Container solutions providing a standard runtime, image format, and build system for Linux containers. These containers are deployable to any Infrastructure as a Service (IaaS) environment. Microservice architectures are inherently horizontally scalable mainly due to REST-based protocols [7]. But, there are almost no specialized tools to figure out performance impacts coming along with this architectural style. These performance impacts might be due to fundamental design decisions

- to use different programming languages for different services
- to use REST-APIs for service composition,
- to decide for few but big or many but small messages in REST-APIs,
- to encapsulate services into containers,
- to use container clusters to handle complexity of deployments,
- to deploy services to different virtual machine types,
- to deploy services to different cloud providers.

This list is likely not complete. Nevertheless, it should be obvious for the reader that network performance is a critical aspect for the overall performance of microservice architectures. And a system architect should be aware of these impacts to ponder what impacts are acceptable or not. Some design decisions, like to use a specific programming language, to design APIs for few but big or many but small messages have to be made very early. For instance: It is simply not feasible to develop a service and run performance benchmarks in the aftermath to find out that the used programming language is known to have problems with specific message sizes. Of course there exist network benchmarks to measure network performance of infrastructures (for example *iperf* [2]) or for specific web applications (for example *httperf* [9]). However, these tools do not support engineers directly in figuring out what the performance impact of specific programming languages, message sizes, containerization, different virtual machine types, cloud provider infrastructures, etc. **in the upfront of a microservice design** might be.

Therefore, we propose a highly automated benchmarking solution called *ppbench* in Sect. 2. Our proposal is intentionally designed for the microservice domain and covers above mentioned performance impacts for upfront design decisions. In Sect. 3 we consider how to visualize benchmark data statistically appropriate to get a better understanding of the above mentioned impacts. We implemented our solution proposal *ppbench* as a software prototype and describe how to download, install and use the benchmark in Sect. 4. We present some exemplary results in Sect. 5. The performed experiments have been derived from above mentioned design questions as examples how to use *ppbench* to answer microservice related performance questions. Section 6 reflects related work and shows how *ppbench* is different compared with already existing network benchmarks. We conclude our findings and summarize some astonishing and sometimes overseen performance impacts on network performances in Sect. 7.

2   Benchmark Design

The proposed benchmark is intentionally designed to support upfront design decision making regarding microservice related performance aspects. To some degree the benchmark might be useful to measure general HTTP performance as well. But this is not the intended purpose. The benchmark is aligned to a general reference model of cloud-native applications (see Fig. 1). This reference model for cloud-native applications is derived and explained in more details in [10]. It assumes that cloud-native applications are composed of services. These services can be deployed via deployment units (often containers) and are operated on an elastic platform (often clustered container runtime environments like Kubernetes, Mesos, Docker Swarm, etc.). The elastic platforms are providing several necessary elasticity features for cloud-native application components (see Fig. 1). The platform hosts (virtual machines) are operated on public or private IaaS services of arbitrary cloud service providers. So, several performance sensitive aspects can be derived from that model and should be covered by a benchmark for microservice-based cloud-native application engineering.

- **Virtual machine types** ❶ can obviously affect processing and network performance.
- **Container** ❷ simplify deployments but can effect network performance due to enlarging the network stack.
- **Software defined networks (SDN)** ❸ are used by elastic platforms, simplify orchestration and networking between containers, but they can affect network performance as well.
- **Programming languages** ❹ can affect processing performance of (containerized) applications and networking libraries of these languages can affect network performance.

To analyze the impact of container, software defined network (SDN) layers and machine types on the network performance of distributed microservice-based systems, several basic experiment settings are provided (see Fig. 2). All experiments rely on a basic *ping-pong* system which provides a REST-based protocol to exchange data. This kind of service coupling is commonly used in microservice architectures. The *ping* and *pong* services are implemented in different languages (see Table 1) to compare programming language and HTTP library impact on network performance ❹. All HTTP libraries are used in a **multithreaded way**, so that services should benefit from multi-core machines. *Ppbench* is designed to be extended for arbitrary programming languages and HTTP libraries, so the list can be easily extended. Currently, provided languages are covering high and medium ranked programming languages like Java (Rank 1) and Ruby (Rank 12) from the TOP 20 and Dart (Rank 28) and Go (Rank 44) from the TOP 50 of the TIOBE[1] programming index.

HTTP requests can be sent to this *ping-pong* system from a *siege* host. Via this request the inner message length between *pong* and *ping* server can be

---

[1] http://www.tiobe.com/index.php/content/paperinfo/tpci/index.html,   September 2015.

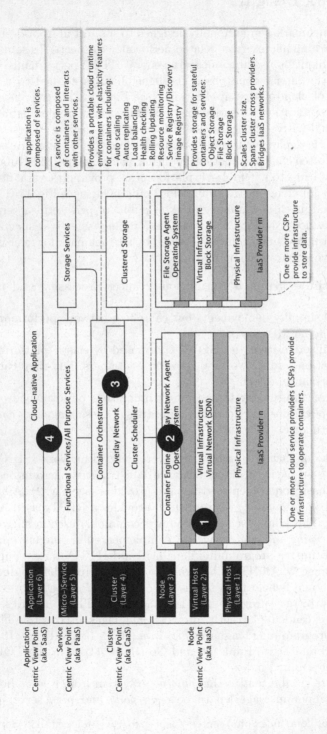

Fig. 1. Cloud-native application reference model, taken from [10].

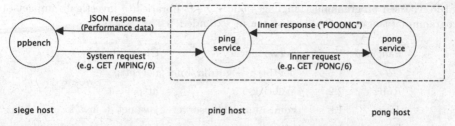

(a) **Bare** experiment to identify reference performance with containerization and SDN applied

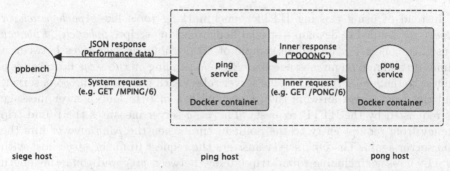

(b) **Container** experiment to identify impact of containers (here *Docker*, other solutions are extendable)

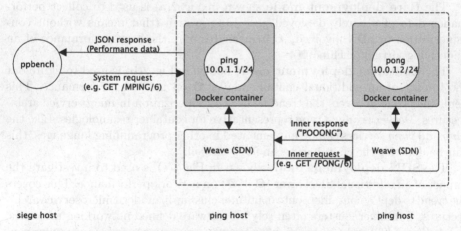

(c) **SDN** experiment to identify impact of SDN (here *Weave*, other solutions are extendable)

Fig. 2. Deployment modes.

defined. So it is possible to measure round-trip latencies between *ping* and *pong* for specific message sizes to cover specific message sizes on application or service layer according to the cloud-native reference model. This can be astonishingly tricky to realize with standard network benchmarks (see Sect. 6).

Table 1. Used programming languages and HTTP libraries to investigate impact of programming languages ❹. This list can be extended.

Language	Version	Server library	Client library
Go	1.5	net/http + gorilla/mux	net/http
Ruby	2.2	Webrick	httpclient
Java	1.8	com.sun.net.httpserver	java.net + java.io
Dart	1.12	http + start	http

Instead of using existing HTTP benchmarking tools like *Apachebench* or *httperf* we decided to develop a special benchmarking script (*ppbench*). *Ppbench* is used to collect a $n\%$ random sample of all possible message sizes between a minimum and maximum message size. *Pbench*, running on the *siege* host, sends a request to the *ping* server. The *ping* server relays each request to the *pong* server. And the *pong* server answers the request with a m byte long answer message (as requested by the HTTP request). The *ping* server measures the round-trip latency from request entry to the point in time when the *pong* answer hits the *ping* server again. The *ping* server answers the request from the *siege* host with a JSON message including round-trip latency between *ping* and *pong*, the length of the returned message, the HTTP status code sent by *pong* and the number of retries to establish a HTTP connection between *ping* and *pong*.

The **Bare deployment mode** shown in Fig. 2(a) is used to collect performance data of a barely deployed *ping-pong* system (that means without containerization or SDN involved, so it measures only the bare performance of an underlying virtual machine ❶).

The **Container deployment mode** shown in Fig. 2(b) is used to figure out the impact of an additional container layer ❷ on network performance. This deployment mode covers the trend to use containerization in microservice architectures. *Docker* is only a type representative for container technologies. Like the *ping* and *pong* services can be implemented in other programming languages, this can be done for the container technology as well.

The **SDN deployment mode** shown in Fig. 2(c) is used to investigate the impact of an additional SDN layer (❷ + ❸) on network performance. This covers the trend to deploy containers onto container clusters in modern microservice architectures. Container clusters often rely on software defined networking under the hood. *Weave* [22] and *Calico* [16] have been taken as type representatives for container focused SDN solutions to connect *ping* and *pong* containers. Because every data transfer must pass the SDN between *ping* and *pong*, the performance impact occurs due to this additional SDN layer. Other SDN solutions like *flannel* [4] are possible and planned in future extensions of the presented prototype (see Sect. 7).

3   Data Visualization

This section focuses mainly on data visualization requirements from a **system architect's** and **performance impact** understandability's point of view.

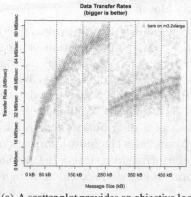

(a) A scatter plot provides an objective look on data

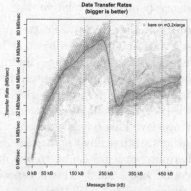

(b) A scatter plot with additional median and 90%/50% confidence bands

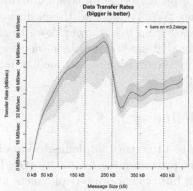

(c) Plot of median and 90%/50% confidence bands only

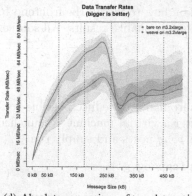

(d) Absolute comparison of two data series

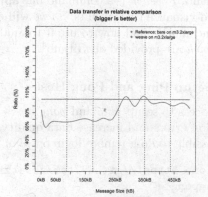

(e) Relative comparison of two data series

Fig. 3. Examples for different visualizations of benchmark data.

Therefore, *ppbench* can present benchmark data in an absolute (scatter plots or confidence bands) or a relative way (comparison plots).

Absolute Scatter Plots. Scatter plots like shown in Fig. 3(a) provide a good, realistic and maybe most objective impression about the distribution of measurements but without guiding descriptive statistical data. To draw general conclusions, we should not be interested in masses of single data points but in trends and confidence intervals [18]. Figure 3(b) shows exactly the same dataset but with a (spline smoothed) median line and 50% and 90% confidence bands as an overlay. This helps to reason the skewness of a distribution and to handle outliers statistically appropriate. So Fig. 3(b) provides the most information. Furthermore, *ppbench* is able to plot only the descriptive statistical data (median and confidence bands). This is especially helpful in cases where three or more data series have to be plotted.

Relative Comparison Plots. It is often more useful to compare data series in a relative way. This is especially true for system architects who are usually interested in questions like: "What would be the performance impact due to a virtual machine type decrease from *m3.2xlarge*[2] down to *m3.xlarge*[3]?". *Ppbench* provides additional plots for that kind of questions shown exemplary in Fig. 3(e). A comparison plot is simply the division of two median lines to compare two or more data series in a relative way.

4   Operating Guideline

On the one hand *ppbench* is a reference implementation of the *ping-* and *pong-*services written in different programming languages (Go, Ruby, Java and Dart) and provided with different but typical cloud deployment forms (bare, containerized, connected via a SDN solution, see Fig. 2). So we explain how to setup hosts to operate *ping* and *pong*-services in the mentioned deployment modes. On the other hand, *Ppbench* is a command line application to run benchmarks and analyze and visualize their results. So we will explain how to install the frontend to run benchmarks, and how to use it to analyze and visualize results. All sources of *ppbench* are provided on Github[4].

4.1   Setup Ping and Pong Hosts

The process to setup *ping* and *pong* hosts is automated to reduce possible configuration errors and increase data quality. After a virtual machine is launched it is possible to do a remote login on *ping* and *pong* hosts and simply run

```
git clone https://github.com/nkratzke/pingpong.git
cd pingpong
./install.sh
```

---

[2] AWS virtual 8-core machine type with 30 GB RAM.
[3] AWS virtual 4-core machine type with 15 GB RAM.
[4] https://github.com/nkratzke/pingpong.

to install all necessary packages and software on these hosts. This will download the latest version of *ppbench* from github.com/nkratzke/pingpong where it is under source control and provided for public use.

4.2   Starting and Stopping Services

The installation will provide a start- and a stop-script on the host, which can be used to start and stop different *ping* and *pong* services in different deployment modes (bare, containerized, SDN, see Fig. 2). In most cases, a benchmark setup will begin by starting a *pong* service on the *pong* host. It is essential to figure out the IP address or the DNS name of this *pong* host (**PONGIP**). This might be a private (IaaS infrastructure internal) or public (worldwide accessible) IP address. The *ping* host must be able to reach this **PONGIP**. To start the Go implementation of the *pong* service provided as a Docker container, we would do the following **on the *pong* host:**

```
./start.sh docker pong-go
```

Starting *ping* services works basically the same. It is important that the *ping* service knows its communicating counterpart (**PONGIP**). To start the Go implementation of the *ping* service provided in its bare deployment mode, we would do the following **on the *ping* host:**

```
./start.sh bare ping-go {PONGIP}
```

By default all services are configured to run on port 8080 for simplicity reasons and to reduce configuration error liability.

4.3   The Command Line Application

Ppbench is written in *Ruby 2.2* and is additionally hosted on RubyGems.org for convenience. So, it can be easily installed (and updated) via the *gem* command provided with *Ruby 2.2* (or higher).

```
gem install ppbench
```

Ppbench can be run on any machine. This machine is called the *siege* host. It is not necessary to deploy the *siege* host to the same cloud infrastructure because *ppbench* is measuring performance between *ping* and *pong*, and not between *siege* and *ping*. In most cases, *ppbench* is installed on a workstation or laptop outside of the cloud. *Ppbench* provides several commands to define and run benchmarks and to analyze data.

```
ppbench.rb help
```

will show all available commands (see Table 2). For this contribution, we will focus on defining and running a benchmark against a cloud deployed *ping-pong* system and on doing some data analytics.

Table 2. Commands of *ppbench*.

Command	Description
run	Runs a ping pong benchmark
latency-comparison-plot	Plot round-trip latencies in a relative way
latency-plot	Plots round-trip latencies in an absolute way
request-comparison-plot	Plots requests per second in a relative way
request-plot	Plots requests per second in an absolute way
transfer-comparison-plot	Plots data transfer rates in a relative way
transfer-plot	Plots data transfer rates in an absolute way
citation	Provides citation information about ppbench
help	Displays help documentation
summary	Summarizes benchmark data
naming-template	Generates a JSON file for naming

4.4   Running a Benchmark

A typical benchmark run with default options can be started as follows:

```
ppbench.rb run
    --host http://1.2.3.4:8080 \
    --machine m3.2xlarge \
    --experiment bare-go \
    benchmark-data.csv
```

The benchmark will send a defined amount of requests to the *ping* service (hosted on IP 1.2.3.4 and listening on port 8080). The *ping* service will forward the request to the *pong* service and will measure the round-trip latency to handle this request. Further command line options to define message sizes, concurrency, repetitions, etc. for running benchmarks against a *ping-pong* system can be found

Table 3. Options for the run command of *ppbench*.

Option	Description	Default	Example
--host	Ping host		--host http://1.2.3.4:8080
--machine	Tag to categorize the machine		--machine m3.2xlarge
--experiment	Tag to categorize the experiment		--experiment bare-go
--min	Minimum message size in bytes	1	--min 10000
--max	Maximum message size in bytes	500000	--max 50000
--coverage	Defines sample size between min and max	0.05	--coverage 0.1
--repetitions	Repetitions for each message	1	--repetitions 10
--concurrency	Concurrent request	1	--concurrency 10
--timeout	Timeout in seconds for a HTTP request	60	--timeout 5

in Table 3. This benchmark data is returned to *ppbench* and stored in a file called
`benchmark-data.csv`. The data is tagged to be run on a *m3.2xlarge* instance,
and the experiment is tagged as *bare-go*. It is up to the operator to select appropriate tags. The tags are mainly used to filter specific data for plotting. This
logfile can be processed with *ppbench* plot commands (see Table 2).

Because tags for machines and experiments are often short and not very
descriptive, there is the option to use more descriptive texts. The following command will generate a JSON template for a more descriptive naming which can
be used with the `--naming` option.

```
ppbench.rb naming-template *.csv > naming.json
```

These naming files look like in Listing 1.1.

Listing 1.1. An exemplary naming file.

```
{
"machines": {
    "m3.xlarge":      "m3.xlarge (AWS, 4 Core)",
    "n1-standard-4":"n1-standard-4 (GCE, 4 Core)"
  },
 "experiments": {
    "bare-go": "Go, bare install",
    "docker-go":"Go, Docker deployed",
    "weave-go": "Go, with Weave SDN networking"
  }
}
```

4.5   Plotting Benchmark Results

Ppbench delegates all relevant statistical data processing and data presentation
to the statistical computing toolsuite R [17]. So, *ppbench* can plot transfer rates,
round-trip latency and requests per second statiscally appropriate. We demonstrate it using transfer rates:

```
ppbench.rb transfer-plot \
    --machines m3.xlarge,m3.2xlarge \
    --experiments bare-go,bare-dart \
    *.csv > plot.R
```

This will generate a R script, which can be used to generate a plot like shown
in Fig. 3(a). The plot will contain only data collected on machines tagged as
m3.xlarge or *m3.2xlarge* and for experiments tagged as *bare-go* or *bare-dart*. The
result is written to standard out. So, it is possible to use *ppbench* in complex
command pipes. Pipes can be used to generate PDF output using the `--pdf`
option (see Table 4).

To add medians and confidence bands like shown in Fig. 3(b), we have to add
the `--withbands` option. To surpress plotting of single measurements like shown
in Fig. 3(c) we have to add the `--nopoints` flag. In most cases, this is the best
option to compare absolute values of two or more data series avoiding the jitter
of thousands of single measurements.

```
ppbench.rb transfer-plot \
    --machines m3.2xlarge \
    --experiments bare,weave \
    --withbands --nopoints \
    *.csv > plot.R
```

Table 4. Options for the plot commands of *ppbench*.

Option	Description	Default	Example
--machines	Select machines	All	--machines m3.xlarge,m3.2xlarge
--experiments	Select experiments	All	--experiments bare,docker
--recwindow	Plot TCP standard receive window	87380	--recwindow 0 (will not plot the window)
--yaxis_max	Max. value on Y-axis	Max.	--yaxis_max 50000
--xaxis_max	Max. value on X-axis	Max.	--xaxis_max 500000
--yaxis_ticks	Ticks to present on Y-axis	10	--yaxis_ticks 20
--xaxis_ticks	Ticks to present on X-axis	10	--xaxis_ticks 30
--withbands	Flag to plot confidence bands	No plot	--withbands
--confidence	Percent value for confidence bands	90	--confidence 75
--nopoints	Flag not to plot single data points	Plot	--nopoints
--alpha	Transparency (alpha) for data points	0.05	--alpha 0.01
--precision	Number of points for medians	1000	--precision 100
--naming	Use user defined names	Not used	--naming description.json
--pdf	Tell R to generate a PDF file	No pdf	--pdf example.pdf
--width	Width of plot (inch, PDF only)	7	--width 8
--height	Height of plot (inch, PDF only)	7	--height 6

Above mentioned commands can be used to show and compare absolute values. However, system architects are usually interested in comparing data series in a relative way. There is a comparison plot command for every metric (latency, transfer rate, request per second, see Table 2). For a relative comparison of the above mentioned example we would do something like that:

```
ppbench.rb transfer-comparison-plot \
    --machines m3.2xlarge \
    --experiments bare,weave \
    *.csv > plot.R
```

This plots a relative comparison like shown in Fig. 3(e). All series are shown relatively to a reference data series. The reference data series is the first combination of the --machines and --experiments entries. In the example shown above, this would be the data series for the *bare* experiment executed on a *m3.2xlarge* machine. All other data series are plotted relatively to this first reference performance. Further command line options for plotting can be found in Table 4.

5   Evaluation

Table 5 shows all experiments which have been performed to evaluate *ppbench*. The experiments had not the intention to cover all microservice related performance aspects but to show exemplarily how to use *ppbench* to answer microservice related performance questions formulated exemplary in Sect. 1. We only present data that was collected in AWS (Amazon Web Services) region *eu-central-1*. We cross checked AWS data with data collected in GCE (Google Compute Engine). Due to page limitations GCE data is not presented but it fully supports our findings. We intentionally worked with a very small set of instance types (*m3.large, m3.xlarge* and *m3.2xlarge*) that show high similarities with other public cloud virtual machines types like GCE machine types (*n1-standard-2, n1-standard-4, n1-standard-8*). These instance types have been selected according to a similarity measure based benchmark described in [11]. Although this covers only a small subset of possible combinations, it is fully sufficient to show how *ppbench* can be used to investigate interesting performance aspects being of interest for system architects designing microservice based systems. We want to evaluate our benchmark by collecting objective data to reflect some often heared cloud-native application performance rules (or myths) that say:

- **Rule 1:** Programming language impact is minor, due to the fact that applications are waiting most of their time on the I/O subsystems (whatever language is selected).
- **Rule 2:** Container impact on network performance is negligible, because that is what Docker is always telling.

- **Rule 3:** SDN impact on network performance is likely severe but hard to figure out precisely.
- **Rule 4:** Therefore, SDN solutions should be operated on machine types with access to large network bandwidth.

We designed the experiments shown in Table 5 to collect data for these cloud-native application performance engineering rules. Experiments P1–P4 are designed to collect data for rule 1, C1–C4 to collect data for rule 2, S1–S4 to collect data for rule 3, and V1–V4 to collect data for rule 4.

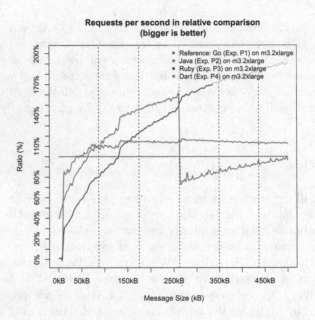

Fig. 4. Exemplary relative programming language impact on requests per second. (Color figure online)

5.1 Impact of Programming Languages (P1–P4)

We used the experiments P1 (Go), P2 (Java), P3 (Ruby) and P4 (Dart) to examine the performance impact of different programming languages (see Table 5). All experiments rely on the bare deployment mode shown in Fig. 2(a). Communication between REST-based services is mainly network I/O. Network I/O is very slow compared with processing. So, programming language impact on network I/O should be minor. Network applications are waiting most of their time on the I/O subsystem. So, a "faster" programming language shall have only limited impact to performance. But in reality is much more complex. We examined that this has nothing to do with the programming languages itself, but the performance (buffering strategies and so on) of the "default" HTTP and TCP libraries delivered with each programming language (see Table 1).

We used requests per second as a metric to visualize language impact on REST-performance (Fig. 4). The most interesting curve is the non-continuous curve for Dart (Fig. 4, lightblue line). It turned out that these **non-continuous effects** are aligned to the standard TCP receive window TCP_{window} on the systems under test [3]. Therefore, we highlighted (throughout complete contribution) TCP_{window} (87380 bytes on the systems under test) as dotted lines to give the reader some visual guidance. At $3 \times TCP_{window}$ we see a very sharp decline for Dart. Something similar is observable for Java and Ruby at $1/10 \times TCP_{window}$.

Due to *ppbench*, we can now recommend specific programming languages for specific ranges of message sizes. **Go** can be recommended for services producing messages fitting in $[0...\frac{1}{2}TCP_{window}]$, **Java** can be recommended for services with messages within $[\frac{1}{2}TCP_{window}...TCP_{window}]$, **Dart** for services producing only messages fitting in $[2 \times TCP_{window}...3 \times TCP_{window}]$, and finally **Ruby** for big messages fitting in $[4 \times TCP_{window}...]$. These detailed insights are astonishingly complex and surprising, because there exist the myth in cloud programming community that Go is one of the most performant languages for network I/O in all cases. We showed that even Ruby can produce better performances, although Ruby is not known to be a very processing performant language.

5.2   Impact of Containers (C1–C4)

Containers are meant to be lightweight. Therefore, the impact on network performance is often ignored. Nevertheless, due to an enlarged network stack there might be some effects on network performance observable. According to our above mentioned insights we used the Go implementations for the *ping-pong* system to investigate the impact of containers for services implemented in languages with **continuous performance behavior**. We used the Dart implementation to figure out the container impact for services implemented in languages with **non-continuous performance behavior**. For both implementations we used the bare deployment mode shown in Fig. 2(a) to figure out the reference performance for each language (C1, C3; see Table 5). And we used the container deployment mode in Fig. 2(b) to investigate the impact of containers on network performance (C2, C4; see Table 5).

We used round-trip latency as a metric to visualize container impact on REST-performance (Fig. 5). The Dart implementation shows distinctive non-continuous network behavior (see Fig. 5(a), blue lines). Containerization of the Dart implementation shows about 10% performance impact for all message sizes (slightly decreasing for bigger messages, see Fig. 5(b)). The containerization impact on the Go implementation is only measurable for small message sizes but around 20% which might be not negligible. For bigger message sizes it is hardly distinguishable from the reference performance. So we see, rule 2, container impact on network performance is negligible, is true for bigger messages but must be handled with care for small messages.

Table 5. Experiments for evaluation, *Docker* has been chosen as container technology. *Weave* has been chosen as SDN technology.

Experiment	Ping service			Pong service		
	Language	Mode	Machine	Language	Mode	Machine
P1/V5	Go	Bare	m3.2xlarge	Go	Bare	m3.2xlarge
P2	Java	Bare	m3.2xlarge	Java	Bare	m3.2xlarge
P3	Ruby	Bare	m3.2xlarge	Ruby	Bare	m3.2xlarge
P4	Dart	Bare	m3.2xlarge	Dart	Bare	m3.2xlarge
C1/V3	Go	Bare	m3.xlarge	Go	Bare	m3.xlarge
C2	Go	Con.	m3.xlarge	Go	Con.	m3.xlarge
C3	Dart	Bare	m3.xlarge	Dart	Bare	m3.xlarge
C4	Dart	Con.	m3.xlarge	Dart	Con.	m3.xlarge
S1/V1	Go	Bare	m3.large	Go	Bare	m3.large
S2/V2	Go	SDN	m3.large	Go	SDN	m3.large
S3	Ruby	Bare	m3.large	Ruby	Bare	m3.large
S4	Ruby	SDN	m3.large	Ruby	SDN	m3.large
V4	Go	SDN	m3.xlarge	Go	SDN	m3.xlarge
V6	Go	SDN	m3.2xlarge	Go	SDN	m3.2xlarge

5.3   Impact of SDN (S1–S4)

SDN solutions contend for the same CPU like payload processes. That is likely the impact why SDN might have noticable performance impacts. According to our above mentioned insights, we used the Go (showing no non-continuous network behavior) implementation for the *ping-pong* system to figure out the impact of SDN. Due to the fact that the Go implementation did not show the best network performance for big message sizes, we decided to measure the performance impact with the Ruby implementation as well (best transfer rates for big message sizes). For both languages we used the bare deployment mode shown in Fig. 2(a) to figure out the reference performance (S1, S3; see Table 5). And we used the SDN deployment mode in Fig. 2(b) to investigate the impact of SDN on network performance (S2, S4; see Table 5). We used intentionally the smallest machine type (m3.large, virtual 2-core system) of our machine types to stress CPU contention effects of the *ping* and *pong* services and the SDN routing processes.

Data transfer rates are used to visualize SDN impact on REST-performances. Figure 6(b) shows relative impact of a SDN layer to REST-performance for Go and Ruby implementations. In second and higher TCP_{window} the SDN impact is clearly measurable for Go and Ruby. The impact for both languages seem to play in the same league (about 60% to 70% of the reference performance). Go seems to be a little less vulnerable for negative performance impacts due to containerization than Ruby. We can even see a positive impact of SDNs for Ruby

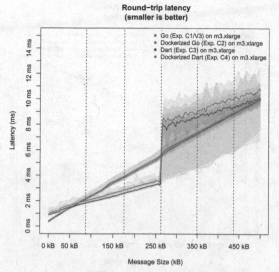

(a) Absolute impact of containers on latencies

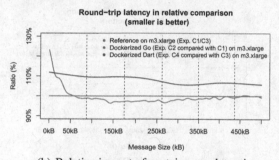

(b) Relative impact of containers on latencies

Fig. 5. Examplary container impact on latencies. (Color figure online)

in the first TCP_{window}. Remember, we identified a non-continuous behavior for Ruby (and for Java as well) at $\frac{1}{10}TCP_{window}$. It turned out that the SDN solution attenuated the non-continuous effect in the first TCP_{window} for the Ruby implementation. This attenuation showed in average a positive effect on network performance in the 1st TCP_{window}.

So, SDN can severely impact performance. But around non-continuous performance points, SDN can even improve network performance.

5.4   Impact of VM Types on SDN Performance (V1–V6)

SDN impact on network performance can be worse (see Fig. 6). This has mainly to do with effects where service processes contend with SDN processes for the same CPU on the same virtual machine. However, these effects should decrease

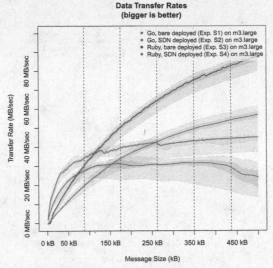

(a) Absolute impact of SDN on transfer rates

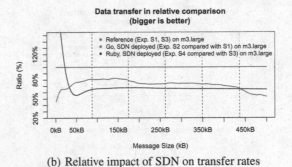

(b) Relative impact of SDN on transfer rates

Fig. 6. Exemplary SDN impact on transfer rates.

on machine types with more virtual cores (independently of the network bandwidth a host has access to). So, we reused the introduced experiments S1 (as V1) and S2 (as V2) to investigate the performance impact of SDNs on a virtual 2-core system (m3.large). We reused the bare deployed experiment C1 (as V3) to investigate the reference performance on a virtual 4-core system (m3.xlarge) and compared it with the same but SDN deployed experiment V4. We did exactly the same with V5 (reuse of P1) und V6 on a virtual 8-core system (m3.2xlarge). All experiments (see Table 5) used the Go implementation for the *ping-pong* system due to Go's continuous network behavior. So, all experiments just measure the impact of an additional SDN layer and not the impact of increased network bandwidth.

Figure 7 compares the performance impact of the SDN deployment mode shown in Fig. 2(c) with the bare deployment mode shown in Fig. 2(a) on different

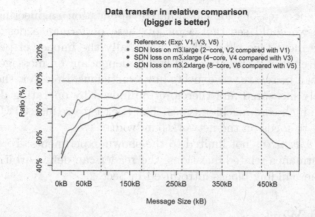

Fig. 7. How different virtual machines can decrease SDN impact on transfer rates.

VM types (2-core, 4-core and 8-core). The SDN impact on 8-core machine types is less distinguishable than on 4- or 2-core machine types. While 2- and 4-core machine types show similar performance impacts in first and second TCP_{window}, the 2-core machine type looses significantly in the third and higher TCP_{window}. High core machine types can effectively attenuate the negative impact on network performance of SDN solutions. This has nothing to do with the network bandwidth of the corresponding machine types. Hence, *ppbench* provided a much more detailed insight into SDN performance behavior.

5.5   Summary

It was very astonishing for us to see that **programming languages** (or their standard HTTP and TCP libraries) might have a substantial impact on REST-performance. Furthermore, three out of four analyzed languages showed non-continuous network behavior, so that messages being only some bytes larger or smaller may show completely different latencies or transfer rates. We identified such effects at $\frac{1}{10}TCP_{window}$ (Ruby, Java) and $3 \times TCP_{window}$ (Dart). According to our measurements, there is no ideal programming language showing best results for all message sizes. Containers are stated to be lightweight. The performance impact of **containerization** might not be severe but is not always negligible. Small message performance may be more vulnerable to performance impacts of containers than big message performance. As expected, the impact of **SDN** can be severe. This is especially true for small core machine types. However, due to attenuation effects, SDN can even show positive effects in case of non-continuous network behavior [12]. Machine types with more cores decrease the performance impact of SDN because CPU contention effects are reduced. SDN impacts can be decreased by a virtual 8-core machine type to a level comparable to containerization. In that case, the performance decrease due to a "wrong" programming language can be even more severe!

There are good reasons for cloud-native application engineering rules. By using *ppbench* we could see that some of these performance design rules by practicians are more true than others. Especially the impact of programming languages is often underestimated and highly dependant on message sizes. And SDN impacts can be severe - as expected by practicians. However, these impacts can be precisely measured and therefore minimized by operating deployments on appropriate high core machine types. The trick is to reduce CPU contention effects and not to focus on the network bandwidth.

Of course, *ppbench* is not limited to the shown experiments. To collect data for other performance related questions, the reader can define arbitrary experiments using the building blocks introduced in Sect. 2.

6   Related Work

There exist several **TCP/UDP networking benchmarks** like *iperf* [2], *uperf* [19], *netperf* [14] and so on. [20] provide a much more complete and comparative analysis of network benchmarking tools. [11] present a detailled list on cloud computing related (network) benchmarks. Most of these benchmarks focus TCP/UDP performance [20] and rely on a specific server component used to generate network load. These benchmarks are valuable to compare principal network performance of different (cloud) infrastructures by comparing what kind of maximum network performances can be expected for specific (cloud) infrastructures. However, in most cases maximum expectable network performances is not very realistic for REST-based protocols. Other **HTTP related benchmarks** like *httperf* [9] or *ab* [1] are obviously much more relevant for REST-based microservice approaches. These tools can benchmark arbitrary web applications. But because the applications under test are not under direct control of the benchmark, these tools can hardly define precise loads within a specific frame of interest. Therefore HTTP related benchmarks are mainly used to run benchmarks against specific test resources (e.g. a HTML test page). This makes it hard to identify trends or non-continuous network behavior.

Ppbench is more a mix of tools like *iperf* (TCP/UDP benchmarks) and *httperf* (HTTP benchmarks) due to the fact that *ppbench* provides a benchmark frontend (which is conceptually similar to *httperf* or *ab*) and a reference implementation under test (*ping-pong* system which is conceptually similar to a *iperf* server). Most of the above mentioned benchmarks focus on data measurement and do not provide appropriate visualizations of collected data. This may hide trends or even non-continuous network behavior. That is why *ppbench* focus on data visualization as well.

7   Conclusion

Companies like Netflix, Google, Amazon, Twitter successfully exemplified elastic and scalable microservice architectures for very large systems. Modern cloud-native application architectures are often realized as microservice architectures and in a way that services are deployed as containerized deployment units on elastic platforms (Apache Mesos, Kubernetes, Docker Swarm, etc.). Furthermore, microservice architectures often use lightweight and REST-based mechanisms for service communication. According to the microservice architecture approach, services are often implemented in different programming languages adding complexity to a system. All this points might lead to decreased performance. Astonishingly, it is quite complex to figure out these impacts in the upfront of a microservice design process due to missing and specialized benchmarks. So, this contribution proposed and evaluated a benchmark intentionally designed for this challenging cloud-native application engineering task. We advocate that it is more useful to reflect fundamental design decisions and their performance impacts in the upfront of a microservice architecture development and not in the aftermath. We showed exemplary how to setup systematic tests to figure out possible performance impacts coming along with design decisions (using different programming languages, using REST-APIs for service composition, encapsulating services into containers, deploying services to different virtual machine types, and more). Finally, our contribution can be used as reference by other researchers to show how new approaches in microservice design can improve performance.

Acknowledgements. This study was funded by German Federal Ministry of Education and Research (03FH021PX4). We thank René Peinl and his research group for their valuable feedback and for their contribution to integrate Calico SDN into *ppbench*. We thank all reviewers for their valuable feedback on our initial conference paper [13], especially Bryan Boreham from Weaveworks.

References

1. Apache Software Foundation: ab - Apache HTTP server benchmarking tool (2015). http://httpd.apache.org/docs/2.2/programs/ab.html
2. Berkley Lab: iPerf - The network bandwidth measurement tool (2015). https://iperf.fr
3. Bormann, D., Braden, B., Jacobsen, V., Scheffenegger, R.: RFC 7323, TCP Extensions for High Performance (2014). https://tools.ietf.org/html/rfc7323
4. CoreOS: Flannel (2015). https://github.com/coreos/flannel
5. Docker Inc.: Docker (2015). https://www.docker.com
6. Docker Inc.: Docker Swarm (2016). https://docs.docker.com/swarm/
7. Fielding, R.T.: Architectural styles and the design of network-based software architectures. Ph.D. thesis (2000)
8. Hindman, B., Konwinski, A., Zaharia, M., Ghodsi, A., Joseph, A.D., Katz, R.H., Shenker, S., Stoica, I.: Mesos: a platform for fine-grained resource sharing in the data center. In: NSDI, vol. 11 (2011)
9. HP Labs: httperf - a tool for measuring web server performance (2008). http://www.hpl.hp.com/research/linux/httperf/
10. Kratzke, N., Peinl, R.: ClouNS - a reference model for cloud-native applications. In: Proceedings of 20th International Conference on Enterprise Distributed Object Computing Workshops (EDOCW 2016) (2016)

11. Kratzke, N., Quint, P.C.: About automatic benchmarking of IaaS cloud service providers for a world of container clusters. J. Cloud Comput. Res. **1**(1), 16–34 (2015)
12. Kratzke, N., Quint, P.C.: How to operate container clusters more efficiently? Some insights concerning containers, software-defined-networks, and their sometimes counterintuitive impact on network performance. Int. J. Adv. Netw. Serv. **8**(3&4), 203–214 (2015)
13. Kratzke, N., Quint, P.C.: ppbench - a visualizing network benchmark for microservices. In: Proceedings of 6th International Conference on Cloud Computing and Service Sciences (CLOSER 2016), vol. 2, pp. 223–231 (2016)
14. netperf.org: The Public Netperf Homepage (2012). http://www.netperf.org
15. Newman, S.: Building Microservices. O'Reilly Media, Incorporated, San Francisco (2015)
16. Project Calico: Calico (2016). https://www.projectcalico.org/
17. R Core Team: R: a language and environment for statistical computing. R Foundation for Statistical Computing, Vienna, Austria (2014). http://www.R-project.org/
18. Schmid, H., Huber, A.: Measuring a small number of samples, and the 3v fallacy: shedding light on confidence and error intervals. IEEE Solid-State Circ. Mag. **6**(2), 52–58 (2014)
19. Sun Microsystems: uperf - a network performance tool (2012). http://www.uperf.org
20. Velásquez, K., Gamess, E.: A comparative analysis of network benchmarking tools. In: Proceedings of the World Congress on Engineering and Computer Science 2009 (WCOES 2009) (2009)
21. Verma, A., Pedrosa, L., Korupolu, M.R., Oppenheimer, D., Tune, E., Wilkes, J.: Large-scale cluster management at Google with Borg. In: Proceedings of the European Conference on Computer Systems (EuroSys), Bordeaux, France (2015)
22. Weave Works: Weave (2015). https://github.com/weaveworks/weave

An Universal Approach for Compliance Management Using Compliance Descriptors

Falko Koetter[1(✉)], Maximilien Kintz[1], Monika Kochanowski[1],
Thatchanok Wiriyarattanakul[1], Christoph Fehling[2], Philipp Gildein[2],
Sebastian Wagner[2], Frank Leymann[2], and Anette Weisbecker[1]

[1] University of Stuttgart IAT and Fraunhofer IAO, Nobelstr. 12, Stuttgart, Germany
{falko.koetter,maximilien.kintz,monika.kochanowski,
thatchanok.wiriyarattanakul,anette.weisbecker}@iao.fraunhofer.de
[2] University of Stuttgart IAAS, Universitätsstr. 38, Stuttgart, Germany
{christoph.fehling,philipp.gildein,sebastian.wagner,
frank.leymann}@iaas.uni-stuttgart.de

Abstract. Trends like outsourcing and cloud computing have led to a
distribution of business processes among different IT systems and organizations. Still, businesses need to ensure compliance regarding laws and
regulations of these distributed processes. This need gave way to many
new solutions for compliance management and checking. Compliance
requirements arise from legal documents and are implemented in all parts
of enterprise IT, creating a business IT gap between legal texts and software implementation. Compliance solutions must bridge this gap as well
as support a wide variety of compliance requirements. To achieve these
goals, we developed an integrating compliance descriptor for compliance
modeling on the legal, requirement and technical level, incorporating
arbitrary rule languages for specific types of requirements. Using a modeled descriptor a compliance checking architecture can be configured,
including specific rule checking implementations. The graphical notation of the compliance descriptor and the formalism it's based on are
described and evaluated using a prototype as well as expert interviews.
Based on evaluation results, an extension for compliance management in
unstructured processes is outlined.

Keywords: Business process management · Compliance modeling ·
Model-driven architecture · Business process compliance · Process
mining

1 Introduction

Cloud computing is both a chance and a challenge for many companies [43],
especially in the field of security, privacy, [40] and - consequently - compliance.
The possibility to acquire services over the cloud in order to better perform their
business processes gives companies new possibilities to manage their business.
However, this adds to the complexity of the involved IT systems as well as

© Springer International Publishing AG 2017
M. Helfert et al. (Eds.): CLOSER 2016, CCIS 740, pp. 209–231, 2017.
DOI: 10.1007/978-3-319-62594-2_11

poses new challenges in managing these. Due to growing regulatory requirements stemming from new laws like the Sarbanes-Oxley Act there is an increasing demand for business process compliance solutions in the industry [35].

However, there is a gap between compliance management and business process management, as one is driven by legal requirements and the other by business needs as well as the new cloud technologies. Additionally, compliance management spans not only business processes but also the process environment consisting of software systems, physical hardware and personnel, as described in [22]. Here, also compliance applications and their needs are mentioned. This requires communication between legal specialists, business users and IT personnel. All this makes keeping processes compliant a cumbersome task.

As business process models and their implementation increase in complexity [6], manual compliance checking is not feasible for large organizations. Existing IT-supported compliance management solutions focus on specific process execution environments (e.g. process engines) and only support specific kinds of compliance rules [17]. As far as processes are executed within such an environment, they can support compliance enforcement with strict process models, transition rules, double-checks, etc. However, not all parts of a business process are usually contained within such solutions, and not all of the compliance requirements can be enforced in such a way. We found the factor preventing effective compliance management is not a lack of tools, but rather a lack of integration, between different kinds of compliance checking as well as between business and IT.

To alleviate this problem, in previous work we proposed an integrating compliance descriptor [26], which bridges legal, business and IT levels by separating laws, compliance requirements that stem from them and compliance rules implementing these requirements. As the approach encapsulates compliance rules, multiple rule languages can be used, providing integration and coverage of all kinds of compliance requirements. We described how to use this compliance descriptor for compliance checking, gathering results from different rules and aggregating them to determine requirement fulfillment and law compliance [27].

In this work, we will build on this approach by developing a conceptual modeling language for compliance descriptors and integrating it with other modeling languages for processes and compliance rules. This is the main contribution of this paper, as it proved necessary to develop a formalism and graphical modeling notation in order to facilitate creating compliance descriptors. Using a model-driven approach, we show how a conceptual model of compliance can be transformed to different artifacts necessary for compliance management in a reference architecture, which is extended from previous work. Evaluating the work using a prototype and real-life example, we show our approach is a feasible solution for bridging business and IT views of process compliance in a rule-language-independent and extensible fashion. Compared to our previous work, we evaluated the approach not only in a prototype but also with a real-life system and real users.

The remainder of this work is structured as follows. In Sect. 2 an overview of work in compliance modeling and compliance integration is given. In Sect. 3

we describe conceptual compliance modeling using the compliance descriptor. In Sect. 4 we describe the extensible architecture for compliance management as well as the model transformation. Section 5 describes the prototype and evaluation. Section 6 describes a detailed outlook on future work. Finally, Sect. 7 gives a conclusion.

2   Related Work

Achieving Business Process Compliance is not a one-time task, but a continuous activity. Different compliance checks are performed at different phases in the business process lifecycle. [17] gives an overview of compliance checking methods and distinguishes design-time, run-time and ex-post compliance checking. In particular, [17] notes a lack of an universal approach, supporting all phases of the lifecycle as well as continuous change. To address this, we introduced an integrating compliance descriptor in [26]. Rather than designing yet another compliance rule modeling language, this compliance descriptor connects laws, compliance requirements and rules in different compliance rule languages. Thus, a link between the business and IT view of compliance requirements is preserved. Using this link, the impact of changes in laws, requirements or implementation to overall compliance can be assessed at any time, enabling maintenance of compliance in the face of change [27].

Similarly to the structure of the compliance descriptor, in [10] three levels of regulatory compliance are defined. Regulations define measures and directives which are implemented by policies, internal controls, and procedures. Furthermore, eight requirements for a compliance management framework are defined, among others enforcement, change management, traceability, and impact analysis. As in other previous work, different types of compliance checking are identified. To tackle these challenges, an architecture for a compliance checking framework based on semantic business process models is proposed. In this architecture, regulations are modeled as semantic policies which are monitored by a policy monitoring component. From these, semantic business rules are generated to be enforced at design-time and run-time by an inference engine [11].

The *SeaFlows Toolset* [28] is a framework for compliance verification of business processes. Using *compliance rule graphs*, rules can be modeled by imposing patterns of process activities and/or conditions on process data at specific points in process execution. Patterns are then checked at design time, while data conditions are checked at run-time in a BPM suite. In further work, the resource perspective, i.e. who performs tasks, has been added in an extended rule graph [39]. This approach is interesting both in combining run-time and design-time rules in the same rule graph as well as combining multiple rules in a single graph and separating it from the process model. However, due to the implementation techniques used, it is dependent of specific modeling and execution environments and doesn't explicitly offer extensibility for other types of compliance rules.

Reference [34, Chap. 10] investigates design-time, run-time and ex-post compliance rules based on process traces, on which Linear Temporal Logic (LTL) expression or compliance rule graphs are tested. Also the impact of process change on compliance in models and running process instances is tackled by investigating the changes in compliance rules and their effects. However, only process models and instances in a workflow engine are in the scope of compliance checking.

Reference [38] defines a graphical modeling language for LTL rules which can be used for design-time compliance checking. A *compliance domain* can be used to attach an LTL rule to a process or a part of a process. In [3] BPMN-Q, a graphical query language for business process models is used for design time compliance checking. Using compliance patterns and anti-patterns of violations, this approach is used to visualize reasons for compliance violations in the process model [4].

aPro [23] is a solution for model-driven process monitoring. Based on a monitoring model of a business process, including data to measure as well as Key Performance Indicators (KPIs) and Goals to monitor, a Complex Event Processing (CEP) based monitoring software is automatically created. aPro is used to monitor compliance rules at run-time (e.g. timing restrictions) [24] and to counteract compliance violations [25].

Various other approaches cover compliance checking at design-time, checking the process model by rules or constraints [14,15].

Compliance of business processes reaches beyond the scope of process models and execution, encompassing aspects like hosting, maintenance, encryption and physical access control. Business process management builds on these aspects, which need to be covered in order to guarantee overall compliance. The Topology and Orchestration Specification for Cloud Applications (*TOSCA*[1].) is an OASIS standard which allows modeling the topology of an application, including the implementation and physical deployment of components like web services and databases. In [42] *Policy4TOSCA* is described, extending TOSCA models by so-called policies to describe non-functional requirements, which may stem from compliance rules. These can be checked during process deployment in order to guarantee a compliant hardware and software stack [27].

While checking compliance rules for a business process can be looked at in a single-system fashion, i.e. the process is executed in a single, homogenous environment, in which compliance is checked, in practice process compliance needs to be managed in a multi-system fashion. One reason for this are heterogeneous, grown IT infrastructures found in practice [31], another reason are the different scopes compliance rules may encompass [17].

Reference [21] describes an approach for integrating compliance checking across multiple companies in cross-organizational business processes. Challenges arise because parts of an organizational process model may not be globally known and because local compliance requirements may contradict global compliance requirements. Thus, a notion of *compliability* is defined, meaning the interaction between process partners conforms to global compliance rules, even though private processes are not known. While this approach allows integrated design-time

---

[1] www.oasis-open.org/committees/tosca/ (accessed 12.3.2015).

compliance across multiple public and private process models, it does not address the challenge of heterogeneous process environments and other kinds of compliance checking, though this shall be addressed in future work.

Reference [33] advocates a separation of compliance management and process management by introducing a separate *compliance engine*, which checks an implementation system for compliance and interrupts it in case of risks. Compliance rules are defined on a business vocabulary from a conceptual model of processes. By separating process and compliance, any information system may be checked, similar to aPro, the model-driven solution for process monitoring used in our approach [23]. Similarly, any compliance rule language may be used. This in general solves the integration problem, but requires a high degree of manual implementation in the compliance engine and in the integrated information system. To lessen the required effort, [32] lists methods of assisting domain experts in the selection of applicable existing rules from a rule repository, e.g. by question trees.

In [8] the problem of aligning compliance rules and their implementation across partners in a business network is investigated. Compliance rules are defined on a conceptual level and then concretized to a specific scenario and process modeling environment, which in turn are concretized in a specific implementing technology. This resembles a compliance descriptors separation of requirements and rules. However, even after concretization, manual implementation and integration needs to be performed to achieve compliance monitoring.

Reference [16] describes a generic compliance evaluation method extending existing enterprise modeling frameworks to encompass compliance evaluation. Similar to this work, partial compliance can be evaluated and is visualized in a heatmap. Compliance is evaluated for individual architecture artifacts aggregated through architecture levels. For example, the compliance of a business process is calculated form the compliance of its activities. In comparison, our approach has a specific process focus and allows definition of emergent compliance rules encompassing multiple activities. Additionally, [16] gives no techniques how to automatically determine compliance at the bottom level, while our approach integrates different compliance rule languages.

Reference [44] proposes a model-driven method for auditing using an ontology, determining accountability and authorization and in turn defining controls to address identified risks. While an approach like this provides the possibility to discover requirements and necessary controls, it does not cover the actual compliance checking. A gap between definition and implementation remains.

A framework for defining and managing compliance requirements is presented in [30]. Requirements are defined using a declarative language and LTL. Design-time compliance checking is supported while run-time compliance checking is only conceptually described. Similarly to the reusable compliance rules in this approach, patterns are used for easier implementation of common requirement types. Compared to this approach, extensibility of checks, e.g. for software deployments, is not covered.

Reference [20] describes a visual language for modeling compliance rules, focusing on the different perspectives involved with compliance management, i.e. control flow, interaction, time, data and resource perspectives. Similar to the integrating approach in this work, the goal is to integrate the modeling of different kinds of rules, e.g. regarding order of activity, mutual exclusion, user authorization and timing. While this approach provides a unified modeling environment for a diverse set of rules, currently only ex-post compliance checking on log files is supported. In comparison the approach in this work, while less homogenous in modeling, provides a greater range of compliance checks and higher extensibility.

Compliance management has fostered many works in the last years, starting in 2003 [1] - however focusing mainly on the US market. In the German insurance industry compliance is a strategic challenge today, as it is organized in the upper management hierarchy [41]. Regulatory compliance comprises large parts of the IT budgets of insurance companies [2]. In the future, costs for compliance management are predicted to further raise, as existing solutions do not cover current requirements [41] and an increase in regulations is seen as realistic [37].

The most relevant topics for insurance companies are Solvency II and data privacy, although national rulings also have been mentioned. 93% of the participants mentioned that their compliance activities are not or only partially supported by IT. The authors of [41] suggest that one of the main challenges in future compliance management is to handle multiple different compliance sources (international and national laws, code of conducts, etc.) and provide IT support thereof. Additional potentials are identified in a higher process efficiency and consolidation of compliance reports. [5] states that compliance needs to be better integrated into business processes. Broadening the view on the USA, a multitude of regulations has been noted as a top priority [36]. The authors recommend using software applications for automation and leveraging the governance efforts.

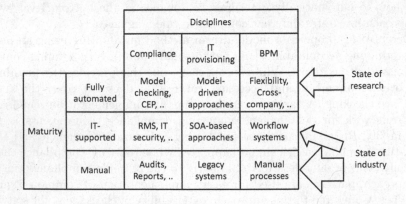

Fig. 1. Compliance in research and industry.

Figure 1 gives a view on the maturity of compliance management and the underlying layers - IT and BPM - in industry and in state of research. Whereas companies - especially in insurance industry - still perform compliance management mostly manually and do not have overall workflow systems in place, the state of research is much more sophisticated. Authors discuss integration of automatic configuration of processes, model-driven approaches for application management, and model checking technology. In order to bridge the gap between state of the industry and research, a (1) unifying approach for management is necessary which allows for (2) change management across different platforms and compliance measures providing an (3) overall view on the state of compliance. This is necessary to support the usage of state of the art technology where it is employed, but also to support legacy systems and manual tasks.

Overall, the related work in research shows first steps for solving the problem of heterogeneous environments and the need for compliance checking to cover the whole business process lifecycle and associated artifacts. However, approaches try to fit all compliance checking rules in homogenous modeling languages or require manual rule implementation. In contrast, this work contributes an integrating compliance management approach, which separates business and IT view of process compliance and encapsulates compliance rules in any rule modeling language.

3   Conceptual Compliance Modeling

As business process compliance needs to cover many aspects of the enterprise, multiple artifacts need to be modeled for compliance management. On the business level, *laws* impose compliance *requirements* on a business *process*. On the IT level, requirements are implemented in *rules* which are checked throughout the process lifecycle, e.g. at design-time, run-time and during deployment.

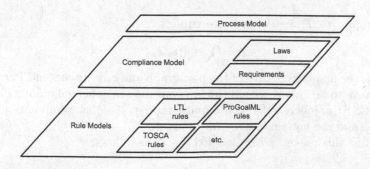

Fig. 2. Compliance-related models.

Figure 2 gives an overview of compliance-related models. A *process model* describes the process on a conceptual level. This process model may be prescriptive or descriptive, i.e. it may be executed or describe the execution by other

IT systems. On an implementation level, different types of compliance rules can be used for compliance checking. LTL rules [38] are used for model checking at design-time, e.g. for verifying the order of activities. ProGoalmML rules [23] are used for compliance monitoring at run-time, e.g. to check if timing restrictions are kept. TOSCA policies [42] are used to verify the physical deployment of process infrastructure, e.g. to satisfy data protection requirements.

For connecting the business and IT levels the *compliance descriptor* is used. It contains all *laws* (and other regulatory documents) applicable to the process. Compliance requirements from these laws are contained as well, described in natural speech as well as in an expression referencing implementing compliance rules. As the compliance descriptor serves to integrate the business and IT levels of compliance management, a modeling language needs to be understood both by business and IT experts. In previous work we defined the structure of the compliance descriptor on an implementation level using XML schema [12]. While this proved sufficient for implementing compliance checking, creating compliance descriptors proved to be difficult. To close this gap, we will define a graphical modeling language for compliance descriptors.

3.1 Laws

In the compliance descriptor, a *law* represents a regulatory document from which compliance requirements stem. A law $l \in L$ is defined as a tuple:

$$l \in L := (n_l, P_l, v_l) \tag{1}$$

where n_l is the name of the law, v_l identifies the version of the law and P_l is a set of paragraphs the law consists of. On a technical level the law is stored as an XHTML document, providing both human readability due to html formatting as well as a well-defined structure for automatic processing due to xml validity. Laws need to be structured in order to make references to paragraphs of the law possible.

Paragraphs $p \in P_l$ are defined as a tuple:

$$p \in P_l := (n_p, c_p) \tag{2}$$

where n_p is the number or name of the paragraph and c_p is its content. Paragraph numbers have to be unique within the law, but may be repeated among different laws. Thus, to uniquely reference a paragraph $p \in P_l$, a combination of its number n_p and the law's name n_l can be used.

We define this reference as a law url $u \in U$ as follows:

$$u \in U := (n_{l_u}, n_{p_u}) \tag{3}$$

with n_{l_u} as the name of a law $l \in L$ and n_{p_u} as the name of a paragraph $p \in P_l$ within that law. Note that if a law url refers to a law, it may only refer to a paragraph within that same law:

$$(n_{l_u}, n_{p_u}) \in U \rightarrow p \in P_l \tag{4}$$

3.2 Entity

Entities within the process context (e.g. a process activity or a server) are modeled as *entities*, so they can be referenced by the compliance descriptor. An entity $e \in E$ is defined as follows:

$$e \in E := (n_e, ref_e) \tag{5}$$

where n_e is the name of the referenced entity and ref_e a unique reference to the entity (e.g. a reference to an activity in a process model).

3.3 Compliance Rules

A compliance rule is an implementation level artifact used to enforce or monitor a fact, e.g. if a variable has a certain value, if an activity is always followed by another activity or if a database is encrypted. Thus, the concrete implementation of a rule is not stipulated by the compliance descriptor. To provide a wide support in regards to compliance rule languages, a rule is described with generic attributes, allowing handling it during compliance management without knowledge of implementation semantics. Only during model transformation is the concrete implementation used.

A compliance rule is described as follows:

$$r \in R := (n_r, d_r, il_r, ph_r, rex_r, VD_r, B_r)$$

where n_r is the name of the compliance rule, d_r its description in natural language. These are written by IT personnel to give business users an understanding of the semantics of the rule. il_r is a unique identifier for the implementation language of the rule, e.g. *LTL*. ph_r identifies the phase of the process lifecycle in which the rule is applied, e.g. design-time or run-time.

The concrete implementation is called a *rule expression* and stored in rex_r. Depending on the type of rule this may either be a formal description of the rule (e.g. an XML file) or a reference to the full implementation (e.g. a reference to a graphical rule model). In any case, a compliance rule must provide a suitable result for evaluation. Depending on the lifecycle phase, a compliance rule may either be fulfilled (*true*), not fulfilled (*false*) or not yet evaluated (*unknown*). Each rule must provide these results.

To facilitate reuse of compliance rules among different process models and requirements, rules are variable. This means certain parts of the rule can be modified in order to adjust the rule to its concrete use case. For example the name and location of a database can be chosen. For this purpose, a so-called variability descriptor is used [29], an XML format which allows variability in arbitrary documents by referencing variability points and possible values. The variability descriptor of a rule is stored or referenced in VD_r and is a set of variability points vp_r:

$$VD_r := \{vp_r\} \tag{6}$$

A set of *bindings* B_r describes which concrete values are chosen for each variability point. A binding $b \in B_r$ is defined as follows:

$$b \in B_r := (vp_b, val_b, type_b)$$

where vp_b identifies the variability point that is bound, val_b is the value the variability point is bound to, which depends on the type of the binding $type_b$.

There are three types $type \in BTYPE$ of bindings [29]. A *constant* value val_b is bound to the variability point. Depending on the variability point, this may for example be an integer or string value. An *entity* e in the compliance descriptor is bound to the variability point. val_b identifies the entity by its name n_e. A *parameter* indicates the variability point is not yet bound in the rule, but will be bound later. val_b indicates the number vp_b shall have in the order of parameters.

The bindings B_r are called a *complete binding*, if:

$$complete(B_r) := \forall vp \in VD_r : \exists b \in B_r : vp_b = vp \wedge type_b \neq parameter$$

A complete binding thus provides a concrete value for each variability point. No further information is necessary to create a concrete rule.

On the other hand the bindings B_r are called a *partial binding*, if:

$$partial(B_r) := (\forall vp \in VD_r : \exists b \in B_r : vp_b = vp) \wedge (\exists b \in B_r : type_b = parameter)$$

Note that even a partial binding must provide a binding for each variability point. Bindings which fail both criteria are invalid.

3.4 Compliance Requirements

A compliance requirement is a single compliance-related requirement to the business process or the process environment stemming from a law. It is used to link laws and implementation. A requirement $q \in Q$ is defined as follows:

$$q \in Q := (n_q, d_q, u_q, cex_r)$$

where n_q is the unique name of the requirement, d_q is a description of the requirement in natural language, provided by business users. $u_q \in U$ is a law url referencing the paragraph the requirement stems from. Note multiple requirements may stem from the same law or even paragraph.

Aside from a description d_q in natural language, it contains a formal compliance expression cex_r, which describes the requirement by referencing compliance rules. The compliance expression $cex \in CEX$ is defined as follows:

$$cex \in CEX := (f_{cex}, R_{cex}, B_{cex})$$

A compliance expression links multiple rules $r \in R_{cex}$ in a formula f_{cex}. This formula uses Boolean operators to relate rules to each other. As rules may be variable, additional binding information may be necessary to create the rule.

This is the case if $partial(B_r)$ is true. Then, additional bindings $B_{r_{add}}$ need to be specified for each binding $b \in B_r$, where $type_b = parameter$. These additional bindings are stored in B_{cex} and are defined as follows:

$$B_{r_{add}} \in B_{cex} := \{(vp_b, val_b, type_b)|type_b \neq parameter \wedge$$
$$(\exists b_r \in B_r : type_{b_r} = parameter \wedge vp_{b_r} = vp_b)\}$$

The bindings are then combined to create are so-called final binding:

$$B_{r_{final}} := B_{r_{add}} \cup \{b \in B_r|type_b \neq parameter\}$$

For the final binding $complete(B_{r_{final}})$ must be true, as otherwise no concrete rule may be created. Then no deployment of rules can take place. Note that there may be multiple final bindings if a rule is used multiple times within compliance expressions.

The formula f_x defines a Boolean expression linking the rules R_{cex}. The additional bindings B_{cex} are specified using parentheses and parameter order. Quotation marks are used to bind constants, names without quotation marks are used to bind entities. The formula language provides the operators AND and OR, as well as defining precedence using parentheses. Note the absence of a negation. The reason for this is twofold. First, the negation of a rule may be counterintuitive. For example, the negation of a rule imposing activity A is always followed by activity B is not that B never follows A, but rather that in at least one possible case B does not follow A. We found that safely using these negations requires proficiency in predicate logic as well as in the implementation language which average business users do not process. Second, as rules may evaluate to *unknown*, ternary Kleene logic [19] is used to evaluate compliance expressions rather than binary Boolean logic. Thus, negations are not used to approximate the behavior of the rule language to the intuitive understanding.

3.5 Graphical Model

Based on the formal description of the compliance descriptor, a graphical modeling language is designed. The modeling elements are shown in Fig. 3, corresponding to the elements defined above.

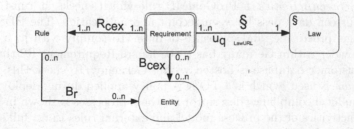

Fig. 3. Modeling elements of the compliance descriptor with cardinalities and formal equivalents for all relationships.

A *law* is modeled as a rounded square with a section sign indicating it contains multiple paragraphs. An *entity* is modeled as an oval. All entities referenced in bindings must be explicitly added to the compliance descriptor. Modeled entities may reference their counterparts in other models, e.g. an activity in a process model.

Rules are modeled as rounded squares without any further decoration. For each entity bound in B_r, a reference to the entity has to be modeled. This explicit modeling of entities serves to visualize the impact of compliance rules and requirements on the process, which would otherwise be hidden in an attribute.

A *requirement* is modeled as a double rounded square, indicating it may consist of multiple rules. Like the rule it has to reference each entity bound in B_{cex}. Additionally, requirements have to reference all rules R_{cex} used in the compliance expression. Requirements reference the law they stem from using their *law url* u_q. In comparison to the other references, a *law url* contains an XPATH expression indicating the law l_u and the paragraph p_u. A law url is decorated with a section sign to distinguish it from the other edges.

3.6   Example

Figure 4 shows a simplified example process from our work with insurance companies (for more detail see [27]). This claim management is used to automatically process damage claims from a customer, e.g. in case of car damages. A claim is received either digitally or in paper and stored in a customer database. The claim is then processed to find additional information, e.g. if the claim is covered by the insurance policy. Based on this data, the claim is decided to be either accepted in full, partially accepted or rejected. Finally, a notification of the result is sent to the customer as a letter.

Two laws relevant for this process are the Code of Conduct of the German Insurance Association (GDV) [13] and the German Federal Data Protection Act (BDSG) [7]. From these laws two requirements are derived as an example. The GDV Code of Conduct states a customer who provides personal data has to be asked in a timely fashion if this data can be used for marketing purposes. Requirement R1 is thus that such a notification takes place after data is received. This requirement is realized using two rules, *followedBy*, an LTL rule checking during design-time if an activity is always followed by another activity, and *maxTimeBetweenActivities*, a ProGoalML rule which checks at run-time that at most fourteen days pass between receipt and notification. The BDSG regulates storage, processing, and exposure of data. To comply with it, a suitable hosting provider within Germany has to be found. Requirement R2 thus states that the customer database is hosted within Germany. To check this, the rule *hostingRegion* is used, which is a TOSCA policy applied during deployment.

The graphical compliance descriptor for these processes is shown in Fig. 4. It references activities of the process model and external rules (for a full specification see [12]).

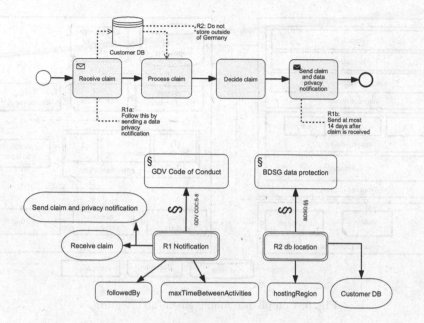

Fig. 4. Example claim management process (top) and compliance descriptor (bottom).

4   Compliance Management Architecture

Figure 5 gives an overview of the compliance management reference architecture. An *editor* in the frontend allows the user to graphically model a compliance descriptor. Additionally, existing capabilities for process and rule modeling can be used within the editor. All models are stored in a *model repository* in the backend. From this repository, a compliance descriptor in XML can be exported. To use a compliance descriptor for compliance management, it needs to be transformed into implementation specific artifacts. For this, the XML compliance descriptor is read by the *model transformation*, which creates rule expressions as well as a so-called VisML file [18], a dashboard description language which describes how rule checking results are to be visualized.

The created rule expressions are deployed to their specific *rule checking implementations* by a *rule deployment* component. For each deployed rule a *deployment descriptor* is created. The details of the deployment descriptor depend on the type of rule, but contain a unique identifier of the rule as well as all details necessary to undeploy it. During *results gathering* the deployment descriptor is used to get all results from the rule checking implementations. These results are then aggregated to determine requirement fulfillment and law compliance, thus translating compliance checking results from an IT level to a business level. All kinds of results are provided in an implementation-independent way to reporting and to a dashboard (via a *value provider*). Encapsulating rule implementations makes the architecture extensible, as only interfaces for rule deployment and result gathering need to be added for each rule language. The deployment

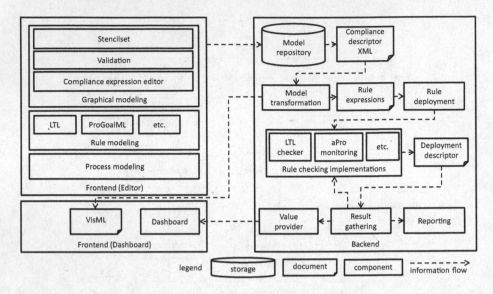

Fig. 5. Compliance management architecture.

descriptor handles implementation specific data in a generic fashion throughout the process lifecycle.

After giving an overview of the architecture, we will describe model transformation used to create concrete rule expressions and a visualization schema.

To create deployable concrete rule expressions, rules R_{final} are created for each final binding $B_{r_{final}}$ resulting from a compliance expression cex.

$$R_{final} = \{(n_r, d_r, il_r, ph_r, rex_r, VD_r, B_{r_{final}})|B_{r_{final}} \in \bigcup_{cex \in CEX} \{B_{r_{final}} \in B_{cex}\}\}$$

Using VD_r and $B_{r_{final}}$, a concrete rule expression $rex_{r_{final}}$ is created from rex_r. This rule expression can then be deployed to its implementation as indicated by il_r. The rule creation and deployment process is described in detail in [27].

Automatically creating process monitoring dashboards using VisML has been described in previous work [18]. Using a ProGoalML file as input, for each KPI or goal the appropriate visualization is selected using a visualization mapping file, and configured with the necessary data source and parameters. For the visualization of compliance requirements, a new visualization for the monitoring of compliance requirements was designed and documented in VisML, and the mapping file was extended. On the back-end, a new data value provider type was implemented to provide the data structure required by the new visualization.

The new visualization for compliance (see example in Fig. 6) was conceived as follows: The dashboard presents a box for each law. The box is labelled with the laws' name and colored in green if all underlying requirements are met, in red if one or more requirement is broken, and in yellow if the status is unknown.

A dashboard user has the possibility to click on a law box. The box is then replaced by a box for each underlying requirement, colored as mentioned above. Thus, the user can immediately see which requirements are fulfilled. The user can go one step further and click on a requirement box, to show the underlying rules in the compliance expression.

Tooltips provide additional information at every stage of the drill-down process from law to requirement to rules. The visualization mapping file was extended to indicate that a law mentioned in a compliance descriptor should be rendered as a "law box" compliance visualization. The VisML generation algorithm was extended as follows:

```
L_gen := ∅
for each  q ∈ Q
   if  l_{u_q} ∉ L_gen
      L_gen := L_gen ∪ {l_{u_q}}
   end if
end for
if  L_gen ≠ ∅
   add compliance value provider
       to data sources
   for each  l ∈ L_gen
      add law box visualization for l
          to dashboard
      add law data set for l
          to data sets
   end for
end if
```

In this algorithm $q \in Q$ are the compliance requirements, L_{gen} the laws on the top level of visualization, which are found by following the law urls u_q of the requirements.

The visualized data is obtained querying a data source with the appropriate parameters to obtain a data set. For compliance visualizations, a new data source *ComplianceValueProvider* was designed. It supports data sets using a parameter *law*, indicating the law for which the data is requested. It supports one or more parameters for each requirement, indicating the level of drill down.

5 Prototype and Evaluation

The architecture has been implemented in a prototype based on the Oryx editor[2], a web-based modeling tool. The prototype contains modeling capabilities for processes, compliance descriptors, and rules (ProGoalML and LTL).

When modeling a compliance descriptor, rules and entities may reference other models, which can be edited in another editor window. The compliance descriptor is automatically converted to XML, which is then used as a basis

---

[2] http://bpt.hpi.uni-potsdam.de/Oryx (accessed 18.3.2015).

for model transformation. During model transformation, rules expressions are automatically created and deployed to their respective rule implementations. A VisML file is generated automatically and used with the configurable dashboard to visualize compliance checking results.

Figure 6 shows the modeling as well as the configured dashboard.

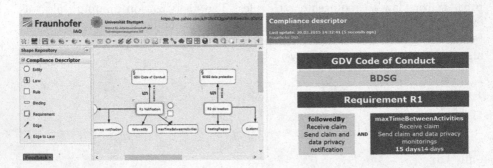

Fig. 6. Prototype compliance descriptor modeling (left) and dashboard (right).

We evaluated the prototype with the example (see Sect. 3.6) and synthetic execution data for run-time rules. The process was modeled together with domain experts of the example process. We found the compliance, process, and rules monitoring to work as an integrated workflow. However, cross-model validation and other usability features like a graphical compliance expression editor may further increase usability for business users. We found the concept of encapsulating implementing rules feasible. However, some checks require additional information of the process implementation, e.g. run-time data to check timings. Currently, these need to be supplied manually to the rule checking implementations. This could be partially automated by allowing IT users to give global configuration files for each type of rule. The created dashboard shows compliance on the rule, requirement, and law levels. Using drilldown, the cause for a lack of law compliance or legal fulfillment can easily be found. Further implementation details can be found in a technical report [12].

Using these preliminary results, we further evaluated the prototype in a real-life process in the German insurance industry and in interviews with compliance experts. Implementing compliance checking with the real process encompassed design-time and run-time. For design-time checking, rules for a claim management process were created from laws and industry guidelines and validated within the process model, proving it to be compliant. For real-time checking, compliance goals on process KPIs and timing restrictions were modeled as compliance rules, which were deployed to a model-driven process monitoring solution [23]. Monitoring data was acquired from the live system using existing business monitoring information, which was manually integrated with the new monitoring solution. Run-time compliance checking was performed on a test system with real instance data. By modifying the running system, timing rules and business rules could be broken, resulting in a compliance violation, shown on the dashboard and as an alert. Deployment rules in TOSCA were not evaluated, as access to the deployment of real-life live systems was not possible. We discussed this solution with IT personnel as well as compliance specialists.

While IT personnel and specialists were generally interested in the approach, it was judged as only feasible with new or overhauled process. Efforts for integration were seen as high, even though partial automation is available, especially in existing infrastructures which cannot easily be changed and don't provide access to data in real-time. Additionally, even if rules are established, it is not possible to check past instances, as monitoring data is gathered only at run-time and not from existing sources (e.g. historical data). Especially for audits compliance checking of past process instances is necessary.

Another challenge in practice is the lack of process automation. Knowledge-intensive processes neither fully automated nor fully structured. Employees have a degree of freedom not captured in the process model. Compliance rules, even if communicated clearly to employees, are not always fulfilled. Examples named by interviewees are the omission of checks to serve customers faster and the use of non-approved or homegrown IT tools like Excel Sheets and cloud services. This suggests the need for further compliance rule types outside of the business process lifecycle not covered in this work yet. Examples could be process mining tools to compare the actual process to the prescribed process and infrastructure assessment tools to find unauthorized applications, files, and communications.

Compliance experts also noted that being notified about each compliance violation can be overwhelming in large processes. Compliance as a cross-sectional task depends on cooperation of all departments. Thus, investigating each violation independently will not be feasible, because it will require too many resources from other departments. Rather, compliance analytics should be able to find the root cause of a violation beyond the violated rule.

Considering this feedback, we find the general approach of using a compliance descriptor and different rule languages to be feasible for automated, new or overhauled business processes. To achieve compliance in real-life, large, unstructured or partially automated business processes, the current range of functionality is not sufficient yet and needs to be extended considering the deficits outlined above.

6   Outlook

As shown in the evaluation, while the chosen approach provides a basis for compliance management, additional aspects need to be covered and incorporated into the solution. In this section we will give an outline of current and future work to address these shortcomings.

Large, unstructured partially automated processes prove a challenge in compliance management, as there is neither a descriptive process model documenting the process nor a prescriptive process model which guides process execution. If either of these models should exist, they are often out-of-date or not adhered to by employees taking shortcuts.

In order to allow compliance checking of these processes, a discovery phase is necessary, in which the as-is process is investigated. The goal of this phase is not only to discover violations, but to also aquire a process model to work on as well as define rules for existing and likely compliance violations so they are detected in the future.

Process mining is a technique for generating process models from process execution logs. The Process Mining Framework (*ProM*) [9] is a widely adopted tool for process mining. We plan to incorporate ProM into our solution to allow support for structured processes.

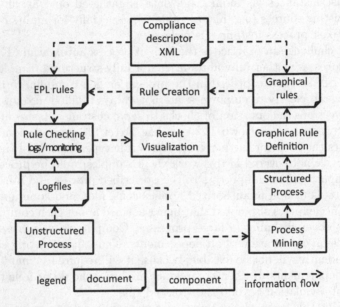

Fig. 7. Preliminary components of compliance checking in unstructured processes using process mining.

Figure 7 shows the preliminary components of a compliance checking solution for unstructured processes. In a first discovery step, process mining is used to create a structured process model from the log file of the unstructured process.

This process model can be used in a graphical editor (Oryx) to find compliance violations and define compliance rules. A future research question is what compliance rules are suitable for unstructured processes and how are they formalized, defined and visualized. Existing rules are only partially applicable. While LTL rules allow model checking on a structured process, they can only check the extracted process as a whole and do not detect which instance causes a violation. While aPro monitoring rules can detect compliance violations in non-executable processes, they are created on the basis of an accurate descriptive process model. In comparison, a structured process model from process mining is only an approximation, as it only contains existing process execution but cannot contain possible process branches not included in the execution logs. Some aPro rules (e.g. a timing goal) still work on unstructured processes, but the majority does not, as the generated Complex Event Processing patterns may not match if the process execution does not follow the process model [23].

Rules for unstructured processes can be checked in two manners. One is a bulk check on past execution logs to find violations in past process instances. The other is a real-time check during process execution. Events are monitored as they are logged to discover violations as soon as they occur. Compliance rules for unstructured processes must support both manners to support design-time, run-time and ex-post rule checking.

To investigate the feasibility of rule languages, we built a preliminary prototype encompassing the components shown in Fig. 8. A process model is mined in ProM (A) and imported as a structured process model to Oryx (B). On this structured model, graphical compliance rules can be defined. In this preliminary prototype two patterns can be defined (C): the antecedence occurence pattern (stipulating an activity has to be followed by another activity) and antecedence absence pattern (stipulating an activity must not be followed by another activity) [20]. From these patterns, CEP rules are generated, which detect pattern violations on an event stream. During design-time and ex-post checking, this event stream is fed to a CEP engine from log files (D). During run-time checking, the event stream will originate from the systems executing the process.

Based on the graphical definitions, rules can be integrated in the compliance descriptor using the same modeling environment, adding rules to requirements and in turn relevant laws.

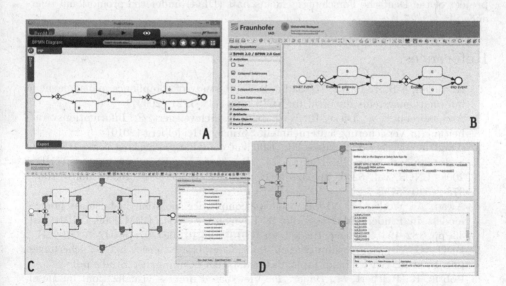

Fig. 8. Preliminary prototype for compliance checking of unstructure processes.

Work on the preliminary prototype is ongoing. After proving general feasibility, future work will investigate which rules are necessary for unstructured process and if existing modeling languages (e.g. [20]) can be integrated. Additionally, usability tests with logs from unstructured insurance processes are planned.

7   Conclusion

In this work we described the compliance descriptor, an integrating approach for compliance management. It allows connected modeling of laws, compliance requirements and technical compliance rules, thus bridging the gap between different compliance levels. In a model transformation step, a compliance descriptor is used to configure a compliance management solution, including different implementations for rule checking. Results of different rule checking components can be aggregated on a single dashboard. We evaluated this approach using a prototype and example from practice, as well as in interviews with experts. We have shown the practical applicability of our approach for compliance modeling, rule modeling, compliance checking and result visualization. However, the evaluation also showed gaps in the proposed solution. We have outlined how in future work we will adress these results, increasing usability of the prototype and adding further rule types, encompassing checks for unstructured processes and historical data.

Acknowledgements. The work published in this article was funded by the Co.M.B. project of the Deutsche Forschungsgemeinschaft (DFG) under the promotional reference SP 448/27-1.

References

1. Abdullah, N.S., Indulska, M., Sadiq, S.W.: A study of compliance management in information systems research. In:,ECIS, pp. 1711–1721 (2009)
2. Aschenbrenner, M., Dicke, R., Karnarski, B., Schweiggert, F.: Informationsverarbeitung in Versicherungsunternehmen. Springer, Heidelberg (2010)
3. Awad, A., Decker, G., Weske, M.: Efficient compliance checking using BPMN-Q and temporal logic. In: Dumas, M., Reichert, M., Shan, M.-C. (eds.) BPM 2008. LNCS, vol. 5240, pp. 326–341. Springer, Heidelberg (2008). doi:10.1007/978-3-540-85758-7_24
4. Awad, A., Weske, M.: Visualization of compliance violation in business process models. In: Rinderle-Ma, S., Sadiq, S., Leymann, F. (eds.) BPM 2009. LNBIP, vol. 43, pp. 182–193. Springer, Heidelberg (2010). doi:10.1007/978-3-642-12186-9_17
5. BDO AG Wirtschaftsprüfungsgesellschaft: Compliance Survey bei Versicherungen (2010). http://www.bdo.de/uploads/media/BDO_Compliance_Studie.pdf
6. Bobrik, R., Reichert, M., Bauer, T.: View-based process visualization. In: Desel, J., Pernici, B., Weske, M. (eds.) BPM 2004. LNCS, vol. 3080. Springer, Heidelberg (2004). doi:10.1007/978-3-540-75183-0_7
7. Bundesdatenschutzgesetz (BDSG): Gesetze im Internet - Bundesdatenschutzgesetz (BDSG) (1990). http://www.gesetze-im-internet.de/bundesrecht/bdsg_1990/gesamt.pdf. Accessed 19 Jan 2016

8. Comuzzi, M.: Aligning monitoring and compliance requirements in evolving business networks. In: Meersman, R., Panetto, H., Dillon, T., Missikoff, M., Liu, L., Pastor, O., Cuzzocrea, A., Sellis, T. (eds.) OTM 2014. LNCS, vol. 8841, pp. 166–183. Springer, Heidelberg (2014). doi:10.1007/978-3-662-45563-0_10
9. Dongen, B.F., Medeiros, A.K.A., Verbeek, H.M.W., Weijters, A.J.M.M., Aalst, W.M.P.: The ProM framework: a new era in process mining tool support. In: Ciardo, G., Darondeau, P. (eds.) ICATPN 2005. LNCS, vol. 3536, pp. 444–454. Springer, Heidelberg (2005). doi:10.1007/11494744_25
10. El Kharbili, M., Stein, S., Markovic, I., Pulvermüller, E.: Towards a framework for semantic business process compliance management. In: Proceedings of the 1st GRCIS, pp. 1–15 (2008)
11. El Kharbili, M., Stein, S., Pulvermüller, E.: Policy-based semantic compliance checking for business process management. In: MobIS Workshops, vol. 420, pp. 178–192. Citeseer (2008)
12. Fehling, C., Koetter, F., Leymann, F.: Compliance Modeling - Formal Descriptors and Tools (2014). http://www.iaas.uni-stuttgart.de/institut/mitarbeiter/fehling/TR-2014-Compliance-Modeling.pdf
13. German Insurance Association (GDV): Verhaltensregeln fuer den Umgang mit personenbezogenen Daten durch die deutsche Versicherungswirtschaft (2012). http://www.gdv.de/wp-content/uploads/2013/03/GDV_Code-of-Conduct_Datenschutz_2012.pdf. Accessed 19 Jan 2016
14. Ghose, A., Koliadis, G.: Auditing business process compliance. In: Krämer, B.J., Lin, K.-J., Narasimhan, P. (eds.) ICSOC 2007. LNCS, vol. 4749, pp. 169–180. Springer, Heidelberg (2007). doi:10.1007/978-3-540-74974-5_14
15. Goedertier, S., Vanthienen, J.: Designing compliant business processes with obligations and permissions. In: Eder, J., Dustdar, S. (eds.) BPM 2006. LNCS, vol. 4103, pp. 5–14. Springer, Heidelberg (2006). doi:10.1007/11837862_2
16. Karagiannis, D., Moser, C., Mostashari, A.: Compliance evaluation featuring heat maps (CE-HM): a meta-modeling-based approach. In: Ralyté, J., Franch, X., Brinkkemper, S., Wrycza, S. (eds.) CAiSE 2012. LNCS, vol. 7328, pp. 414–428. Springer, Heidelberg (2012). doi:10.1007/978-3-642-31095-9_27
17. Kharbili, M.E., de Medeiros, A.K.A., Stein, S., van der Aalst, W.M.P.: Business process compliance checking: current state and future challenges. In: MobIS, LNI, vol. 141, pp. 107–113. GI (2008)
18. Kintz, M.: A semantic dashboard description language for a process-oriented dashboard design methodology. In: Proceedings of 2nd MODIQUITOUS 2012, Copenhagen, Denmark (2012)
19. Kleene, S.C.: Introduction to Metamathematics. North-Holland Publishing Co., Amsterdam (1952)
20. Knuplesch, D., Reichert, M.: A visual language for modeling multiple perspectives of business process compliance rules. In: Software and Systems Modeling, pp. 1–22. Springer, Heidelberg (2016)
21. Knuplesch, D., Reichert, M., Pryss, R., Fdhila, W., Rinderle-Ma, S.: Ensuring compliance of distributed and collaborative workflows. In: 9th Collaboratecom, pp. 133–142. IEEE (2013)
22. Kochanowski, M., Fehling, C., Koetter, F., Leymann, F., Weisbecker, A.: Compliance in BPM today - an insight into experts' views and industry challenges. In: Proceedings of INFORMATIK 2014, GI (2014)
23. Koetter, F., Kochanowski, M.: A model-driven approach for event-based business process monitoring. In: Rosa, M., Soffer, P. (eds.) BPM 2012. LNBIP, vol. 132, pp. 378–389. Springer, Heidelberg (2013). doi:10.1007/978-3-642-36285-9_41

24. Koetter, F., Kochanowski, M.: A model-driven approach for event-based business process monitoring. In: Information Systems and e-Business Management, pp. 1–32 (2014)

25. Koetter, F., Kochanowski, M., Kintz, M.: Leveraging model-driven monitoring for event-driven business process control. In: Workshop zur Ereignismodellierung und -verarbeitung im Geschaeftsprozessmanagement (EMOV) (2014, to appear)

26. Koetter, F., Kochanowski, M., Renner, T., Fehling, C., Leymann, F.: Unifying compliance management in adaptive environments through variability descriptors (short paper). In: IEEE SOCA 2013, pp. 214–219. IEEE (2013)

27. Koetter, F., Kochanowski, M., Weisbecker, A., Fehling, C., Leymann, F.: Integrating compliance requirements across business and IT. In: 18th EDOC, pp. 218–225. IEEE (2014)

28. Ly, L.T., Knuplesch, D., Rinderle-Ma, S., Göser, K., Pfeifer, H., Reichert, M., Dadam, P.: SeaFlows toolset – compliance verification made easy for process-aware information systems. In: Soffer, P., Proper, E. (eds.) CAiSE Forum 2010. LNBIP, vol. 72, pp. 76–91. Springer, Heidelberg (2011). doi:10.1007/978-3-642-17722-4_6

29. Mietzner, R., Metzger, A., Leymann, F., Pohl, K.: Variability modeling to support customization and deployment of multi-tenant-aware software as a service applications. In: Proceedings of PESOS 2009, pp. 18–25. IEEE Computer Society, Washington, DC (2009)

30. Papazoglou, M.: Making business processes compliant to standards and regulations. In: 2011 15th IEEE International Enterprise Distributed Object Computing Conference (EDOC), pp. 3–13, August 2011

31. Patig, S., Casanova-Brito, V., Vögeli, B.: IT requirements of business process management in practice – an empirical study. In: Hull, R., Mendling, J., Tai, S. (eds.) BPM 2010. LNCS, vol. 6336, pp. 13–28. Springer, Heidelberg (2010). doi:10.1007/978-3-642-15618-2_4

32. Ramezani, E., Fahland, D., Aalst, W.M.P.: Supporting domain experts to select and configure precise compliance rules. In: Lohmann, N., Song, M., Wohed, P. (eds.) BPM 2013. LNBIP, vol. 171, pp. 498–512. Springer, Cham (2014). doi:10.1007/978-3-319-06257-0_39

33. Ramezani, E., Fahland, D., Werf, J.M., Mattheis, P.: Separating compliance management and business process management. In: Daniel, F., Barkaoui, K., Dustdar, S. (eds.) BPM 2011. LNBIP, vol. 100, pp. 459–464. Springer, Heidelberg (2012). doi:10.1007/978-3-642-28115-0_43

34. Reichert, M., Weber, B.: Enabling Flexibility in Process-aware Information Systems: Challenges, Methods, Technologies. Springer, Heidelberg (2012)

35. Sadiq, S., Governatori, G., Namiri, K.: Modeling control objectives for business process compliance. In: Alonso, G., Dadam, P., Rosemann, M. (eds.) BPM 2007. LNCS, vol. 4714, pp. 149–164. Springer, Heidelberg (2007). doi:10.1007/978-3-540-75183-0_12

36. SAI Global: 2013 Insurance Industry Compliance Benchmark Study (2013). http://compliance.saiglobal.com/community/resources/-whitepapers

37. Scherer, G.S.H.: Assekuranz 2015 - Eine Standortbestimmung. Universität Sankt Gallen - Institut für Versicherungswirtschaft, Sankt Gallen, Schweiz (2015)

38. Schleicher, D., Fehling, C., Grohe, S., Leymann, F., Nowak, A., Schneider, P., Schumm, D.: Compliance domains: a means to model data-restrictions in cloud environments. In: 15th EDOC, pp. 257–266. IEEE (2011)

39. Semmelrodt, F., Knuplesch, D., Reichert, M.: Modeling the resource perspective of business process compliance rules with the extended compliance rule graph. In: Bider, I., Gaaloul, K., Krogstie, J., Nurcan, S., Proper, H.A., Schmidt, R., Soffer, P. (eds.) BPMDS/EMMSAD -2014. LNBIP, vol. 175, pp. 48–63. Springer, Heidelberg (2014). doi:10.1007/978-3-662-43745-2_4
40. Takabi, H., Joshi, J.B., Ahn, G.J.: Security and privacy challenges in cloud computing environments. IEEE Secur. Priv. **8**(6), 24–31 (2010)
41. Wagner, R., Steinhüser, D., Engelbrefcht, O., Meinherz, A.: Agenda 2015: Compliance Management als stetig wachsende Herausforderung für Versicherungen (2010)
42. Waizenegger, T., et al.: Policy4TOSCA: a policy-aware cloud service provisioning approach to enable secure cloud computing. In: Meersman, R., Panetto, H., Dillon, T., Eder, J., Bellahsene, Z., Ritter, N., Leenheer, P., Dou, D. (eds.) OTM 2013. LNCS, vol. 8185, pp. 360–376. Springer, Heidelberg (2013). doi:10.1007/978-3-642-41030-7_26
43. Wei, Y., Blake, M.B.: Service-oriented computing and cloud computing: challenges and opportunities. IEEE Internet Comput. **14**(6), 72–75 (2010)
44. Weigand, H., Elsas, P.: Model-based auditing using REA. Int. J. Account. Inf. Syst. **13**(3), 287–310 (2011). Research Symposium on Information Integrity and Information Systems Assurance (2012)

Fostering the Reuse of TOSCA-based Applications by Merging BPEL Management Plans

Sebastian Wagner[1]([⊠]), Uwe Breitenbücher[1], Oliver Kopp[2], Andreas Weiß[1], and Frank Leymann[1]

[1] IAAS, University of Stuttgart, Universitätsstraße 38, 70569 Stuttgart, Germany
{wagner,breitenbucher,Weib,leymann}@informatik.uni-stuttgart.de
[2] IPVS, University of Stuttgart, Universitätsstraße 38, 70569 Stuttgart, Germany
kopp@informatik.uni-stuttgart.de

Abstract. Complex Cloud applications consist of a variety of individual components that need to be provisioned and managed in a holistic manner to setup the overall application. The Cloud standard TOSCA can be used to describe these components, their dependencies, and their management functions. To provision or manage the Cloud application, the execution of these individual management functions can be orchestrated by executable management plans, which are workflows being able to deal with heterogeneity of the functions. Unfortunately, creating TOSCA application descriptions and management plans from scratch is time-consuming, error-prone, and needs a lot of expert knowledge. Hence, to save the amount of time and resources needed to setup the management capabilities of new Cloud applications, existing TOSCA description and plans should be reused. To enable the systematic reuse of these artifacts, we proposed in a previous paper a method for combining existing TOSCA descriptions and plans or buildings blocks thereof. One important aspect of this method is the creation of BPEL4Chor-based management choreographies for coordinating different plans and how these choreographies can be automatically consolidated back into executable plans. This paper extends the previous one by providing a much more formal description about the choreography consolidation. Therefore, a set of new algorithms is introduced describing the different steps required to consolidate the management choreography into an executable management plan. The method and the algorithms are validated by a set of tools from the Cloud application management ecosystem OpenTOSCA.

1 Introduction

Due to the steadily increasing use of information technology in enterprises, accurate development, provisioning, and management of applications becomes of crucial importance to align business and IT. While developing application components and modelling application architectures and designs is supported by sophisticated tools, application management still presents major challenges: Especially

M. Helfert et al. (Eds.): CLOSER 2016, CCIS 740, pp. 232–254, 2017.
DOI: 10.1007/978-3-319-62594-2_12

in Cloud Computing, management automation is a key prerequisite since manual management is (i) too slow to preserve Cloud properties such as elasticity and (ii) too error-prone as human operator errors account for the largest fraction of failures in distributed systems [1,2]. Thus, management automation is a key incentive in modern IT.

While various management technologies[1] exist that are capable of automating *generic* management tasks, such as automatically scaling application components or installing single software components, the automation of *complex, holistic, and application-specific management processes* is an open issue. Automating complex management processes, e. g., migrating an application component from one Cloud to another while avoiding downtime or acquiring new licenses for employed software components, typically requires the orchestration of multiple heterogeneous management technologies. Therefore, such management processes are mostly implemented using workflows languages [6], e. g., BPEL [7] or BPMN [8], since other approaches such as scripts are not capable of providing the reliability and robustness of the workflow technology [9].

Creating management processes, however, requires integrating the different invocation mechanisms, data formats, and transport protocols of each employed technology, which needs enormous time and expertise on the conceptual as well as on the technical implementation level [10].

To avoid continually reinventing the wheel for problems that have been already solved multiple times for other applications, developing new applications by reusing and combining proven (i) structural application fragments as well as (ii) the corresponding available management processes would pave the way to increase the efficiency and quality of new developments. However, while automatically combining and merging individual application structures is resolved [11], integrating the associated management processes is a highly non-trivial task that still has to be done manually. Unfortunately, similarly to manually authoring such processes, this leads to error-prone, time-consuming, and costly efforts, which is not appropriate for modern software development and operation.

In this paper, we tackle these issues. We first present a method that describes how to employ choreographies to systematically reuse existing management workflows. Choreography models enable coordinating the distributed execution of individual workflows without the need to adapt their implementation. Thus, they provide a suitable integration basis to combine different management workflows without the need to dive into or change their technical implementation.

Since choreographies are not intended to be executed on a single workflow engine—which is a mandatory requirement in application management as typically sensitive data such as credentials or certificates have to be exchanged between the coordinated workflows—they must be transformed into an executable workflow model. Therefore, we introduced in our former work [12], that was presented at the *6th International Conference on Cloud Computing and Services Science (CLOSER)*, a process consolidation approach that transforms

---

[1] E.g., configuration management technologies such as Chef [3] or Puppet [4], or Cloud management platforms such as Heroku [5].

a choreography including all coordinated workflow models into one single executable workflow model.

The consolidation results also in a faster execution due to reduced communication over the wire. It also simplifies deployment as only a single workflow has to be deployed instead of various interacting workflows along with the choreography specification itself. Thus, reusing management workflows following this approach leads to significant time and cost savings when developing new applications out of existing building blocks. In this paper, which is an extended version of the original paper [12], we a provide in Sect. 4 a more detailed and comprehensive description of the consolidation approach. Therefore, the new contributions are an additional set of algorithms describing the consolidation steps in a much more formal way than in the original paper. To discuss these algorithms also the choreography meta-model was extended.

To validate the presented approach, we apply the developed concepts to the choreography modelling language BPEL4CHOR [13] and the Cloud standard TOSCA [14,15]. For this purpose, we developed a standard-based, open-source Cloud application management prototype by extending the OpenTOSCA ecosystem [16–18] in order to support managing applications based on choreographies, that are transparently transformed into executable workflows behind the scenes.

The remainder is structured as follows. Section 2 provides background and related work information along with a motivating scenario. In Sect. 3, we conceptually describe the method for reusing TOSCA-based applications and their management plans by introducing management choreographies. In Sect. 4 we formally discuss the step of the method, that transforms a choreography into an executable management plan. Section 5 validates the method proposed in Sect. 3 and Sect. 6 concludes the work.

2   Background and Related Work

This section discusses background and related work about (i) the Cloud standard TOSCA, (ii) management workflows, and (iii) the transformation and consolidation of choreographies. In Sect. 2.3, we introduce a motivating scenario that is used throughout the paper to explain the approach.

2.1   TOSCA and Management Plans

In this section, we introduce the *Topology and Orchestration Specification for Cloud Applications (TOSCA)*, which is an emerging standard to describe Cloud applications and their management. We explain the fundamental concepts of TOSCA that are required to understand the contributions of this paper and simplify constructs, where possible, for the sake of comprehension. For more details, we refer interested readers to the TOSCA Specification [14] and the TOSCA Primer [15]. TOSCA defines a meta-model for describing (i) the structure of an application, and (ii) their management processes. In addition, the standard introduces an archive format that enables packaging applications and all required files, e.g., installables, as portable archive that can be consumed by TOSCA runtimes to provision and manage new instances of the described application. The structure of the application is described in the form of an *application topology*, a directed graph that consists of vertices representing the components of the application and edges that describe the relationships between these components, e. g., that a *Webserver* component is *installed* on an *operating system*. Components and relationships are typed and may specify properties and management operations to be invoked. For example, a component of type *ApacheWebserver* may specify its *IP-address* as well as the *HTTP-port* and provides an operation to *deploy* new applications. In addition, required artifacts, e. g., installation scripts or binaries implementing the application functionality, may be associated with the corresponding components, relationships, and operations. Thereby, TOSCA enables describing the entire structure of an application in the form of a self-contained model, which also contains all information about employed types, properties, files, and operations. These models can be used by a TOSCA runtime to fully automatically provision instances of the application by interpreting the semantics of the modeled structure [15,19].

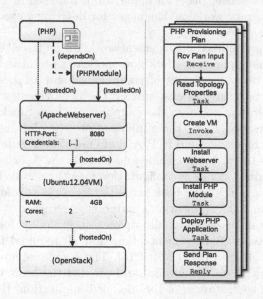

Fig. 1. TOSCA example: topology (left) and provisioning plan (right).

Figure 1 shows an example on the left rendered using VINO4TOSCA [20]. The shown topology describes a deployment consisting of a *PHP* application that is hosted on an *ApacheWebserver* running on a virtual machine (VM) of type *Ubuntu12.04VM*. This VM is operated by the Cloud management system *OpenStack*. To run the PHP application on an Apache Webserver, a *PHPModule* needs to be installed. In the topology the component types and relationship types, e. g., the desired *hostedOn*, of the VM, are put in brackets. The component properties, e. g., the desired RAM of the VM, are depicted below the component types. The actual application implementation, i. e., the PHP files implementing the functionality, is attached to the PHP component.

While the provisioning of simple applications can be described *implicitly* by such topology models, TOSCA also enables describing complex provisioning and management processes in the form of *explicitly* modeled *management plans*. Management plans are executable workflows that specify the (i) activities to be executed, (ii) the control flow between them, i. e., their execution order, as well as (iii) the data flow, e. g., that one activity produces data to be consumed by a subsequent activity [6]. There exists standardized workflow languages and corresponding engines, for example, BPEL [21] or BPMN [22], that enable describing workflows in a portable manner. Standard-compliant workflow engines can be employed to automatically execute these workflow models. The workflow technology is well-known for features such as reliability and robustness [6], thus, providing an ideal basis to automate management processes [7]. In addition, there are extensions of workflow standards which are explicitly tailored to the management of applications. For example, BPMN4TOSCA [8] is an extension to easily describe management plans for applications modeled in TOSCA. TOSCA supports using arbitrary workflow languages for describing executable management plans [14].

Figure 1 shows a simplified management workflow on the right that automatically provisions the application (data flow modeling is omitted for simplicity). The first activity reads properties of components and relationships from the topology model, which enables customizing the deployment without adapting the plan. Other information, e. g., the endpoint of Open Stack, are passed via the plan's start message. Using these information, the plan instantiates a new virtual machine by invoking the HTTP-API of Open Stack. Afterwards, the plan uses SSH to access the virtual machine and installs the Apache Webserver and the PHP module using Chef [3], a configuration management technology. Finally, the application files, which have been extracted from the topology, are deployed on the Webserver and the application's endpoint is returned.

The TOSCA standard additionally defines an exchange format to package topology models, types, management plans and all required files in the form of a *Cloud Service Archive (CSAR)* [14,15]. These archives are portable across standards-compliant TOSCA runtimes and provide the basis to automatically provision and manage instances of the modeled application. Runtimes such as OpenTOSCA [16] also enable automatically executing the associated management workflows, thereby, enabling the automation of the entire lifecycle of Cloud applications described in TOSCA. Thus, TOSCA provides an ideal basis for systematically reusing (i) proven application structures as well as their (ii) management processes as both can be described and linked using the standard.

2.2 Choreography Transformation

There exist manual approaches for transforming choreographies to executable processes (plans). Hofreiter et al. [23] suggest for instance a top-down approach where business partners agree on a global choreography by specifying the inter-action behavior the processes of the partners have to comply with. The chore-ography and the corresponding processes have to be modeled in UML and the authors propose a manual transformation to BPEL. Mendling et al. [24] use the Web Service Choreography Description Language (WS-CDL) [25] to model choreographies and to generate BPEL process stubs out of it. However, these process stubs have to be also completed manually. Another drawback of WS-CDL is that it is an interaction choreography which is less expressive than intercon-nection models as we will briefly discuss in Sect. 3.3.

In Sect. 4 a process consolidation algorithm is presented to generate an exe-cutable process from a choreography. Existing process consolidation techniques, e. g., from Küster et al. [26] or Mendling and Simon [27], focus on merging seman-tically equivalent processes, which is different from the proposed consolidation algorithm that merges *complementing* processes of a choreography into a single process.

In contrast to our approach Herry et al. [28] aim to execute a former central-ized management workflow in a decentralized fashion. To accomplish that they are describing an approach to decompose the management workflow into a set of different interacting agents coordinating its execution.

2.3 Motivating Scenario

This section describes a motivation scenario based on the previous example to explain the difficulties of implementing executable management plans and the significant advantage that would be enabled by an approach that facilitates systematically reusing and combining existing workflows. As described before, for provisioning the PHP-based example application several management tasks have to be performed: Open Stack's HTTP-API has to be invoked for instanti-ating the VM while SSH and Chef are used to install the Webserver. However, already this simple example impressively shows the difficulties: Two low-level management technologies including their invocation mechanisms, data formats, and transport protocols have to be (i) understood and (ii) orchestrated by a workflow. This requires complex data format transformations, building integra-tion wrappers to invoke the technologies, and results in many lines of complex workflow code [10]. Thus, implementing such management plans from scratch is a labor-intensive, error-prone, and complex task that requires a lot of expertise in very different fields of technologies - reaching from *high-level* orchestration to *low-level* application management. Therefore, systematically reusing existing plans and combining them and coordinating them would significantly improve these deficiencies.

Figure 2 shows an example how TOSCA may support this vision. On the left, the provisioning plan and the topology of the TOSCA example introduced in Sect. 2.1 is shown. On the right, a topology is shown that describes the deploy-ment of a MySQL database including the corresponding provisioning plan. This plan automatically provisions a new VM, installs the MySQL database manage-ment system, creates a new database, and inserts a specified schema, which is

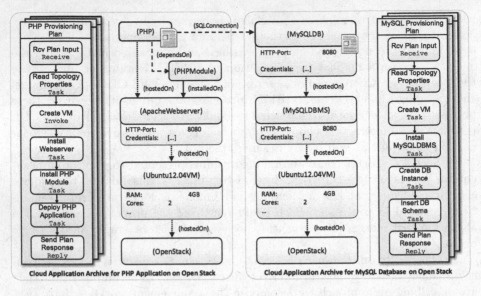

Fig. 2. Motivating scenario showing that management plans have to be combined to reuse existing topology models and management processes.

attached to the MySQL component. Thus, if a LAMP-application[2] has to be developed, the two topologies could be merged and connected with a new relationship of type *SQLConnection*. Obviously, to provision the combined stack, also their provisioning plans have to be combined. However, while merging TOSCA topology models can be done easily using tools such as Winery [17], manually combining workflow models is a crucial and error-prone task since (i) the individual control flows and possible violations have to be considered, (ii) low-level artifacts, e. g., XML schema definitions, have to be imported, and (iii) typically hundreds of lines of workflow code have to be integrated. Handling these issues manually is neither efficient nor reliable. Therefore, a systematic approach for combining TOSCA topologies and management plans is required that enables combining plans without the need to deal with their actual implementation.

3 A Method to Reuse TOSCA-based Applications

This section presents a generic method to systematically reuse existing TOSCA-based topology models and their management plans as building blocks for the development of new applications. The method is subdivided in two phases and shown in Fig. 3: (i) a *manual modeling phase*, which describes how application developers and manager model new applications by reusing existing topology

---

[2] An application consisting of Linux, Apache, MySQL, PHP components.

Fig. 3. Steps of the method to systematically reuse TOSCA-based (i) application topologies and (ii) their corresponding management plans.

models and plans, and (ii) an *automated execution phase*, which enables automatically deploying and managing the modeled application. The five steps of the method are explained in detail in the following.

3.1 Select and Merge TOSCA Topology Models

In the first step, the application developer sketches the desired deployment and selects appropriate TOSCA topology models from a repository to be used for its realization. The selected topologies are merged by copying them into a new topology model, which provides a *recursive aggregation model* as the result is also a topology that can be combined with others again. This is a manual step that may be supported by TOSCA modeling tools such as the open-source implementation Winery [17]. In previous works, we showed how multiple application topologies can be merged automatically while preserving their functional semantics [11] and how valid implementations for custom component types can be derived automatically from a repository of validated cloud application topologies [29]. These works support technically merging individual topologies, but the general decisions which topologies to be used are of manual nature as only developers are aware of the desired overall functionality of the application to be developed.

3.2 Connect Merged Parts of the Application

The resulting topology model contains isolated topology fragments that may have to be connected with each other. For example, the motivating scenario requires the insertion of a *SQLConnection* relationship to syntactically connect the merged topology models. Using well-defined relationship types enables specifying the respective semantics. This is also a manual step as these connections exclusively depend on the desired functionality. Moreover, TOSCA enables specifying *requirements* and *capabilities* of components, which can be used to automatically derive possible connections [15]. Modeling tools may use these specifications to support combining the individual fragments, but in many cases the final decisions must be made manually by the application developers. For example, if multiple business components and databases exist, in general, a modeling tool cannot derive with certainty which component has to connect to which database.

3.3   Coordinate Management Plans by Choreographies

Similarly to connecting isolated topology fragments, their management plans need to be combined for realizing holistic management processes that affect larger parts of the merged application at once, for example, to terminate the whole application. However, as discussed in Sect. 2.3, manually merging workflow models is a highly non-trivial and technically error-prone task. Therefore, we propose using *interconnection choreographies* to coordinate the individual workflows without changing their actual implementation. Interconnection choreographies define interaction specifications for collaborating processes by interconnecting *communication activities*, i. e., *send* and *receive* activities, of these processes via set of *message links*[3]. This enables modeling different interaction styles between the individual management workflows, e. g., asynchronous and synchronous interactions. Thus, in this step, (i) application managers analyse required management processes, (ii) select appropriate management workflows of the individual topology models, and (iii) coordinate them by modeling choreographies. In addition, (iv) depending on required input and output parameters of the individual workflows, the data flow between the workflow invocations has to be specified. For example, the MySQL provisioning workflow of the motivating scenario outputs the endpoint and credentials of the database, which are required to invoke a management plan of the PHP model that connects the

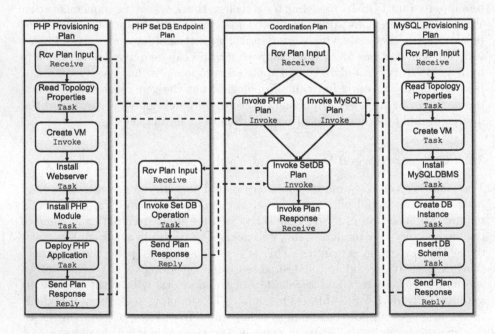

Fig. 4. Provisioning choreography coordinating management plans.

---

[3] In contrast to interaction choreographies that model message exchanges as abstract interactions not considering the workflow implementation.

PHP frontend to this database[4]. This is a manual step as the desired functionality, in general, cannot be derived automatically for application-specific tasks.
For example, the individual provisioning plans of the motivating scenario can
be used to model the overall provisioning of the entire application as well as to
implement management plans that scale out parts of the application to handle
changing workloads.

Figure 4 shows an example choreography that coordinates three management workflows from the motivating scenario. The coordination plan invokes
in parallel the provisioning workflows of the PHP and MySQL topology models,
respectively, by specifying message links to their receive activities. After their
execution, messages are sent back to the coordination plan, which continues with
invoking the aforementioned management workflow for transferring the database
information (endpoint, database name, and credentials) to the PHP application
by invoking the corresponding management operation.

3.4 Transform Choreographies into Executable Workflows

After manually modeling the choreography, the resulting model has to be transformed into an executable workflow. This has to be done as choreographies are
not suited to be executed on a single workflow machine: unnecessary communication effort between the different workflows would slow down the execution
time [30] and passing sensitive data over the wire, e. g., the database credentials,
is not appropriate. Therefore, in this step, the choreography is automatically
translated into an executable workflow model. This is described in detail in the
next section and implemented by our prototype.

3.5 Deploy and Execute Resulting Workflows

In the last step, the generated workflow model is deployed on an appropriate
workflow engine. Afterwards, the plan can be triggered by sending the start
message to the workflow's endpoint. TOSCA runtimes such as OpenTOSCA [16]
explicitly support management by executing such workflows.

4   Process Consolidation

To transform the management choreography into an executable workflow we
provide a set of algorithms in Sect. 4.2 implementing the process consolidation
approach described in [31] and [32]. Compared to the original paper, which is
language-agnostic, the algorithms and the corresponding meta-model are focused
on BPEL. This is because BPEL supports block-structured and graph-based

---

[4] Such management workflows can be realized in a generic manner by binding them
exclusively to operations defined by the respective component type. TOSCA enables
exchanging the implementations of these operations on the topology layer to implement application-specific management logic.

modeling [33]. Hence, the consolidation can be also applied to other block-structured languages such as BPMN. Moreover, BPEL has a well-defined operational semantics and it is still one of the most prominent executable workflow languages [34].

4.1   Choreography Meta-model

The algorithms base on the interconnection choreography meta-model that is defined in the following. The choreography meta-model uses the process meta-model introduced in [6] and the formalization introduced in [35] as foundation. For simplicity reasons BPEL constructs such as compensation handlers, event handlers, termination handlers or loops are omitted in this meta-model and left for future works.

Definition 1 (Process). *A process $P = (A, E, \mathcal{E}^f, \mathcal{H}, \mathcal{V}, Cond)$ is a directed acyclic graph where A is the set of activities within the process. $E \subseteq A \times A \times Cond$ is the set of directed control links between two activities. $\mathcal{E}^f$ denotes the set of fault events that may occur during the execution of a process. $\mathcal{H} \subseteq A_{scope} \times \{\perp, \mathcal{E}^f\} \times A$ denotes the set of hierarchy relations between a scope activity and its child activities. The nesting $\perp$ indicates that an activity is a direct child of a scope activity and the nesting $\mathcal{E}^f$ indicates that an activity is a fault handling activity of a scope. $Cond$ denotes the set of transition conditions and join conditions used within P that can be evaluated to* true *or* false.

The transition condition of a control link is evaluated if the execution of the source activity of the link (i) completed, (ii) faulted, or (iii) if it is affected by dead path elimination [6,36].

Definition 2 (Activity). *The operational semantics of an activity $a \in A$ is implied by its type which can be assigned to an activity with the function* type $: A \to \mathcal{T}$. *The following sets of activities are distinguished:*
$A = A_{invoke} \cup A_{receive} \cup A_{reply} \cup A_{task} \cup A_{empty} \cup A_{opaque} \cup A_{assign} \cup A_{scope}$
An activity with incoming control links has a join condition referring to the states of its incoming control links. It can be assigned with the function joinCond $: A \to Cond$. *The condition must evaluate to* true *to start the execution of the activity. If no join condition is defined explicitly, the activity execution is started when one of its in incoming control links is activated.*

Invoke activities A_{invoke} within a process send messages to another process over a message link. These messages can be received by receive activities $A_{receive}$. An invoke activity supports either the asynchronous or synchronous one-to-one interaction[5] pattern [37]. The asynchronous invoke activity sends a message to the receive within the other process in a "fire and forget" manner, i. e., after the

---

[5] In one-to-one interactions an invoke activity sends a message to exactly one receive activity, while in one-to-many interactions an invoke activity communicates with multiple receives, e. g., via loops.

message was sent the invoke completes and its successor activities are performed. A synchronous interaction is modeled at the sender side with an invoke activity, that waits until it receives a response from the called partner process before it completes, i.e., it "blocks" until the response is received. At the receiver's side a synchronous interaction is modeled with a receive activity that is followed in the control flow by a set of one or more reply activities A_{reply} sending the response back to the calling invoke. During runtime just one reply activity must be executed, hence, the reply activities for a single receive must reside on mutual exclusive branches in the control flow.

Task activities A_{task} implement the actual management logic, such as executing human tasks, calling scripts etc. This activity type is not part of BPEL specification but introduced here to indicate that a management operation is performed. Empty activities A_{empty} do not perform any business functions but act a synchronization point for control links. Assign activities A_{assign} perform data assignments, such as copying a value from one variable to another. As data flow is out of scope of this work, we will not further discuss how these assignments are performed in detail. Opaque activities A_{opaque} act as placeholder for other activities.

A scope activity $s \in A_{scope}$ defines an execution context for its direct and indirect child activities and defines a common fault handling behavior on them. Therefore, a scope activity defines a set of fault handlers. Each handler contains one fault handling activity to process the thrown fault. A fault handler reacts on a certain fault event $fault \in \mathcal{E}^f$ thrown by the direct or indirect child activities of s. A scope activity has exactly one standard fault handler that reacts on all faults not being caught by the other fault handlers. This standard fault handler can be defined with the function catchAll : $\mathcal{E}^f \rightarrow \{true, false\}$. Scopes can be arbitrarily nested where the root scope is the process, i.e., a process is just a special type of scope. In contrast to scopes whose fault handlers may rethrow faults to their parent scopes, the process scope must not rethrow any faults.

The following further functions are used in the algorithms in Sect. 4.2. The function incoming : $A \rightarrow \wp(E)$ returns the incoming control links and the function outgoing : $A \rightarrow \wp(E)$ the outgoing control links of an activity a. The function parentHR : $A \rightarrow \mathcal{H}$ returns the hierarchy relation between the given activity and its direct parent activity. To denote the projection to the ith component of a tuple π_i is used.

Definition 3 (Choreography). *A choreography $C \in \mathcal{C}$ is defined by the tuple $C = (\mathcal{P}, \mathcal{ML})$, i.e., it consists of a set of interacting processes $\mathcal{P}$ and the message links $\mathcal{ML}$ between them. A message link ml connects a sending and a receiving activity: $\mathcal{ML} \subset A_{invoke} \cup A_{reply} \times A_{invoke} \cup A_{receive}$. $\forall ml \in \mathcal{ML} : P_1 \neq P_2 \wedge P_1, P_2 \in \mathcal{P}$ where $\pi_1(ml) \in \pi_1(P_1) \wedge \pi_2(ml) \in \pi_1(P_2)$. A message link is activated when the sending activity is started. A receiving activity cannot complete until its incoming message link was activated.*

In a choreography the processes interact just via message exchanges. Hence, activities originating form different processes are isolated form each other, i.e., state changes of activities within one process, such as faults, are not affecting activities originating from other processes directly.

4.2   Choreography-Based Process Consolidation

The process consolidation operation gets a choreography as input and returns a single process P_μ. The operation ensures that P_μ contains all activities A_{task} defined within the processes of C and that the execution order between these activities is preserved. P_μ is able to generate the same set of activity traces of A_{task} during runtime as C [31,32]. Since the consolidation was just described on a conceptual level, we provide a more formal description in the following. Algorithm 1 acts as entry point for the consolidation. It creates the consolidated process P_μ and calls the algorithms implementing the consolidation steps. Note that we only address the control flow aspects of the consolidation here. Data flow aspects are just briefly discussed.

Algorithm 1. Process Consolidation.

```
1:  Pμ ← new Process
2:  A_Pμ ← π₁(Pμ), E_Pμ ← π₂(Pμ), ℰ^f_Pμ ← π₃(Pμ)
3:  ℋ_Pμ ← π₄(Pμ), 𝒱_Pμ ← π₅(Pμ), Cond_Pμ ← π₆(Pμ)
4:  procedure CONSOLIDATE(C)
5:      ADDPROCESSELEMENTS(C)
6:      MATERIALIZECONTROLFLOW(C)
7:      A_s = {a ∈ A_Pμ | a ∈ A_scope}
8:      RESOLVEVIOLATIONS(A_s)
9:  end procedure
```

4.3   Adding Process Elements

Algorithm 2 adds the activities, control links, variables etc. being defined in the processes of choreography C to P_μ (line 3). The activities originating from different processes in C have to be isolated from each other in P_μ. This ensures that the original property of a choreography is preserved, that faults occurring in one process are not directly propagated to activities in another processes. The isolation is guaranteed by adding a scope s for each process P to be merged (lines 4 to 9). The attached fault handler catches and suppresses all faults that may be thrown from the activities within the scope.

Algorithm 2. Add elements of process to be merged to P_μ.

```
1:  procedure ADDPROCESSELEMENTS(C)
2:      for all P ∈ π₁(C) do
3:          A_Pμ ← A_Pμ ∪ π₁(P),…,Cond_Pμ ← Cond_Pμ ∪ π₆(P)
4:          s ← new scope
5:          fault ← new ℰ^f
6:          catchAll(fault) ← true
7:          a_fh ← new empty
8:          ℋ_Pμ ← ℋ_Pμ ∪ {(s,fault,a_fh),(Pμ,⊥,s)}
9:          A_Pμ ← A_Pμ ∪ {s,a_fh}
10:     end for
11: end procedure
```

4.4   Control Flow Materialization

The control flow materialization shown in Algorithm 3 derives the control flow between activities originating from different processes from the interaction patterns defined in C. Here the materialization of asynchronous and synchronous one-to-one interactions is discussed. The materialization of one-to-many interactions is described in [32].

Algorithm 3. Control Flow Materialization.

 1: **procedure** MATERIALIZECONTROLFLOW(C)
 2:     $\mathcal{ML}_{inv} = \{ml \in \pi_2(C) \mid \pi_1(ml) \in A_{invoke}\}$
 3:     **for all** $ml_{inv} \in \mathcal{ML}_{inv}$ **do**
 4:         $inv \leftarrow \pi_1(ml_{inv})$
 5:         $rcv \leftarrow \pi_2(ml_{inv})$
 6:         $ML_{rp} \leftarrow \{ml \in \pi_2(C) \mid \pi_1(ml) \in A_{reply}\}$
 7:         **if** $ML_{rp} = \emptyset$ **then**
 8:             MATERIALIZEASYN(inv, rcv)
 9:         **else**
10:             MATERIALIZESYN($inv, rcv, \pi_1(ML_{rp})$)
11:         **end if**
12:     **end for**
13: **end procedure**

To determine the interaction pattern Algorithm 3 checks in line 7 for each invoke activity if it is also a target of one or more message links ML_{rp} originating from reply activities. If this is the case the synchronous control flow materialization is called, otherwise the asynchronous materialization is called.

The materialization for asynchronous interactions is implemented by Algorithm 4. The algorithm replaces the invoke activity inv and receive activity rcv with the *synchronization activities* syn_{inv} and syn_{rcv}. Activity syn_{inv} serves as synchronization point for the control links of the former invoke activity inv. Thus, it inherits the control links and join condition of the invoke inv. The activity also emulates the message transfer by assigning the data that were transported in message before to the variable, where the message content was copied to by the receive activity. The new activity syn_{rcv} gets the control links and join condition of rcv assigned. This preserves the control flow order between the predecessor and successor activities of the former rcv. To emulate the control flow constraint implied by an asynchronous interaction, i. e., that successor activities of the former activity rcv are not started before the message was sent over the message link, a new control link $e_{inv2Rcv}$ is created between syn_{inv} and syn_{rcv}.

To perform the synchronous consolidation, beside invoke inv and receive rcv, also the set $A_{replyInv}$ of possible reply activities for inv is passed to Algorithm 5. An example for a synchronous interaction with two reply activities is depicted in Fig. 5. If a fault occurs during the execution of $b4$ reply activity $b5'$ within fault handler fh is executed. In the standard faultless flow $b5$ is performed.

Algorithm 4. Asynchronous Control Flow Materialization.

1: **procedure** MATERIALIZEASYN(inv, rcv)
2: $\quad syn_{inv} \leftarrow new$ **assign**
3: $\quad$ REPLACEACTIVITY(inv, syn_{inv})
4: $\quad syn_{rcv} \leftarrow new$ **empty**
5: $\quad$ REPLACEACTIVITY(rcv, syn_{rcv})
6: $\quad e_{inv2Rcv} \leftarrow new$ **link**(syn_{inv}, syn_{rcv}, **true**)
7: $\quad$ joinCond(syn_{rcv}) $\leftarrow$ joinCond(rcv) AND $e_{inv2Rcv} =$ **true**
8: **end procedure**
9: **procedure** REPLACEACTIVITY($oldAct, newAct$)
10: $\quad A_{P_\mu} \leftarrow (A_{P_\mu} \cup \{newAct\}) \setminus \{oldAct\}$
11: $\quad \forall e \in$ incoming($oldAct$) : $\pi_2(e) \leftarrow newAct$
12: $\quad \forall e \in$ outgoing($oldAct$) : $\pi_1(e) \leftarrow newAct$
13: $\quad$ joinCond($newAct$) $\leftarrow$ joinCond($oldAct$)
14: $\quad hr_{parent} \leftarrow$ parentHR($oldAct$)
15: $\quad \mathcal{H}_{P_\mu} \leftarrow (\mathcal{H}_{P_\mu} \cup \{(\pi_1(hr_{parent}), \pi_2(hr_{parent}), newAct)\}) \setminus \{hr_{parent}\}$
16: **end procedure**

Algorithm 5. Synchronous Control Flow Materialization.

1: **procedure** MATERIALIZESYN($inv, rcv, A_{replyInv}$)
2: $\quad syn_{inv} \leftarrow new$ **assign**
3: $\quad$ REPLACEACTIVITY(inv, syn_{inv})
4: $\quad syn_{rcv} \leftarrow new$ **empty**
5: $\quad$ REPLACEACTIVITY(rcv, syn_{rcv})
6: $\quad e_{inv2Rcv} \leftarrow new$ **link**(syn_{inv}, syn_{rcv}, **true**)
7: $\quad syn_{rcRp} \leftarrow new$ **empty**
8: $\quad \forall e \in$ outgoing(inv) : $\pi_1(e) \leftarrow syn_{rcRp}$
9: $\quad e_{inv2RcRp} = new$ **link**(syn_{inv}, syn_{rcRp}, **true**)
10: $\quad$ **for all** $rp \in A_{reply}$ **do**
11: $\quad\quad syn_{rp} \leftarrow new$ **assign**
12: $\quad\quad$ REPLACEACTIVITY(rp, syn_{rp})
13: $\quad\quad e_{rp2RcRp} \leftarrow new$ **link**(syn_{rp}, syn_{rcRp}, **true**)
14: $\quad\quad E_{P_\mu} \leftarrow E_{P_\mu} \cup \{e_{rp2RcRp}\}$
15: $\quad$ **end for**
16: $\quad$ joinCond(syn_{rcRp}) $\leftarrow e_{rp2RcRp_i} =$ **true** $OR \ldots OR\ e_{rp2RcRp_n} =$ **true** $\triangleright$
    ($n = |A_{reply}|$)
17: **end procedure**

In a first step Algorithm 5 replaces inv and rcv with synchronization activities and creates a control link between these activities. This is implemented in the same way as for asynchronous interactions (lines 2 to 6). In the second step, the loop in line 10 replaces each reply activity rp being an origin of a message link targeting the former activity inv with another synchronization activity syn_{rp}. The created reply activities syn_{rp} assign the response data to the same variable, where the content of the reply message (transported over message link ml_{rp}) was copied to by activity inv. In the example in Fig. 5 two synchronization

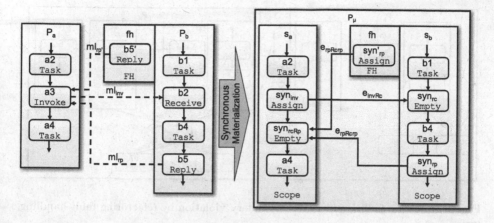

Fig. 5. Synchronous control-flow materialization.

activities syn_{rp} and syn'_{rp} are created from the two reply activities. To emulate the control flow constraint that the successor activities of a synchronous invoke are not started before the response message is sent from exactly one of the reply activities, a control link is created from each synchronization activity syn_{rp} to the new activity syn_{rcRp} (created in line 7). The join condition of syn_{rcRp} set in line 16 ensures that syn_{rcRp} is executed when one of the synchronization activities syn_{rp} completed.

4.5   Resolving Cross-Boundary Link Violations

Workflow languages supporting block-structured and graph-based modeling impose certain restrictions on control links crossing the boundaries of block-constructs such as loops, subprocesses, error handlers etc. For instance, in BPMN it is forbidden that the source activity of a control link lays outside of a BPMN subprocess while the target of this link is pointing to an activity inside the subprocess. In BPEL control links must not pass the boundary of loops, event handlers or compensation handlers. For fault handlers just control links pointing outside the handler are allowed. The control flow materialization, however, may create control links crossing the boundary of block-constructs if the sending or receiving activity, where control link is created from, is located in such a construct. In [38] we proposed algorithms for resolving these *cross-boundary violations* for links crossing BPMN and BPEL loops. Here an algorithm is provided for solving these violations for fault handlers.

An example scenario for a cross-boundary violation involving a fault handler is shown on the left side in Fig. 6. There activity a_{fhRoot} is the fault handling activity of fault handler fh. It contains the fault handling logic including synchronization activities (not depicted in Fig. 6). The control flow materialization created the invalid control link e_{cbl} pointing to activity a_{fhRoot} and causing a cross-boundary violation.

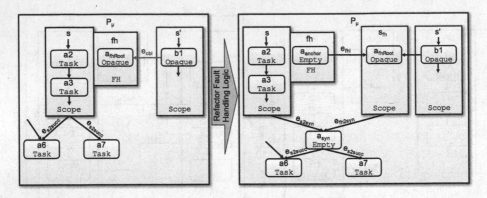

Fig. 6. Resolving fault handler cross-boundary violation by refactoring fault handling logic.

An informal discussion about resolving the violations for fault handlers was provided in [12]. Based on this discussion Algorithm 6 is introduced. The algorithm needs to resolve the violations while preserving the control flow semantics of BPEL scopes and fault handlers. Before the algorithm is introduced this semantics is briefly discussed.

If the execution of a scope completes successfully without throwing a fault, all of its fault handlers are uninstalled. However, if a scope throws a fault the normal processing within the scope stops. All control links originating from activities within the scope which were not activated before the faulted occurred are deactivated (dead path elimination). Then the fault handler catching the fault is installed, i. e., the fault handling activities within the fault handler are performed. All other fault handlers are uninstalled as well. After fault handling completed the control flow continues at the scope boundaries and the transition conditions of all control links originating from the scope are evaluated as usual. The deactivation of a fault handler implies that the root activity of the fault handler and also its child activities cannot be executed anymore or, spoken in terms of BPEL, are marked as *dead*. Consequently, all control links originating from within an uninstalled fault handler are deactivated.

Algorithm 6 performs two steps. In a first step procedure RESOLVEVIOLATIONS checks for each of the provided scopes if there exist cross-boundary violations caused by control links pointing into one or more fault handlers. Therefore, all fault handlers of scope s are determined in line 6. In line 9 the actual violation check is performed. In the example in Fig. 6 only fault handler fh violates the cross-boundary constraint.

In the second step procedure REFACTORFAULHANDLERLOGIC resolves the violation by moving the fault handling logic a_{fhRoot} out of the fault handler into a new *fault handling scope* s_{fh}, such as the one depicted at the right side of Fig. 6. Thereby, for each violated fault handler a separate fault handling scope s_{fh} is created. All fault handlers not being affected by a cross-boundary violation remain unchanged. To ensure that the fault handling logic a_{fhRoot} is executed if the

Algorithm 6. Removing Control Links Pointing into Fault Handlers.

1: **procedure** RESOLVEVIOLATIONS(A_s)
2:     **for all** $s \in A_s$ **do**
3:         $hr_{parentS} \leftarrow$ parentHR(s)
4:         $a_{parentS} \leftarrow \pi_1(hr_{parent})$
5:         $a_{syn} \leftarrow \varnothing$
6:         $FH = \{hr \in \mathcal{H}_{P_\mu} \mid s = \pi_1(hr) \wedge \pi_2(hr) \in \mathcal{E}^f\}$
7:         **for all** $fh \in FH$ **do**
8:             $A_{FH} = \pi_3(FH)$
9:             **if** $\exists e \in E : \pi_1(e) \notin A_{FH} \wedge \pi_2(e) \in A_{FH}$ **then**
10:                 **if** $a_{syn} = \varnothing$ **then**
11:                     $a_{syn} \leftarrow new$ **empty**
12:                     $\mathcal{H}_{P_\mu} \leftarrow \mathcal{H}_{P_\mu} \cup \{(a_{parentS}, \perp, a_{syn})\}$
13:                     $e_{s2syn} \leftarrow new$ **link**($s, a_{syn},$ **true**)
14:                     $\forall e \in$ outgoing(s) : $\pi_1(e) \leftarrow a_{syn}$
15:                 **end if**
16:                 REFACTORFAULTHANDLERLOGIC($s, fh, a_{parentS}, a_{syn}$)
17:             **end if**
18:         **end for**
19:     **end for**
20: **end procedure**
21: **procedure** REFACTORFAULHANDLERLOGIC($s, fh, a_{parentS}, a_{syn}$)
22:     $s_{fh} \leftarrow new$ **scope**
23:     $fault \leftarrow \pi_2(fh)$
24:     $a_{fhRoot} \leftarrow \pi_3(fh)$
25:     $a_{anchor} \leftarrow new$ **empty**
26:     $e_{fhl} \leftarrow new$ **link**($a_{anchor}, s_{fh},$ **true**)
27:     $e_{fh2syn} \leftarrow new$ **link**($s_{fh}, a_{syn},$ **true**)
28:     $E_{P_\mu} \leftarrow E_{P_\mu} \cup \{e_{fhl}, e_{fh2syn}\}$
29:     $\mathcal{H}_{P_\mu} \leftarrow (\mathcal{H}_{P_\mu} \cup \{(a_{parentS}, \perp, s_{fh}), (s_{fh}, \perp, a_{fhRoot}), (s, fault, a_{anchor})\}) \setminus (s, fault, a_{fhRoot})$
30:     joinCond(a_{syn}) $\leftarrow$ joinCond(a_{syn}) OR $e_{fh2syn} =$ **true**
31: **end procedure**

corresponding fault is caught, a new control link e_{fhl} is created in line 26. It connects the newly created empty activity a_{anchor} (line 25) within the fault handler with the fault handling logic a_{fhRoot}. The additional activity a_{syn} created in line 11 becomes the source of the outgoing control links of scope s (line 14) to ensure that the successor activities of s are not started before either s or its fault handling scopes completed. Activity a_{syn} is just created once per scope s. To preserve the property that control links originating from scope boundaries are always performed no matter whether the scope s completed successfully or not, a_{syn} becomes the target for the set of control links originating from each created fault handling scope. The join condition of a_{syn} created in line 30 ensures that a_{syn} and its successors (i. e., the former successors of s) are performed if one these links is activated.

The resulting example process P_μ is depicted at the right side of Fig. 6. Moving the fault handling logic out of the fault handlers of scope s resolves the violations but also preserves the original control flow. If scope s completes successfully, all of its fault handlers are uninstalled and the dead path elimination marks the activities within the fault handlers as dead. Hence, also activity a_{anchor} is marked as dead and its successor activity a_{fhRoot}. The the successors of scope s can be executed as link e_{s2syn} is activated.

If a fault occurs during the execution of scope s the execution of s is interrupted and the matching fault handler is installed. If the matching fault handler was not changed by Algorithm 6 its fault handling logic is directly performed. After the completion of the fault handling logic link e_{s2syn} is activated and thus also the successors of s. In case a fault handler is called whose fault handling logic was moved out of the fault handler, activity a_{anchor} is executed. As a_{anchor} is an empty activity, it completes directly and its outgoing control link causes the execution of the actual fault handling logic a_{fhRoot}. Since the BPEL control link semantics requires that all incoming control links of an activity are evaluated, it is guaranteed that the former successor activities (in the example $a6$ and $a7$) of scope s are not started before either s or its fault handling logic completed.

4.6   Consolidation Example

The single process *LAMP Provisioning Plan* shown in Fig. 7 results from the application of Algorithm 1 on the provisioning choreography. As all plans interact

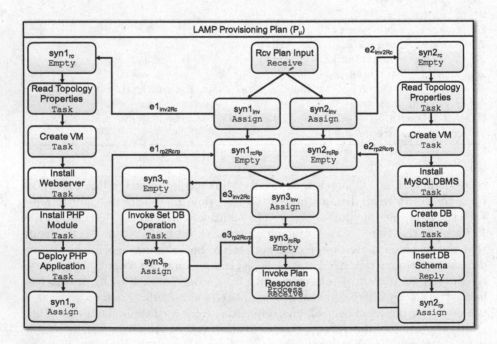

Fig. 7. Consolidation of a provisioning choreography into single LAMP provisioning plan.

synchronously only the synchronous control flow materialization is applied. Thus, the sending and receiving activities related to each message link are replaced with synchronization activities as described in Sect. 4.4.

5 Validation

In this section, we validate the practical feasibility of the presented method by a prototypical implementation. We applied the method and merge algorithms to the choreography modeling language BPEL4CHOR and extended the OpenTOSCA Cloud management ecosystem to support choreographies. This ecosystem consists of (i) the graphical TOSCA modelling tool *Winery* [17], the (ii) *OpenTOSCA container* [16], and (iii) the self-service portal *Vinothek* [18]. An overview of the entire prototype is shown in Fig. 8: Application developers use Winery to merge existing topology models, while application managers use the choreography modelling tool ChorDesigner [39] to coordinate the associated management workflows.

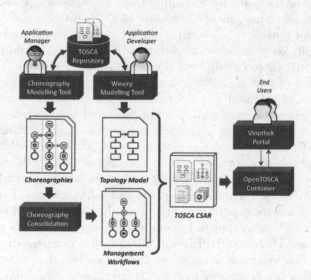

Fig. 8. Architecture of the open-source Cloud management prototype.

Based on the merge algorithm described in Sect. 4 a process consolidation tool was developed for generating a single executable BPEL processes out of a choreography[6]. Therefore the algorithm was extended to accommodate the language idiosyncrasies of BPEL. This includes the emulation of the choreography's data flow in the merged process and the elimination of cross-boundary violations. Beside asynchronous and synchronous one-to-one interactions the tool does also support the consolidation of one-to-many interactions [32,37].

---

[6] The prototype is available as Open-source: https://github.com/wagnerse/chormerge.

The merged topology model as well as the generated management plans can be packaged as CSAR using Winery. The resulting CSAR can be installed on the OpenTOSCA container, which internally deploys the workflows and, thereby, makes them executable. To ease the invocation of provisioning and management workflows, we employ our TOSCA self-service portal Vinothek, which wraps the invocation of workflows by a simple user interface for end users. All tools are available as open-source implementations, thus, the developed prototype provides an end-to-end Cloud application management system supporting choreographies for modelling coordinated management processes.

6   Conclusion and Future Work

In this work, we proposed a method for developing new TOSCA-based applications in a more efficient way by reusing topologies and management plans from existing applications. This method describes the steps to be performed to combine existing application topology artifacts or parts thereof into a single topology and how the plans for managing these artifacts can be coordinated by BPEL4Chor-based management choreographies. To provide the choreography again as plan for reusing it in other topologies and for the efficient execution of the choreography on a single workflow engine, this method encompasses a step to consolidate the choreography into a single management plan. Therefore, we introduced a set of new algorithms describing the consolidation of interacting BPEL processes in a formal way. To validate the method different tools of the OpenTOSCA ecosystem were used. Each of these tools enables one or more steps of the proposed method to be performed in a semi-automatic manner.

As BPEL4Chor has the same modeling capabilities as BPMN collaborations [40] the proposed algorithms can be also applied on BPMN collaborations. In the near future also the OpenTOSCA ecosystem will be extended to model and enact BPMN-based management plans and collaborations. Failures during the execution of a management plan may need already completed tasks in other plans of the same choreography to be compensated, i. e., the effects of these tasks must be undone. Therefore, BPEL and BPMN offer compensation constructs to implement this behavior. To support the consolidation of plans interacting via compensation constructs, we plan to extend process consolidation approach accordingly. To provide a comprehensive set of reusable management plans, we also plan to transform low-level management scripts into plans.

Acknowledgements. This work was partially funded by the BMWi projects SmartOrchestra (01MD16001F), NEMAR (03ET40188), and by the DFG project SitOPT (610872).

References

1. Brown, A.B., Patterson, D.A.: To Err is Human. In: EASY, p. 5 (2001)
2. Oppenheimer, D., Ganapathi, A., Patterson, D.A.: Why do internet services fail, and what can be done about it? In: USITS (2003)

3. Opscode, Inc.: Chef official site (2015). http://www.opscode.com/chef
4. Puppet Labs Inc.: Puppet official site (2015). http://puppetlabs.com/puppet/what-is-puppet
5. Coutermarsh, M.: Heroku Cookbook. Packt Publishing Ltd., Birmingham (2014)
6. Leymann, F., Roller, D.: Production Workflow: Concepts and Techniques. Prentice Hall PTR, New Jersey (2000)
7. Keller, A., Badonnel, R.: Automating the provisioning of application services with the BPEL4WS workflow language. In: DSOM (2004)
8. Kopp, O., Binz, T., Breitenbücher, U., Leymann, F.: BPMN4TOSCA: a domain-specific language to model management plans for composite applications. In: BPMN (2012)
9. Herry, H., Anderson, P., Wickler, G.: Automated planning for configuration changes. In: The Past, Present, and Future of System Administration, Proceedings of the 25th Large Installation System Administration Conference, LISA, USENIX Association (2011)
10. Breitenbücher, U., Binz, T., Kopp, O., Leymann, F., Wettinger, J.: Integrated cloud application provisioning: interconnecting service-centric and script-centric management technologies. In: CoopIS (2013)
11. Binz, T., Breitenbücher, U., Kopp, O., Leymann, F., Weiß, A.: Improve resource-sharing through functionality-preserving merge of cloud application topologies. In: CLOSER. SciTePress (2013)
12. Wagner, S., Kopp, O., Leymann, F.: Consolidation of interacting bpel process models with fault handlers. In: Proceedings of the 5th Central-European Workshop on Services and their Composition (ZEUS), CEUR (2013)
13. Decker, G., Kopp, O., Leymann, F., Weske, M.: BPEL4Chor: Extending BPEL for modeling choreographies. In: ICWS. IEEE (2007)
14. OASIS: TOSCA v1.0 (2013). http://docs.oasis-open.org/tosca/TOSCA/v1.0/os/TOSCA-v1.0-os.html
15. OASIS: TOSCA Primer v1.0 (2013). http://docs.oasis-open.org/tosca/tosca-primer/v1.0/tosca-primer-v1.0.html
16. Binz, T., Breitenbücher, U., Haupt, F., Kopp, O., Leymann, F., Nowak, A., Wagner, S.: OpenTOSCA – a runtime for TOSCA-based cloud applications. In: Basu, S., Pautasso, C., Zhang, L., Fu, X. (eds.) ICSOC 2013. LNCS, vol. 8274, pp. 692–695. Springer, Heidelberg (2013). doi:10.1007/978-3-642-45005-1_62
17. Kopp, O., Binz, T., Breitenbücher, U., Leymann, F.: Winery – a modeling tool for TOSCA-based cloud applications. In: Basu, S., Pautasso, C., Zhang, L., Fu, X. (eds.) ICSOC 2013. LNCS, vol. 8274, pp. 700–704. Springer, Heidelberg (2013). doi:10.1007/978-3-642-45005-1_64
18. Breitenbücher, U., et al.: Vinothek - a self-service portal for TOSCA. In: Proceedings of the 6nd Central-European Workshop on Services and their Composition (ZEUS), CEUR (2014)
19. Breitenbücher, U., et al.: Combining declarative and imperative cloud application provisioning based on TOSCA. In: IC2E (2014)
20. Breitenbücher, U., et al.: Vino4TOSCA: a visual notation for application topologies based on TOSCA. In: CoopIS (2012)
21. OASIS: Web Services Business Process Execution Language (WS-BPEL) Version 2.0. OASIS (2007)
22. OMG: Business Process Model and Notation (BPMN), Version 2.0 (2011)
23. Hofreiter, B., Huemer, C.: A model-driven top-down approach to inter-organizational systems: from global choreography models to executable BPEL. In: CEC (2008)

24. Mendling, J., Hafner, M.: From WS-CDL choreography to BPEL process orchestration. J. Enterp. Inf. Manage. **21**, 525–542 (2008)
25. Kavantzas, N., Burdett, D., Ritzinger, G., Fletcher, T., Lafon, Y., Barreto, C.: Web Services Choreography Description Language Version 1.0 (2005)
26. Küster, J., Gerth, C., Förster, A., Engels, G.: A tool for process merging in business-driven development. In: Proceedings of the Forum at the CAiSE (2008)
27. Mendling, J., Simon, C.: Business process design by view integration. In: Eder, J., Dustdar, S. (eds.) BPM 2006. LNCS, vol. 4103, pp. 55–64. Springer, Heidelberg (2006). doi:10.1007/11837862_7
28. Herry, H., Anderson, P., Rovatsos, M.: Choreographing configuration changes. In: Proceedings of the 9th International Conference on Network and Service Management, CNSM (2013)
29. Soldani, J., Binz, T., Breitenbücher, U., Leymann, F., Brogi, A.: TOSCA-MART: a method for adapting and reusing cloud applications. Technical report, University of Pisa (2015)
30. Wagner, S., Roller, D., Kopp, O., Unger, T., Leymann, F.: Performance optimizations for interacting business processes. In: IC2E. IEEE (2013)
31. Wagner, S., Kopp, O., Leymann, F.: Towards verification of process merge patterns with Allen's Interval Algebra. In: Proceedings of the 4th Central-European Workshop on Services and their Composition (ZEUS). CEUR (2012)
32. Wagner, S., Kopp, O., Leymann, F.: Choreography-based consolidation of multi-instance BPEL processes. In: Proceedings of the 4th International Conference on Cloud Computing and Service Science (CLOSER). SciTePress (2014)
33. Kopp, O., Martin, D., Wutke, D., Leymann, F.: The difference between graph-based and block-structured business process modelling languages. Enterp. Model. Inf. Syst. **4**, 3–13 (2009)
34. Leymann, F.: BPEL vs. BPMN 2.0: should you care? In: BPMN (2010)
35. Kopp, O., Mietzner, R., Leymann, F.: Abstract syntax of WS-BPEL 2.0. Technical report computer science 2008/06, University of Stuttgart, Faculty of Computer Science, Electrical Engineering, and Information Technology, Germany (2008)
36. OASIS: Web Services Business Process Execution Language Version 2.0 - OASIS Standard (2007)
37. Barros, A., Dumas, M., Hofstede, A.H.M.: Service interaction patterns. In: Aalst, W.M.P., Benatallah, B., Casati, F., Curbera, F. (eds.) BPM 2005. LNCS, vol. 3649, pp. 302–318. Springer, Heidelberg (2005). doi:10.1007/11538394_20
38. Wagner, S., Kopp, O., Leymann, F.: Choreography-based consolidation of interacting processes having activity-based loops. In: Proceedings of the 5th International Conference on Cloud Computing and Service Science (CLOSER). SciTePress (2015)
39. Weiß, A., Karastoyanova, D.: Enabling coupled multi-scale, multi-field experiments through choreographies of data-driven scientific simulations. Computing **98**, 439–467 (2014)
40. Kopp, O., Leymann, F., Wagner, S.: Modeling choreographies: BPMN 2.0 versus BPEL-based approaches. In: EMISA (2011)

Experimental Study on Performance and Energy Consumption of Hadoop in Cloud Environments

Aymen Jlassi[✉] and Patrick Martineau

Université François-Rabelais de Tours, CNRS, LI EA 6300, OC ERL CNRS 6305,
Tours, France
{aymen.jlassi,patrick.martineau}@univ-tours.fr

Abstract. The big data applications are a resource and energy intensive applications. Cloud providers wish to better utilize the technologies of virtualization in order to solve the evolving needs of infrastructures, alongside the growing demand. The virtualization technology based on container is increasingly popular in the high performance domain, this work is the evaluation of this technology in the context of big data and cloud computing domains. It focuses on the software Hadoop, as a big data application, it evaluates the performance impact and energy consumption. The objective is to understand the tradeoff between performance and energy efficiency depending on the technology of virtualization. The outcomes of this paper are: Firstly, the evaluation of the technology of virtualization based on containers on the cloud using Hadoop as a big data application. Secondly, the comparison of the traditional virtualization with the merging container technology. We analyze the impact of the coexistence of virtual machines (or containers) on the CPU, memory, hard disk throughput and network bandwidth. Thirdly, the reduction of the big data application deployment cost using the cloud. Fourthly, the Hadoop community finds an in-depth study of the resource consumption depending on the deployment environment. Our evaluation shows that: *(i)* The container (Docker) technology is a performance enhancement and energy saving technology compared to the traditional technology of virtualization. *(ii)* Performance of Hadoop cluster based on containers is significantly better than the traditional virtualization technology. *(iii)* Data replication rate influences the completion date of job. *(vi)* Coexisting containers (or virtual machines) influence the energy consumption and the completion time of the applications.

Keywords: Cloud computing · Virtualization · Hadoop MapReduce · Power consumption · Performance

1 Introduction

The cloud-computing domain is based on the quality of the services offered to customers and the capacity of providers to ensure performances and security. The virtualization is the most important technology that has the capacity to

© Springer International Publishing AG 2017
M. Helfert et al. (Eds.): CLOSER 2016, CCIS 740, pp. 255–272, 2017.
DOI: 10.1007/978-3-319-62594-2_13

hide the complexity of the infrastructure and to optimize resource exploitation. It helps providers to reduce costs. The virtualization technology was introduced in 1960 by IBM [25]. It transparently enables time-sharing and resource-sharing on servers. It aims at improving the overall productivity by enabling many virtual machines to run on the same physical support. Many categories of virtualization tools [25] are used in data centers. In this paper, we classify them on the full and light virtualization: The full virtualization is based on the management of virtual machine (VM). The VM is guest operating system (OS) that runs in parallel over physical hosts. A hypervisor ensures the interpretation of instruction from the guest OS to the host OS. The light virtualization is based on the management of containers on a physical host, the containers share functions from the kernel of the host OS and have direct access to its library. In the last decade, the light technology of virtualization has been shifting quickly, it allows to obtain a cost-effective clusters of servers. Docker is the most sophisticated tool in its category; it offers a large and more intensive range of capability to manage hardware resources. This classification is also used in [22]. Traditionally, the cloud computing and big data [29] environments are mainly based on the heavy virtualization tools. The main reason are in the following points: (i) the benefits and convenience of the heavy virtualization, (ii) the complete isolation of the environment between guests and host OS. Nowadays, the Docker technology offers multiple capabilities of resource isolation. It reaches an adequate level of maturity and it can be tested with big data tools. Hadoop software is a big data environment, it is based on the MapReduce Model, which was introduced by Google in 2004 [16] as a parallel and distributed computation model. It is largely adopted in companies and data centers; for example Facebook [15] and Amazon [15] use it to answer the computation needs.

In this work, we study and compare the two categories of virtualization. The experiments must be made using the Docker technology, the VMware technology and the Hadoop software. We therefore analyze the influence of the platform's resource variation. During the evaluation, we consider (i) the completion time of the workload, (ii) the quantity of hardware resources and (iii) the energy consumption criteria. We prove, then, that this technology gives a cost-effective cluster with a better efficiency in most cases.

The remainder is as follows. In Sect. 2, the previous studies on literature are presented. In Sect. 3, concept and terms used in this paper are reminded. In Sect. 4, the methodology used in the experiments is presented. In Sect. 5, the results are presented and discussed. The conclusion is presented in Sect. 7.

2   Related Works

Since the last decade, Hadoop has been interesting the scientists community. Many benchmarks and evaluation tests have been done in order to evaluate its performances or to compare it to other softwares. There are mainly two levels in the benchmark of the software Hadoop.

The first one focuses on the comparison of Hadoop with the existing engine in the big data computing. For example, Fadika et al. [1] benchmark Hadoop with three data-intensive operations to evaluate the impact of the file system, network and programming model on performances. Stonebraker et al. [2] compare Mapreduce model to parallel database, they focus on the performance aspect. Pavlo et al. [3] prove that Hadoop is slower than two state-of-the-art parallel database systems, in performing a variety of analytical tasks, by a factor of 3.1 to 6.5. Jiang et al. [8] give an in-depth study of MapReduce performance to identify bottleneck factors that affect the performance on Hadoop, they show that the best tuning of these factors improves the same benchmark used in [3] and [10] by a factor of 2.5 to 3.5. Zechariah et al. [5] compare Hadoop, LEMO-MR and twister (three implementations of the MapReduce model). Gu et al. [4] compare Hadoop software (HDFS/MapReduce) to the softwares Sector/Sphere. Jefrey et al. [6] focus on the file system to identify bottlenecks, they identify the weaknesses of the Hadoop file system to solve and the best practices to follow in the cluster deployment.

The second one focuses on the performances and the energetic consumption of Hadoop using different deployment architectures. For example, Kontagora et al. [7] benchmark Hadoop performances using full-virtualization (using VMware Workstation). The paper [20] evaluates Hadoop's performances using openStack, KVM and XEN. It compares performances using openStack deployment with the physical deployment. Reference [30] compare the Hadoop software using different tools of container technology, however, neither Docker technology nor heavy technology are considered in the comparison.

The Docker technology is benchmarked in other contexts as the HPC technology. For example, Xavier et al. [24] present an in-depth performance evaluation of the containers based on the virtualization for HPC. They present the evaluation of the tradeoff between performance and isolation. In the same context and compares the job executions using containers with executions using physical infrastructure deployment, it confirms that the overload due to the use of the container and the time completion are about 5%. Reference [22] analyzes the resource isolation in the context of Docker technology and confirms that container isolation is less secure than isolation offered by traditional tools of virtualization (heavy). For an accurate study, [21] presents the state of the art of all open source projects, which adapt Docker technology to the context of the Cloud.

We conclude that the Docker technology has been evaluated in the context of HPC technology, which has its specificity. In most cases, the big data and HPC are two divergent fields of technologies. Each one has its own scheduling policies, resources requirements workloads affinities. The topic of this paper focuses on the use of Hadoop software with the Docker technology as a light virtualization tool. It compares this emerging technology with the traditional virtualization technology. It focuses on the resources exploitation, the time completion of the benchmarks and the energetic consumption. It analyze and discuss assumptions acquired from experiments performed in the HPC context.

3 Background

This section contains definitions of various concepts and terms used in this work. It presents the Hadoop software characteristics and it defines the heavy and light virtualization technologies.

Google introduced the model MapReduce as a distributed and parallel Model for data intensive computing. Every job generates a set of "map" and "reduce" tasks, which is executed in a distributed fashion over a cluster of machines. "Map" tasks have to be executed before "reduce" tasks. Tasks have to be executed the nearest to the needed data input. Data outputs of tasks map are transferred from the machine where tasks "map" run to the machines where "reduce" tasks run using the network.

3.1 The Hadoop Implementation

Hadoop implements the MapReduce model; the computation level is named "Yarn" and is composed of three elements, which manage job execution. At first, the Resource Manager (RM) is the master daemon; it assures synchronization over different elements and distributes resources between jobs. On a second point, the Node Manager (NM) is the responsible for the resource exploitation per slave machine. The Application Master (AM) is responsible for managing the lifecycle of a job. The scheduler in the RM is responsible for the management of the resources. The scheduling policies are based on these assumptions:

1. The scheduler considers the homogeneity criteria of the cluster thus slave machines run jobs at the same rate.
2. The tasks progress linearly during a they tend to finish in waves, thus tasks having a low progress rate are considered as slow tasks.
3. The tasks in the same category require the same amount of resources.

The storage level is named Hadoop file system (DFS) and is composed of the NameNode (NN) as a server, which contains the cartography of block's file. The datanode is the second element of the storage architecture: it is responsible for maintaining data blocks and communicates with namenode to perform operations like adding, moving, deleting. It also applies a number of NN decisions like ensuring data replication and load balancing operations. The sizes of the files in DFS are from megabytes up to terabytes. They are partitioned into data blocks. The size of a block is a decisive point to reduce the duration of the workload execution. When the scheduler cannot assign tasks to machines where data are stored, network bandwidth is allocated to migrate blocks.

3.2 The Light or Container Technology

In big companies like Facebook and Yahoo, a cluster of Hadoop contains a large number of machines. The optimization of the resource exploitation offers the opportunity to reduce costs and increase benefits. The companies profit from the virtualization in the cloud to improve resource exploitation. As the energy management presents an important field, much research over the Cloud aim minimize the electric consumption of the data center. This paper analyze the effect of the use of virtualization tools over the energetic consumption. It presents proportional relations between different kinds of resources and the consumption of energy. It is important to mention that the energetic gain over a cluster of four machines will be weak. The idea is to detect the variation of the consumption as small as it is. In a large cluster scale, the variation in energy consumption is not negligible and has an important impact on the overall cost.

4   Methodology

The experiments are repeated with both types of virtualization tools. The first topic of this work is to compare the performance variation using the two technologies of virtualization; we compare the time completion of the used benchmarks. The second topic is to focus on the Docker technology and give and in-depth study of this technology; we aim to identify the inadequate or badly spent resources. The third topic analyses the variation of the energy consumption according to the experiments and tests. Two sets of experiments have been carried out, the first set uses an homogeneous cluster of machine. In the second set, we vary resource capacities to thoroughly analyze performance variations between heterogeneous and homogeneous platforms. CPU, memory, hard disk and total load over physical machine are recuperated during the experimentation. We consider the time execution of the job. In all these works, the physical host has 12 CPU cores, 31.5 GB of RAM and 500 GB of hard disk (Table 1). The experiments are partitioned on two main parts: when we study Hadoop in homogeneous cluster, the virtual machines and the containers have the same configuration; they have 2 cores (and 2 thread per core), 5 GB of RAM and 80 GB of hard disk (Table 1). To ensure the best evaluation of the platform, some configuration parameter could be fixed. For example, the rate of data replication is two (this number is depending on the size of the cluster) and the capacity of node manager is set to 3 GB of RAM and 3 cores. When we address the problems in the heterogeneous cluster, we use another configuration of virtual machines (VMs and containers) depending on the resource we are studying. The experiments are based on two levels, the first one considers two slave machines and

Table 1. Configuration of machines (physical or virtual) used in the experiments.

	Host machine	Client machine
Processor	Intel ®Xeon (R) CPU E5-26200 @ 2.00 GHz	
CPU cores	12	2 cores (4 threads)
RAM (GB)	31.5	5
HDD (GB)	500	80
OS	Ubuntu 14.10	

the second one considers four slave machines. The slave machines can be virtual machines or containers. All experiments are repeated 5 times. The Ganglia software is used to recuperate LOAD, CPU, RAM metrics. It overloads the Hadoop cluster with 2% [12]. The software hsflow is combined to Ganglia to retrieve I/O bound of the hard disk access on the slave machines. These metrics offer the possibility of an in-depth study in the variation in resource utilization during experiments. In order to measure energetic consumption, we use a specific engine mounted to the electrical outlet, it measures overall the energy consumption of the cluster machines every 2 s and save it on an external memory card. Four workloads (TestDFSIO-read, TestDFSIO-write, TeraGen, TeraSort) are used in our benchmarks. In order to reach the topic of this work; we use the benchmarks TeraGen and TeraSort and TestDFSIO. They are used by Vmware organisation; intel [9] and AMD [28] to evaluate their products. They are considered as a reference and are used in many other works like [1]. The first kind of workloads is TeraGen and TestDFSIO. They stress the hard disk and I/O resources, they are based on a set of "map" tasks which writes random data in HDFS in the a sequential manner. In these works; they generate three sizes of data 10, 15 and 20 GB using 2 then 4 slave machines.

The second one is TeraSort, this benchmark stresses: memory, network and compute resources. Each data generated with TeraGen is sorted with TeraSort. TeraSort is known for the capacity to aggregate output of the TeraGen workload. It is based on a set of "map" tasks and "reduce" tasks. The job TeraSort is forced to use four reduce tasks, we aim to dispatch the compute on many slave machines. The four workloads (TestDFSIO-read, TestDFSIO-write, TeraGen, TeraSort) used in the evaluation are based on the MapReduce model; each of them has the capacity to stress specific resource thus the evaluation results will be more accurate.

Hadoop is designed to work on a homogeneous cluster. Defined policies (configuration of files and the default schedulers) don't consider the configuration of machines when they schedule tasks, however, the clusters and technologies grow up continuously and companies don't have guarantee to supply the cluster with the same machine's configurations. Thus, we study the influence of the variation of the machine configuration on the performance of the workloads executions. We vary the quantity of the resources of the Hadoop slave machines and we analyze experiments results. We double the RAM and CPU resources of slave machines in the cluster and the resource capacities of the slave node. The slave

Table 2. Configuration of slave machines and (NM) used in heterogenuous context.

Resources	Slave machine 1	Slave machines 2 and 3	Slave node configuration
CPU (cores/Vcores)	4 cores (4 threads)	2 cores (4 threads)	6 Virtual cores
Memory (GB)	10	5	6
HDD space (GB)	80	80	-

node is the Node Manager (NN) of the Hadoop's cluster. Table 2 introduces the configuration of the slave machines (VM or container) and nodes (which give the configuration of slaves in the Hadoop) considered at this part of experiences.

5   Experimental Results and Discussion

In this section, we provide the results of the experiments. The first subsection discusses the influence of the execution workloads on the performance of the two types of virtual clusters. In the second subsection, we focus on the variation in the resources i.e. CPU and I/O bounds. In the third subsection, we consider a heterogeneous cluster to analyze the variation of resource utilization during experiments. In the fourth subsection, we study the influence of overload of the energy consumption.

5.1   Evaluation of the Machine's Overload Capacity

The overload of a machine can be defined as the difference between load of a physical machine (without any slave machine) and load after the start of slave machines on it. Figure 1 presents (i) the overload of the physical machines without the running of any slave machines (ii) the overload of the physical machine with slave VM when they are idle (iii) the overload of the physical machine with Docker containers when they are idle. We have noticed that the virtual machines reserve total configured memory since its start: thus 17 GB of memory is booked (3 VM) for cluster with two slaves and 28 GB is booked for the cluster with 4 slaves (5 VM). But the containers use resources only when they need them. The host uses 5 GB of memory with three containers and 10 GB with 5 containers. This is the minimum memory needed to start hosts, guest operating systems and Hadoop daemons. The Fig. 1 shows that the overload is measured between 3–5% for Docker containers and between 10–25% for the commercial tools. During experiments, we record overload of the physical machine. Figure 3 compares the overload capacity using the two

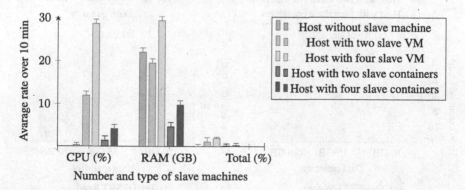

Fig. 1. Resources overload with different number of virtual machines and different types of virtualization tools.

technologies of virtualization and the workload TeraGen. These experiments also consider different size of Hadoop cluster and 20 GB of generated data. We remark in these conditions that container is lighter than traditional virtual machine and for example, the workload TeraGen causes less overhead than VM.

The Fig. 3(a) illustrates the total overload due to the execution of TeraGen. The difference is interesting because it is a large difference in load between the two types of virtualization. The heavy virtualization is characterized by the reservation of the needed memory when these VMs start. It is visible from Fig. 3(b) that the amount of memory reserved by the traditional virtualization increases compared to the results of memory consumption, shown in Fig. 1. The additional memory is used by the hypervisors to interpret the guest operations which are executed by the host operating system.

Docker technology uses only the amount of memory needed to run their process, otherwise memory would be released. This behavior is due to the Docker container policy. The last requires consuming memory not a provisioned memory, thus the memory management in Docker is more flexible than the memory management in traditional virtualization tool. The Docker containers cause less memory overload than traditional VM which reserves 100–200 MB memory per VM for hypervisor. In addition, traditional VM independently reserves a fixed

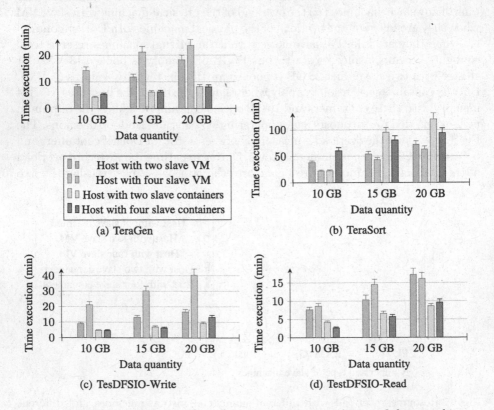

Fig. 2. Time execution in function of the data quantity and type of slave machines.

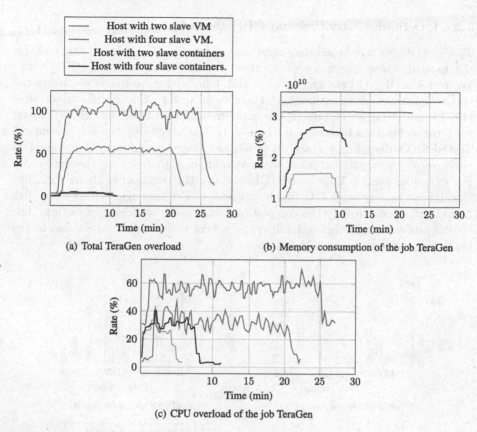

Host with two slave VM
Host with four slave VM.
Host with two slave containers
Host with four slave containers.

(a) Total TeraGen overload

(b) Memory consumption of the job TeraGen

(c) CPU overload of the job TeraGen

Fig. 3. An example of the variation of the overload due to use of different virtualization tools.

amount of memory. The Docker containers offer the possibility to fix a maximum amount of memory a container can use. However, when this memory is not explored by the container, it can be used by another processes. Concerning the variation of the CPU cores and memory, we present analyses and we give an in-depth description. We take as an example the execution of different job with 20 GB of data and we noticed that with the jobs TeraGen and TeraSort, the cluster using Docker technology is more efficient than cluster with traditional virtualization (Fig. 2(a) and (b)). We obtained same results for the same job, executed on the same size of a clusters. The use of two slave machines gives better performances than the use of four slave machines. Thus, the number of slave machines should be correctly chosen to avoid the degradation in performances. We give in next part an in-depth study of Hard disk and CPU bounds exploration.

5.2   I/O-Bound Variation and CPU Bound

The TestDFSIO benchmarks are used to evaluate the HDFS health, they utilize the hard disk resource more than other resources as memory or CPU cores. We run TestDFSIO benchmarks on 2 and 4 hadoop slave machines, using the two technologies of virtualization. Then we also vary the writable data sizes (10, 15 and 20 GB). We present the experimental results of the disk write and read throughput and average IO in Fig. 4. The results of the job execution: TestDFSIO-write, (Fig. 4(a) to (c)) proves that the throughput and average I/O are inversely proportional with the overload measured during the execution. For example, using 4 VMs over 20 GB of data, the overload is about 85%, but throughput and average I/O highly decreases. The management of hard disk bound influences directly the completion time of the workload execution. The results proves that Docker technology use a best policy to manage access to the hard disk compare to the VMware tool.

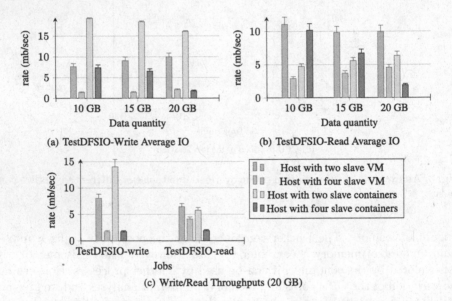

(a) TestDFSIO-Write Average IO

(b) TestDFSIO-Read Avarage IO

(c) Write/Read Throughputs (20 GB)

Fig. 4. Throughput and Average IO of the job TestDFSIO execution with different virtualization tools.

We use the workloads TeraGen and TeraSort to stress the CPU bound. It has a considerable influence over the performance and the energy consumption. In our experiments, we use two cores per slave machines. Docker technology offers two policies to manage CPU resources. The first method is to reserve a specified number of cores per container. The second one uses a relative share rate between containers. It associates a weight to each container and it shares the existing compute resources between them. Please note that all the experiments, explained earlier use the first reservation policy. We will focus on the fair share policy in

Sect. 5.3. We notice that using TeraGen (Fig. 3(a)) and TeraSort; containers cause about the half of the CPU overload than traditional virtualization tool. Figure 2(a) and (b) shows the completion time of the used workflow over a cluster of 2 and 4 slave machines. The cluster with two slave machines gives better performances than the cluster with four. One reason is the architecture, the use of four slave machines increases the competition to access resources and the total overload increases in consequence. For example, when we evaluate the cluster with two slave machines, six cores (CPU) are booked for the virtual cluster so the host OS has the six other cores to use and to run the instructions. However, cluster with four slave machines uses 10 cores (CPU) thus only two cores are used by the host OS. We observe the same behavior for the memory resource use. Thus, we note a performance degradation. As we use the same policy to manage CPU resource in the two cases of study (2 and 4 slave machines), we conclude that the main reason of the performance degradation is the management of throughput and memory policies. We use TestDFSIO read workload to test the read throughput. The results are summarized on Fig. 4(a) to (c). We noticed that the running of multiple slave machines on the same physical host creates a concurrent access to the hard disk. Despite the replication of data used in Hadoop (which is equal to 2), there is a difference in performance between slave machines, depending on the used technology. We give an in-depth description of the memory management policies in Docker technology in Sect. 5.1.

Despite the congestion of resources, when we work with a cluster of four slave machines, in most cases, the Docker container offers better performances in most cases.

In the next subsection, we focus on the execution of the workloads on a heterogeneous cluster and compare the two technologies of CPU management, available in the Docker technology.

5.3 Performance Variation in a Heterogeneous Cluster

We analyze in this subsection the influence of the heterogeneous cluster on the performance of Hadoop. The first step of the experiments uses two different configurations of slave machines. The second step changes the capacity of each resource and analyses the influence of each resource variation in Hadoop performances. In the first step, we use an additional configuration of slave machine, then we obtain new Hadoop cluster with two types of slave machine's configurations: one configuration has 4 cores, 10 GB of RAM and 80 GB of hard disk, the second has 2 cores, 5 GB of RAM and 80 GB of hard disk. The results of the execution of the workloads TeraGen and TeraSort over a heterogeneous and homogeneous cluster of three slave machines is presented in Fig. 5(a). It confirms that Hadoop outperforms better with homogeneous cluster. In the Fig. 5(b), we focus on the Docker technology performances. We run the two jobs in this four use cases: (1) homogeneous cluster of three machines, (2) heterogeneous cluster (described in Table 2), (3) heterogeneous cluster with increasing the memory and (4) heterogeneous cluster with increasing only of the number of cores. Then we conclude from this experiment that varying only RAM or CPU resources is

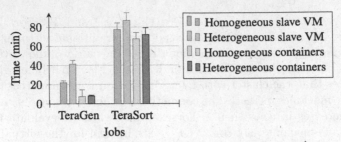

(a) Influence of the heterogeneous cluster (two types of virtualization)

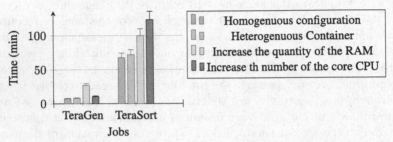

(b) Influence of resources variations of Docker containers on the execution of jobs

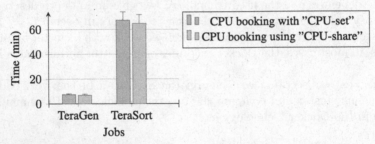

(c) Comparison between the two policies for managing the CPU resource allowed by the Docker technology

Fig. 5. Completion time of the workloads TeraGen and TeraSort with heterogeneous platforms.

not helpful for the Hadoop performances. There are two reasons of this conclusion. The first one is that increasing the capacity of memory in client machines stresses the host operating system and limits its performances. In the same manner, increasing the number of virtual cores per container limits the number of CPU cores used by the host system and decreases its computing capacity. The second one is noted at the scheduling level. After observing the tasks assignment at the Hadoop scheduler level with the two technologies, in the first third of the time execution, we noticed that workloads have a high rate of tasks failures on the slave machine (SM) two and three. However, there is no task failure in the first SM. These results are due to: *(i)* The homogeneity of the cluster, considered by the scheduler (capacity scheduler is used in experiments).

(ii) The mismatch between the resource definition in the configuration files. *(iii)* The available resources on the cluster. During the remaining period of execution of Terasoft and TeraGen, the scheduler has the tendency to re-run the failed tasks with double capacity of resources. The scheduler adapts its behavior and affects the major quantity of tasks, which have double capacity of resources to SM-1. For example, behind the capacity of Nodemanager's resources in Table 2 (slave daemon on the Hadoop cluster), the SM-1 runs three tasks all the time, two tasks have 2 GB of memory and the third has 1 GB of memory. The SM-2 and SM-3 have the tendency to execute tasks with 1 GB of memory, four times more than tasks having 2 GB of RAM. We notice that the scheduler always affects one Vcores (virtual cores) per task. This is caused by the fact that Hadoop considers the criteria for resource homogeneity during scheduling of the tasks.

In Fig. 5(c); we focus on the CPU resources. In previous experiments, we use the reservation policy to affect CPU cores in containers (Sect. 5.3). On the next step, we compare the two policies, given by Docker to explore computing capacity between containers. The results show that there is a thin difference between policy on the described environment of experiments. The sharing method (with CPU-share option) performs better than the affectation method (reservation policy with the CPU-set). The share policy gives the opportunity to share unused compute resources between containers and don't limit them to a specific number of cores. As the Hadoop context is concerned, The share policy increases the capacity of slots on the slave machines. Thus, it increases performances without having the negative aspect on the host OS. When the number of slave machines per physical host is maintained, the share policy gives better performances. It ensure a minimum rate of computational capacity per container. When the host machine has a free computational resources, these resources are shared respecting the relative share between containers. As a result, the compute capacity per Hadoop NM daemon increases and will have a good influence on the completion time of the workloads. When the overload limit is reached, the two methods have the same behavior and the decrease on performances.

5.4   Evaluation of the Energy Consumption

The energy consumption is an important issue in the big data context i.e. Yahoo deploys a Hadoop cluster over more than 2000 servers; Facebook deploys Hadoop over 600 servers; General Electric deploys Hadoop on a cluster of 1700 servers. At this scale, the energy is a critical aspect which influences considerably the cost of cluster exploitation. A research realized by the U.S. Environmental Protection Agency [17] and the Natural Resources Defense Council [18] announced in 2007 that the cost of energy consumption for cluster management was very high. The commission in the European Union defined the code of conduct on datacenter energy efficiency since 2008 [19]. Through the experiences, it is clear that the load and the energy consumption are proportional, when we run 4 slave machines. The overload and energy consumption increase and they are higher than the case of 2 slave machines. We can conclude that when the overload of physical machine increases (more than 85%), the performance degrades and then, energy consumption increases. Installing many virtual machines on the physical host increases the energy consumption and they have negative influence on the job's execution performances. The overload on physical host is proportional to the number of slave machines and the workload running on them. Figure 2(a) and (b) show the completion time of TeraGen and TeraSort workloads over a cluster with different slave machines. The cluster of two slave machines is better performing than the cluster with four machines and has a bit lower consumption than four slave machines cluster. The virtualization technology is used by the server providers to manage the load on the physical machine and to optimize energetic consumption. The overload on the physical machine is the aggregation of all overload of their guest when the VMs run in a higher load. Working with the same type of job, size of cluster and quantity of data (Fig. 6), there is a thin difference between the use of the two virtualization tools. Docker technology consumes less energy than traditional tools. This one is caused by the use of containers instead of the overload due to the use of virtual machines.

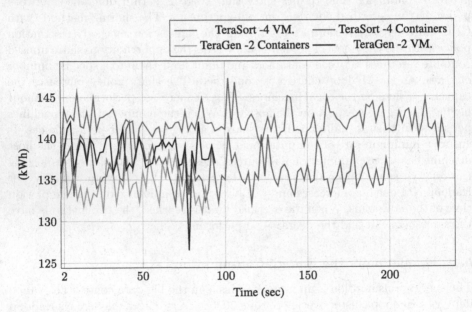

Fig. 6. Energetic consumption of different size of clusters with jobs TeraGen and Tera-Sort.

6 Discussion

Big data applications are known by their capacity to perform processing operations on a huge quantity of data at a respectful time. These applications have a constant challenge faced by the data volume and structure. The hardware architecture and capacity have a direct influence on the application performance.

The main issues of this work are aiming to assess the capacity of the light virtualization to be used for the deployment of big data application. Thus we discuss *(i)* the performance of virtual deployment of Hadoop cluster, *(ii)* the energy management of big data application based on the container technologies and *(iii)* The elasticity.

In our experiment, we focus on the Hadoop software. Scheduling operation depends on the memory and CPU resources, the scheduler considers a homogeneous cluster when it ensures the task scheduling operation. Using the docker technology, our experiment proves that the performance of Hadoop increases compared to VMware tool by a factor of 70% on average. The main reasons are *(i)* the memory and CPU management policies and *(ii)* the new record libraries used in Docker. For example, the first thing to note is that there are two solutions for managing memory resources: *(1)* a container can use all of the memory on the host with the default settings. If Hadoop does not need the memory resource, it will be released and another container can use this quantity to improve its performance. *(2)* If you want, you can limit memory for all of the processes inside of a container. Thus, the memory management in Docker is more flexible than the memory management in traditional virtualization tool, that is based on the second policy, the separation of VM and host physical memory and other techniques like transparent page sharing and memory compression.

In the other side, Docker makes it possible to specify a value of shares of the CPU available to the container or attaching containers to cores. The two policies of CPU management in Docker increase the computation capacity of Hadoop. The Hadoop computation unit is based on virtual cores per slave node and each slave node is deployed in a container. It is based on the sharing of the computing capacity between a number of virtual cores through time. The fair share policy can increase the rate of computing resources. However, it strongly influences the performance of the host operating system since it limits the resources of the clusters. We choose to fix the number of cores per slave machine. Hence, this method offers an accurate report about the resource exploitation and it allows a better comparison between these results.

The start-up time of a container is less than one second and when a container is useless, it can be killed in less than one second. As a consequence, adding a computing node to Hadoop becomes easy; it can help to answer the increase of the overload. Thus Docker influence positively the Hadoop scalability. However the migration of data between physical server remains the massive challenge to be faced for the container and traditional technology of virtualization.

Docker container causes less resource overload than virtual machine. Containers cause about the half overload of the resource CPU compared to the virtual machine. This benefit influences positively the performance and the energy consumption. We only consider the resource isolation; the other kinds of isolation (like user or session isolation) are not realized in this work. The two technologies used in this paper can isolate the CPU resources, memory and Hard disk resources. However, despite of the evolution of the hard disk resource isolation (as blkio controller), the access rate to hard disks remains an open problem. They degrade the total performance and influence the energy consumption.

7 Conclusion

The Hadoop software is adapted to be used with homogeneous platforms. However, hardware technologies are changing continuously and it is not possible to ensure the same Hardware configurations when the cluster is evolving. On the other side, the energy consumption is directly related to the load of resources on the physical host, thus, the performance depends on the number of slave machines per host and it also depends on the execution of workloads. In this paper, four objectives are approached: *(i)* the analysis and the study of many assumptions concerning the configurations of big data platforms. *(ii)* The comparison of performances of the Hadoop platform with the two technologies of virtualization. *(iii)* The study of the variation of the performance in the case of homogeneous and heterogeneous platforms and *(iv)* The deals with the energy consumption on the Hadoop cluster (refer to Sect. 5.4). The containers have an efficient policy to manage memory and CPU and I/O bound resources. They improve considerably the performance of the Hadoop cluster. The energy consumption is directly related to the load of resources on the physical host, i.e. the load of physical host is proportional with the energy consumption. However, the performance depends on the number of slave machines per host and it also depends on the execution of workloads.

In this work, benchmarks argue that the light virtual technology is better to use in the Hadoop context. This work is the first level to assess the possibility to use the Docker technology to improve performance and energy consumption of big data application on the cloud. The deployment of the Hadoop cluster optimizes the resource exploitation and minimizes idle resources either by using traditional virtualization or container technology. However, in the major experiment, the containers cause less overload on resources and offer better performances.

Concerning the security level, in addition to the resource isolation improvement in Docker, Hadoop has its own policy to ensure data security and integrity [30].

In future work, we would like to work on the elasticity and the load balancing to improve reliability of the Hadoop cluster, using the light technology.

Acknowledgements. This work was sponsored in part by the CYRES GROUP in France and French National Research Agency under the grant CIFRE n° 2012/1403.

References

1. Fadika, Z., Govindaraju, M., Canon, R., Ramakrishnan, L.: Evaluating Hadoop for data-intensive scientific operations. In: 5th IEEE International Conference on Cloud Computing, pp. 67–74. IEEE Press, Honolulu (2012)

2. Stonebraker, M., Abadi, D., DeWitt, D.J., Madden, S., Paulson, E., Pavlo, A., Rasin, A.: MapReduce and parallel DBMSs: friends or foes? J. Commun. ACM. **53**, 64–71 (2010)
3. Pavlo, A., Paulson, E., Rasin, A., Abadi, D.J., DeWitt, D.J., Madden, S., Stonebraker, M.: A comparison of approaches to large-scale data analysis. In: International Conference on Management of Data, pp. 165–178. ACM, New York (2009)
4. Yunhong, G., Grossman, R.L.: Lessons learned from a year's worth of benchmarks of large data clouds. In: 2nd Workshop on Many-Task Computing on Grids and Supercomputers, pp. 3:1–3:6. ACM, New York (2009)
5. Fadika, Z., Dede, E., Govindaraju, M., Ramakrishnan, L.: Grid information services for distributed resource sharing. In: 12th International Conference on Grid Computing, pp. 90–97. IEEE Computer Society, Washington, D.C. (2011)
6. Shafer, J., Rixner, S., Cox, A.L.: The Hadoop distributed filesystem: balancing portability and performance. In: IEEE International Symposium on Performance Analysis of Systems and Software (ISPASS), pp. 122–133. IEEE Press, White Plains (2010)
7. Kontagora, M., Gonzalez-Velez, H.: Benchmarking a MapReduce environment on a full virtualisation platform. In: 10th International Conference on Complex Intelligent and Software Intensive Systems, pp. 433–438. IEEE Computer Society, Washington, D.C. (2010)
8. Jiang, D., Ooi, B.C., Shi, L., Wu, S.: The performance of MapReduce: an in-depth study. J. Proc. VLDB Endow. **3**, 472–483 (2010)
9. Huang, S., Huang, J., Dai, J., Xie, T., Huang, B.: The HiBench benchmark suite: characterization of the mapreduce-based data analysis. In: Agrawal, D., Candan, K.S., Li, W.-S. (eds.) New Frontiers in Information and Software as Services. LNBIP, vol. 74, pp. 209–228. Springer, Heidelberg (2011). doi:10.1007/978-3-642-19294-4_9
10. Pavlo, A., Paulson, E., Rasin, A., Abadi, D.J., DeWitt, D.J., Madden, S., Stonebraker, M.: A comparison of approaches to large-scale data analysis. In: International Conference on Management of Data, pp. 165–178. ACM, New York (2009)
11. Understanding Full Virtualization, Paravirtualization and Hardware Assist. http://ww.vmware.com/files/pdf/VMware_paravirtualization.pdf
12. Intel Virtualization Technology (Intel VT). http://www.intel.com/content/www/us/en/virtualization/virtualization-technology/intel-virtualization-technology.html
13. Massie, M., Li, B., Nicholes, B., Vuksan, V., Alexander, R., Buchbinder, J., Costa, F., Dean, A., Josephsen, D., Phaal, P., Pocock, D.: Monitoring with Ganglia. O'Reilly Media Inc., Sebastopol (2012)
14. Baru, C., Bhandarkar, M., Nambiar, R., Poess, M., Rabl, T.: Setting the direction for Big Data benchmark standards. In: Nambiar, R., Poess, M. (eds.) TPCTC 2012. LNCS, vol. 7755, pp. 197–208. Springer, Heidelberg (2013)
15. Hadoop Wiki PowerBy. https://wiki.apache.org/hadoop/PoweredBy
16. Dean, J., Ghemawat, S.: MapReduce: simplified data processing on large clusters. J. Commun. ACM. **51**, 107–113 (2008)
17. Joe, L., Steve, C., Bruce, H., Rebecca, D., Evan, H., Danielle, S., Danielle, S., Andrew, F.: Report to Congress on Server and Data Center Energy Efficiency. U.S. Environmental Protection Agency, New York (2007)
18. Pierre, D.: American Data Centers Are Wasting Huge Amounts of Energy. U.S. Environmental Protection Agency, New York (2014). www.nrdc.org/energy
19. Data Centres Energy Efficiency. http://iet.jrc.ec.europa.eu/energyefficiency/ict-codes-conduct/data-centres-energy-efficiency

20. Xu, G., Xu, F., Ma, H.: Deploying and researching Hadoop in virtual machines. In: IEEE International Conference on Automation and Logistics, pp. 395–399. IEEE Press, Zhengzhou (2012)
21. Peinl, R., Holzschuher, F.: The Docker ecosystem needs consolidation. In: 5th International Conference on Cloud Computing and Services Science, Lisbon, pp. 535–542 (2015)
22. Reshetova, E., Karhunen, J., Nyman, T., Asokan, N.: Security of OS-level virtualization technologies: Technical report. CoRR (2014)
23. Surviving the Zombie Apocalypse Containers, KVM, Xen, and Security. https://archive.fosdem.org/2015/schedule/event/zombieapocalypse/
24. Xavier, M.G., Neves, M.V., Rossi, F.D., Ferreto, T.C., Lange, T., De Rose, C.A.F.: Performance evaluation of container-based virtualization for high performance computing environments. In: 21st IEEE Euromicro International Conference on Parallel, Distributed, and Network-Based Processing, pp. 233–240. IEEE Press, Belfast (2013)
25. Wen, Y., Zhao, J., Zhao, G., Chen, H., Wang, D.: A survey of virtualization technologies focusing on untrusted code execution. In: 6th International Conference on Innovative Mobile and Internet Services in Ubiquitous Computing, pp. 378–383. IEEE Press, Palermo (2012)
26. Jlassi, A., Martineau, P., Tkindt, V.: Offline scheduling of map and reduce tasks on Hadoop systems. In: 5th International Conference on Cloud Computing and Services Science, Lisbon, pp. 178–185 (2015)
27. Getting Started with systemd. https://coreos.com/docs/launching-containers/launching/getting-started-with-systemd/
28. Hadoop Performance Tuning Guide - AMD. http://www.admin-magazine.com/HPC/Vendors/AMD/Whitepaper-Hadoop-Performance-Tuning-Guide
29. Gandomi, A., Haide, M.: Beyond the hype: Big Data concepts, methods, and analytics. J. Int. J. Inf. Manag. **35**, 137–144 (2015)
30. Xavier, M.G., Neves, M.V., De Rose, C. A. F.: A Performance comparison of container-based virtualization systems for MapReduce clusters. In: 22nd Euromicro International Conference on Parallel, Distributed, and Network-Based Processing, pp. 299–306. IEEE Press, Torino (2014)

Towards a Framework for Privacy-Preserving Data Sharing in Portable Clouds

Clemens Zeidler$^{(\boxtimes)}$ and Muhammad Rizwan Asghar

Department of Computer Science, The University of Auckland,
Auckland 1142, New Zealand
{clemens.zeidler,r.asghar}@auckland.ac.nz

Abstract. Cloud storage is a cheap and reliable solution for users to share data with their contacts. However, the lack of standardisation and migration tools makes it difficult for users to migrate to another Cloud Service Provider (CSP) without losing contacts, thus resulting in a vendor lock-in problem. In this work, we aim at providing a generic framework, named *PortableCloud*, that is flexible enough to enable users to migrate seamlessly to a different CSP keeping all their data and contacts. To preserve the privacy of users, the data in the portable cloud is concealed from the CSP by employing encryption techniques. Moreover, we introduce a migration agent that assists users in automatically finding a suitable CSP that can satisfy their needs.

Keywords: Portable cloud · Privacy · Data sharing · Data migration · Migration costs · Migration agent

1 Introduction

Cloud storage is a cheap and reliable alternative to a local storage system. A Cloud Service Provider (CSP) is considered to ensure availability of cloud services so that users can get access to their data from anywhere at any time. Leveraging cloud storage can be an attractive business model for individuals as well as for enterprises that do not have resources to deploy and maintain custom storage solutions. However, data in the cloud is stored at geographically dispersed locations, thus raising serious privacy concerns. Assuming that the CSP is *honest-but-curious* [1], data has to be kept confidential. Technically, the confidentiality of the data can be guaranteed by employing encryption before storing the data in the cloud.

Many CSPs allow their users to share data with each other, which is a great way to collaborate with third parties. For example, Dropbox[1] enables users to share files with each other. However, sharing data with users who do not belong to the same CSP is usually limited and less secure and requires a manual token or key exchange with third parties. Throughout this chapter, we call third parties *contacts, i.e.,* the parties with whom users shares their data.

---

[1] https://www.dropbox.com/.

© Springer International Publishing AG 2017
M. Helfert et al. (Eds.): CLOSER 2016, CCIS 740, pp. 273–293, 2017.
DOI: 10.1007/978-3-319-62594-2_14

There are various reasons why a user may want to migrate her data from one CSP to another one. For example, if there are cheaper CSPs available, the service conditions have changed or the current service is not reliable enough. Furthermore, there are jurisdictional restrictions on the CSP [2]. In the worst case, a CSP might have to shut down its services for financial or legal issues. For instance, if a CSP is used for illegal file sharing, the CSP may face legal issues and its service may get interrupted. For innocent users, this can lead to loss of their personal data.

Having contacts at a certain CSP can be a hindrance for a user to migrate to another CSP since these contacts would then be lost, *i.e.*, a user and a contact would not be able to access and share data with each other anymore. Another problem that makes it hard to migrate a cloud service is that there is often a lack of tools for a seamless migration. For example, there is no simple way to migrate data between CSPs when data needs to be transformed to a different format or encryption scheme. These problems that could stop users from migrating to a different CSP are also known as vendor lock-in [3–5].

In this chapter, we propose *PortableCloud*, a generic framework that addresses the problem of vendor lock-ins and allows users to seamlessly migrate data between CSPs that run *PortableCloud*. If required, data can even be removed from the cloud and migrated to a local service that runs *PortableCloud*. Data can be shared between contacts that reside either at the same or a different CSP (see Fig. 1). To preserve the privacy of users, data is stored encrypted and can only be accessed by authorised parties. When migrating the portable cloud to a new CSP, all contacts are kept and automatically notified about the migration, *i.e.*, the migration is transparent to users and their contacts. We provide a migration cost analysis of the portable cloud migration. Further, we propose an agent that informs users about CSPs with better conditions in order to help them to migrate to a new CSP.

Our contributions can be summarised as follows:

- A proposal of a novel privacy-preserving portable cloud framework *PortableCloud* that enables seamless migration to a new CSP while maintaining all existing contacts.
- A cost analysis of a portable cloud migration.
- A migration agent that assists users in migrating to another CSP.

Section 2 motivates and defines requirements for *PortableCloud*. Section 3 provides an overview of the system model. Section 4 elaborates *PortableCloud*. Section 5 explains the migration process, analyses the migration cost and describes a migration agent. Section 6 discusses privacy aspects of the portable cloud and how the portable cloud can be used by enterprises. Section 7 reviews related work. Section 8 concludes this chapter and gives directions for future work.

2 Scenario and Requirements

2.1 Motivating Scenario

In this section, we briefly explain a scenario that motivates why we need a portable cloud. Let us assume an organisation that has to store data of its users who could share data with their contacts that may or may not belong to the same organisation. A typical example of such an organisation is a university, where data is shared between researchers from the same university as well as with collaborating researchers from other universities and industrial partners.

We consider that the organisation outsources storage services to a CSP. Like a typical storage system in the trusted environment, accounts are created and access rights are defined for users and external contacts for sharing data among them residing on the CSP servers. Since a CSP could compromise the privacy of users when getting cleartext access to the data, the organisation employs encryption techniques before storing any data in the cloud. Consequently, sharing data becomes a matter of sharing secret keys for the encrypted data.

The organisation might decide to migrate from an existing CSP to a new one. There could be various reasons behind such a migration. The most prominent reasons include cost, limitation of (hardware and software) resources and trust. For instance, there is a CSP that (i) offers cheaper services, (ii) does not have sufficient hardware to offer more storage or a particular version of software is not supported or (iii) a CSP is not trusted anymore due to some unexpected event or bad experience.

For migrating from one CSP to another, the organisation could face a number of problems. First, it is not easy to just copy data from one system to another. For instance, a migration process could require downloading a part of encrypted data locally and re-encryption before uploading it to the new CSP. Unfortunately, current CSPs do not support tools for aforementioned operations. Second, all users and their contacts need to be notified about the new CSP. This is not only a tedious task but could also be erroneous and can lead to serious access right issues for users in the system. Third, the new encryption keys and the new tools, which are required to access the data, need to be distributed to the users and contacts. In the case of the second and third problems, any manual change by users or contacts could result in some human errors. Fourth, during the migration process, users and contacts might face issues in accessing the data or consuming services. Even if the old CSP keeps serving till completion of the migration process, there could be inconsistencies in the data if it is modified meanwhile. In other words, there could be serious issues if any data write operation is encountered during the migration process. Last but not least, organisations need to manually explore themselves CSPs that can make competitive offers. More precisely, the existing infrastructure does not take into account automatic discovery of alternative CSPs or their services.

Currently, all these problems result in vendor lock-in issues because an organisation cannot easily choose to migrate to another CSP. An organisation has to consider any migration carefully since it could affect their users as well as their users' contacts.

2.2   Requirements

From the above scenario, we can deduce a set of requirements for a portable cloud architecture. First, the option to migrate to a set of different CSPs can lead to certain security threats since it could be difficult to judge which CSPs keep data private and could be trusted. The data must not only be encrypted, but users and contacts must be able to ensure that the data can only be read by authorised entities, is not modified by unauthorised entities [6] and the origin of the data is authenticated [7]. More precisely, the framework of portable clouds must provide mechanisms to ensure user privacy, integrity and provenance. This can be achieved by using cryptographic methods to encrypt the data to conceal private information from the CSP and sign it. Furthermore, any information that any CSP could infer from the stored or exchanged data must be as minimal as possible.

Second, the migration should be seamless for all stakeholders, in particular for users and contacts. Moreover, the downtime for a migration should be small and contacts need to be notified automatically about the migration. This makes it easy for users to migrate to a new CSP without losing connections to existing contacts, *i.e.*, contacts are able to access data at the new CSP.

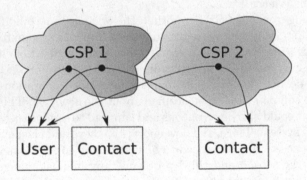

Fig. 1. Data sharing between a user and her contacts who may or may not be at the same CSP. Even after a migration to a new CSP contacts are still able to access data at the new CSP.

Third, users must be able to share data with their contacts that may or may not belong to the users' CSP (see Fig. 1). This is a quite natural requirement since data sharing should stay functional after the migration.

To summarise the requirements:

1. User privacy must be preserved.
2. Data integrity and provenance must be provided.

3. A migration to a new CSP must be transparent and seamless to the user and her contacts.
4. Data needs to be shareable with contacts that may or may not belong to the users' CSP.

3   System Model

A *user* of a *CSP* stores all her data at the CSP. Data at the CSP can be shared with *contacts*. Contacts are users of the same or different CSPs. In our system model, we assume the following entities:

- **Cloud Service Provider (CSP):** It is responsible for storing the data and runs the *PortableCloud* software. The CSP ensures that only authorised parties can access the data. Furthermore, the CSP manages the communication between users and their contacts.
- **User or Organisation:** It is an entity that owns the data managed by the CSP and regulates access to the data by deploying access control policies on the CSP. It is a client of the CSP. In case of an organisation, there could be an administrator. However, in case of individuals, we do not distinguish between an administrator and a user. It is responsible for the encryption of the data before storing it in the cloud.
- **Contact:** A contact is a party that can access data at the user's CSP according to an access control policy defined by the user. Contacts are users of the same or of a different CSP.

Data at the CSP could be stored in any format. It could be managed in files or there could be a database. We assume that the CSP is honest-but-curious [1]. This means the CSP runs PortableCloud correctly but the user cannot trust the CSP to provide confidentiality. For this reason, the user encrypts confidential data locally to prevent the CSP to gain cleartext access to this data. The user encrypts data using a secure symmetric key encryption algorithm, such as AES.

To be able to securely communicate with contacts, each user has a set of (asymmetric) key pairs. These are signing keys and verification keys for digital signatures and public keys and private keys for the public key encryption. We assume that the initial key exchange takes place, typically once, through a secure channel. In this way, information such as data encryption keys or migration notifications can be exchanged in a secure manner.

4   Proposed Framework

In this work, we propose *PortableCloud* a framework that aims at providing cloud portability by seamlessly allowing data migration from one CSP to another one. Using *PortableCloud*, we not only address the vendor lock-in problem but also preserve the privacy of users by encrypting data before storing it in the cloud.

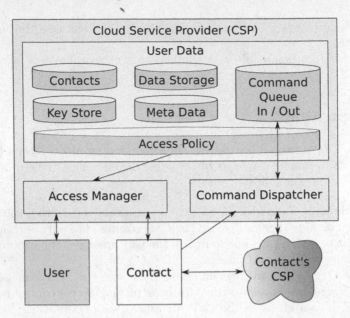

Fig. 2. *PortableCloud*: a user stores data at the CSP in an encrypted way. Only authorised contacts get access to the data at the CSP. The user and her contacts can exchange commands through their CSPs.

PortableCloud enables users to share data with their contacts who may or may not belong to the same CSP.

Figure 2 depicts *PortableCloud*. In *PortableCloud*, the core system entities include a CSP, users, their contacts. A CSP is the core component of *PortableCloud* and the user data is stored at the CSP. The user data contains all information needed by the CSP to operate. For example, the user data contains the access policy and a command queue as a way to communicate with contacts. This encapsulated design of the user data makes it easy to migrate to a new CSP since the user data can be used as it is at the new CSP.

In this section, we provide an overview of the technical details of *PortableCloud*. We point out possible solutions and techniques to implement a portable cloud. After we discuss the CSP (Sect. 4.1), we describe how a new portable cloud account is set up (Sect. 4.2). In *PortableCloud*, users and contacts can communicate over a secure channel (Sect. 4.3). After a contact is established (Sect. 4.4), users and contacts can share and access the data (Sect. 4.5). *PortableCloud* also ensures data integrity and provenance (Sect. 4.6).

4.1   Cloud Service Provider (CSP)

A CSP is responsible for storing the user's data. It also regulates access to the data if a contact satisfies access policy specified by the user. Moreover, the CSP dispatches control commands between the user and her contacts, *e.g.*, commands to notify contacts about the migration. The user interacts with the CSP through a client such as a desktop, mobile app or web page. The CSP mainly consists of three main components including *User Data*, *Access Manager* and *Command Dispatcher*.

User Data. The user data is the core entity at the CSP that holds all the data of a single user. It also contains information the CSP needs to operate, *e.g.,* it includes the access control policy that is deployed to regulate access to the data. As we describe later, this simplifies the migration process since the user data is largely decoupled from the CSP. All sensitive elements of the user data are encrypted using a symmetric key encryption algorithm, where each element uses a separate symmetric key. The user data contains the following sub-components.

Data Storage. The data storage is a repository to store the data. Typically, it could be a database, a file system or a combination of both. Since the data in cleartext could compromise privacy of users, data elements (say files) are encrypted using different symmetric keys. This increases the security since contacts can only decrypt these entries for which they possess the corresponding decryption keys.

Meta Data. The meta data consists of structural information about the data, *e.g.,* directories and file names or table and column names of an encrypted database [8]. Furthermore, the meta data contains integrity and provenance information for the data stored in the data storage. Entries in the meta data are encrypted with encryption keys of the associated data.

Key Store. The key store is used to manage and store cryptographic keys at the CSP. It is important to note that all secret cryptographic keys are stored securely in an encrypted manner. Only the user can access the key store and get access to the keys using a password. This prevents the CSP accessing secret keys [9]. Thus, having access to the keys in the key store enables the user to decrypt all user data stored at the CSP.

We consider that keys in the key store are encrypted with a symmetric master key. This master key, which could be realised as a role key, is stored encrypted as well. The master key can be decrypted by deriving a user key from the user's password using a Key Derivation Function (KDF) [10] (see Fig. 3). This key chain approach is similar to the one used in the Linux Unified Key Setup (LUKS)[2] or adopted by Boxcryptor[3].

The key chain approach makes it easy to change a password for the key store. For changing the password, only the key store master key has to be re-encrypted with a user key derived from a new password. All other keys in the key store do not need to be re-encrypted.

---

[2] https://gitlab.com/groups/cryptsetup.
[3] https://www.boxcryptor.com.

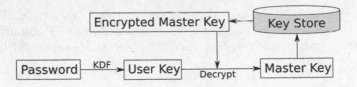

Fig. 3. A user key is generated from the user password using a Key Derivation Function (KDF). The user key is then used to decrypt the master key that allows the user access to the key store.

Note that the security of the key store is limited by the complexity of the user password. The KDF can make it harder for brute-forcing attacks, *i.e.,* by using a salt and a high number of iterations. However, simple passwords can still be guessed easily. If a high degree of security is required, a password in form of a large cryptographic key should be used and stored safely at the user client.

Contacts. This entity contains information about all contacts known to the user. This information contains the CSP location of the contacts as well as contacts' public keys and verification keys. Furthermore, each contact entry contains information to access and decrypt shared data that is located at the contact's CSP. All the contact information is stored encrypted.

Access Policy. The access control policy – or access policy in short – specifies what contacts are eligible to access at the user's CSP. In this work, we consider that access policies are readily available in the cleartext to allow the CSP to regulate access to the data. Without loss of generality, in case of sensitive access policies, we could employ encrypted policy enforcement mechanisms, such as [11, 36–38].

Command Queue. The command queue holds incoming or outgoing commands. Commands are generic messages that can be used to communicate requests between a user and her contacts. As discussed in Sect. 4.3, using commands, users can communicate securely with their contacts.

All commands in the incoming command queue are fetched and handled by the user. Outgoing commands are placed in the outgoing command queue together with a CSP address of the receiving contact. The CSP address is stored in cleartext in order to allow the CSP to dispatch the command.

Access Manager. The access manager is a sub-component of the CSP, which ensures that only authorised entities can get access to the data stored at the CSP. It authenticates users and contacts and provides access to the requested data, given the deployed access policies are satisfied.

Command Dispatcher. The command dispatcher is a sub-component of the CSP that dispatches commands from the outgoing command queue and aims at delivering them to the target CSP. If a command has been delivered successfully, it is removed from the queue. Moreover, this sub-component receives incoming commands and places them in the incoming command queue.

4.2 Account Creation

Once a new user signs up, a signing key pair as well as an encryption key pair are generated. Both key pairs are securely stored in the key store. As already explained in Sect. 4.1, a key chain is used, which starts from the user's password from which the user client derives the user key. Using this key chain, users can get access to the user data.

For authenticating the user to the CSP, we need to present user credentials. As a possible solution, we can use another password (different from one already mentioned above) but it would require the user to manage two passwords: one for authentication and another for decrypting the key chain, thus raising usability concerns. To avoid this usability issue, we propose authentication using the same password for deriving an additional authentication key from it. When signing up, the user chooses a different set of KDF parameters, *i.e.*, number of iterations and salt, to generate the authentication key.

4.3 Command Passing

A user needs a way to communicate with contacts at the same or different CSPs, say to establish a new contact (Sect. 4.4) or to share data with a contact (Sect. 4.5). In general, a direct peer-to-peer connection between the user and a contact is not always possible, in particular when the contact is offline. For this reason, the CSP is used to pass commands between communicating parties.

To send a command to a contact, the command is delivered to the contact's CSP as illustrated in Fig. 4. The contact's CSP puts the command into the contact's incoming command queue. In case the contact's CSP is not available,

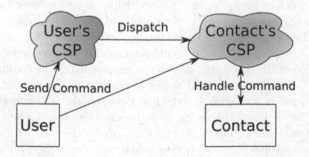

Fig. 4. The user can pass commands to the contact's CSP. These commands are handled by fetching them from the incoming command queue.

the command is placed into the user's outgoing command queue and the user's CSP aims at delivering the command at a later time.

Commands are always signed by the sender and, whenever possible, encrypted with the public key of the receiver, assuming the public key has already been exchanged.

4.4   Contact Establishment

Data can be shared with contacts at the same or a different CSP. To establish a link between a user and an unknown contact, they have to exchange their verification keys and public keys. This key exchange could take place out-of-band, say using PGP [12]. Alternatively, we can rely on Public Key Infrastructure (PKI), where a trust anchor, which is a root Certificate Authority (CA), issues X.509 certificates [13].

After a successful contact establishment, both the user and the new contact create a new contact entry in their contacts databases. Each contact entry contains the contact's keys as well as the CSP location of the contact.

4.5   Data Access and Sharing

Data Access. The user has full access to the user data by logging into the CSP using her user name and password (see Sect. 4.2). However, the access for contacts needs to be restricted and is managed using *access tokens*. The typical frameworks for access tokens include OAuth[4], OpenID Connect[5], where tokens are used to provide access to a service.

Typically, an access token is an identifier that is presented to a service provider to get access to requested services or resources. In the traditional access token model, there is no way to identify the requester. However, we consider special access tokens that are used to allow contacts access to specified resources. In the context of portable clouds, access tokens consist of two parts: a private signing key and a public verification key. The public part of the token, *i.e.,* the verification key, is stored in the user's access policy and is mapped to access rights specified in an Access Control List (ACL) [14] (see Fig. 5). The ACL can also be used to define group access, *i.e.,* multiple access tokens are mapped to the same access rights in the ACL. As described later, an eligible contact is in possession of the private part of the token, *i.e.,* the signing key.

A contact can access data by signing an access request using the private signing key. If the CSP can verify the access request using the public verification key in the ACL then the requested access is granted to the contact.

There are various scenarios in which access tokens are generated and how they are distributed. First, the user generates an access token when sharing data with a contact and sends the private signing key to the contact (see Sect. 4.3). Second, the contact generates an access token and sends the public verification

---

[4] http://oauth.net.
[5] http://openid.net/connect.

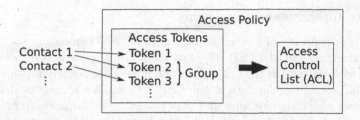

Fig. 5. Contacts can use tokens to gain access to the shared data. Tokens or groups of tokens are then associated with access rights defined in an ACL.

key to the user. Third, one of the contact's signing keys is used in which case the user already knows the public verification key of the contact (see Sect. 4.4). In the following, we assume the first scenario, *i.e.*, the user generates an access token when sharing data.

Data Sharing. A user can share data with contacts located at the same or at a different CSP (see Fig. 1). To access shared data, a contact needs an access token and an encryption key to decrypt the data (see Fig. 6).

Fig. 6. A contact can access or modify data at a user's CSP. Therefore, the contact needs an access token and an encryption key.

To share data with a contact, the user has to send the token as well as the encryption key to the contact. This is done using a secure command as described in Sect. 4.3. Moreover, the access rights in the access policy have to be updated for the used token. If the contact declines the sharing offer, the changes made to the access policy are reverted. When revoking data access for a certain access token, the affected contacts are notified.

In case a contact wants to share data with the user and the user accepts the sharing offer, the user adds a new *shared data entry*. This shared data entry contains information about the shared data; the access token as well as the data encryption key. It is important to note that the contact's CSP location is already stored in the contacts database; thus, all information to access the contact's data is available.

4.6   Data Integrity and Provenance

An important property of a cloud storage is that users can ensure the integrity of their data stored at the CSP [15], *i.e.,* detecting if potential attackers have tampered with the data at the CSP. Moreover, if the data is shared with a contact and the contact writes the data to the CSP, *i.e.,* modifies, adds or deletes the data, the user may want to ensure that the changes really originate from a certain contact [7]. On the other hand, when the user writes the data, the user may want to certify that changes indeed originates from the user. This means not only data integrity is required but also data provenance is needed for a cloud storage.

To verify data integrity, *integrity information* is generated by the writer. One way to generate integrity information is to encrypt the data hash with the data encryption key and use this encrypted hash as integrity information [6]. The integrity information is stored in the meta data entity and can be accessed by users or contacts who can access the associated data. A user or a contact who is able to decrypt the data can also decrypt the integrity information and verify the data integrity by comparing the included hash to the hash of the actual data.

Integrity information does not help to ensure data provenance since multiple different writers may have write access to the same data. The writer also has to provide *provenance information*. The provenance information is stored along with the integrity information in the meta data. Same as the integrity information, the provenance information is encrypted with the data encryption key to prevent the CSP accessing the provenance data. The provenance information contains a hash that identifies the performed write operation, the ID of the writer and the ID of the hosting CSP user. Moreover, a time stamp can help to track when changes were made. Before encrypting the provenance information, the provenance information is signed with the writer's signing key.

The user can verify the data provenance by verifying that the hash of the write operation is compatible with the actual data. The receiver ID contained in the provenance information ensures that the write operation was indeed intended for the user. By verifying the writer's signature on the provenance information, the user can ensure that the changes were performed by the writer.

Unlike integrity information, the user may want to prevent contacts to access the provenance data. The provenance data contains information about with whom the user shares her data and the user may want to conceal this information [7]. For example, if a user shares data with multiple contacts, the user may want to hide who else has access to the data. For that reason, the user can define in the access policy if contacts are allowed to access the provenance data from the meta data entity.

5   Migration

The migration process of *PortableCloud* consists of two main steps. First, the user data has to be copied to the new CSP. Second, all contacts need to be notified about the new CSP. It is important to ensure that the migrations should be transparent for the contacts and there should be a minimal downtime.

Since the user data does not need to be adapted for the new CSP, the migration can take place through a direct data transfer between both CSPs. However, copying a large amount of user data can take a significant amount of time. For that reason, the data should be copied gradually. This can be done by first copying a snapshot of the user data and then successively copy new changes made during the migration. With the assumption that new changes are small, the time to synchronise the data with the new CSP during the migration is small.

One problem that can affect the migration is ongoing write or read transactions that are performed by contacts. A conservative approach is to block new changes at the old CSP and wait unless all the data is migrated. Alternatively, we can imagine more sophisticated approaches that are able to handle ongoing changes during and after the migration.

Once all data is copied to the new CSP, the contacts need to be notified about the migration. This is done by sending them *migration commands* that contain the location of the new CSP. A problem that can occur here is that a contact might not be reached. One reason for that could be the temporary unavailability of the contact's CSP. However, since the migration command is in the outgoing command queue, the new CSP will try to deliver the message at a later point. Another situation when a migration command cannot be delivered is when the user and a contact both migrate to a new CSP at the same time. In this case, there is no easy way to determine the CSP location of the migrating contact. For this reason, the old CSP can be configured to point to the new CSP location when contacts try to communicate with the old CSP. One possible approach is to introduce one or more central name servers where users can register their CSP location.

If a user receives a migration command from a migrated contact, the new contact's CSP location has to be updated in the user's contacts list. Furthermore, the user has to verify that undelivered outgoing commands to the migrated contact are updated to target the contact's new CSP.

5.1   Migration Costs

For enterprises as well as individuals, the costs and services of a CSP are important. If a preferable CSP (say based on various factors such as quality of service or costs) is available, the user may consider migrating to this CSP.

Data Sharing Systems. In the following, we discuss the migration between two different data sharing systems. It also includes the migration to the portable cloud architecture. We assume that the user encrypts data on the client side to prevent the CSP accessing the data.

One of the major costs includes set up costs, such as initial setup fees for the new CSP. There are various sources of costs when transferring data from the old to the new CSP. First, one or both of the involved CSPs may have data transfer fees. Second, it might not be possible to transfer the data directly between both CSPs and the user needs a local data storage to copy the data. For instance, data formats or databases may be incompatible, data requires re-encryption or there is a lack of APIs to transfer data directly.

Once the data is transferred to the new CSP, connections to old contacts have to be re-established and access policies have to be set up. In general, there is no automatic way to convert the old access policy to a new system. For this reason, the access policy has to be verified manually, which can be an expensive and erroneous process, *e.g.*, due to human errors crucial data could accidentally be leaked to wrong contacts.

Portable Clouds. For the migration of the portable clouds, there may be set up and data transfer costs. Even the small migration downtime for the portable clouds could lead to further costs.

PortableCloud minimises the cost described above. Since the data can be transferred directly between the old and the new CSP, expensive data re-encryption and round trips to the user's local storage could be eliminated. Furthermore, *PortableCloud* ensures that contacts can still access the shared data and no new encryption keys have to be exchanged. This not only minimises the service downtime but also is fail-safe against human errors, *i.e.*, the old access policies are re-used at the new CSP.

The user as well as all contacts do not need to update or reconfigure their client software since the migration process is transparent. This eliminates support costs and expensive software adoptions.

5.2   Migration Agent

There are various decision making and other tools that could assist during the migration process [5,16,17]. Like these tools, we use a migration agent in *PortableCloud*. The migration agent calculates costs based on various parameters of interest, which includes, but are not limited to, historical growth pattern, manual input or a combination of both. The migration agent assists users in providing statistics about data usage, forecasting and listing alternative CSPs that can offer similar or even better services. If the agent finds a better CSP, it suggests it to the user as a migration option. For that, the cloud agent maintains a knowledge base of alternative CPSs in real-time. This knowledge base is updated regularly by services that host the migration agent.

A core aspect when considering migration of the portable clouds is cost. The costs identified in Sect. 5.1, *e.g.*, initial set up and data transfer costs, are taken into account. Another interesting parameter is the migration time. The migration agent can estimate how long a migration will take, *e.g.*, how much time the account set up and the data transfer will take. This helps the user in estimating when the new service of the CSP is available.

The migration agent also estimates usage patterns and notifies a user about possible performance problems and issues. These problems could be a result of lacking or surplus of data storage, transfer problems with the users/contacts, or stability and reliability issues with the CSP. For example, if for a certain period of time the user consumes less storage space than she pays for, the migration agent analyses if there are more suitable (*i.e.,* economical) options available.

The migration agent also considers different factors such as customer satisfaction, reputation or legal issues with the CSP. However, these factors are subjective and have to be considered carefully. Furthermore, the migration agent helps users in finding better service plans at the current CSP, if available.

6 Discussion

One goal of *PortableCloud* is to maintain privacy of users. In this section, we discuss what information the CSP can gain about the user and what information is concealed from the CSP. Moreover, we discuss requirements and solutions for an enterprise that uses the portable clouds.

Privacy. There is some general knowledge a CSP has about its users. For instance, when registering with a CSP, information such as the user name/login name, email address, phone numbers, postal address or payment details may be revealed to the CSP.

All data the user stores at the CSP is encrypted and can only be read by the user who has the corresponding key. In *PortableCloud*, the meta data is also protected. Thus, the CSP cannot learn any sensitive user data.

Outgoing commands contain the target CSP in order to deliver a command to a certain contact. This may reveal the identity of contacts. Although all information about contacts is stored encrypted, the CSP can derive information about the number of contacts of a user. To address this issue, Oblivious RAM (ORAM) [18,19] or related techniques may be necessary.

The access policy maps access tokens to an access control map, which may reveal information to the CSP. For example, the CSP can analyse how many access tokens exist for a certain data entry and may derive information about the number of contacts or the importance of the data entry.

Note that the CSP can analyse traffic from/to the user data, which could also reveal information about the stored data as well as about the contacts [20,39].

Enterprises. In *PortableCloud*, as described above, users control their data and manage contacts they share their data with. However, for commercial enterprises, this model might not be an ideal option. In the following, we describe what requirements an enterprise may have concerning portable clouds and how *PortableCloud* can be customised to fulfil these tailored requirements.

An enterprise usually has a number of employees and there are certain restrictions on how data can be shared with internal and external contacts. For this reason, the enterprise needs a way to manage their employees. To do so, the enterprise takes the role of an admin user who can manage a group of users at the CSP (see Fig. 7). The admin user has several privileges such as:

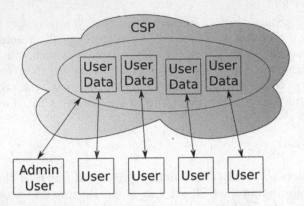

Fig. 7. An enterprise can administrate and manage multiple users (*e.g.,* its employees). The enterprise may have special access rights to its employees' data and key stores.

- Administration of new users, *i.e.,* creation and deletion.
- Control data access among users of the enterprise and external contacts.
- Prevent users to migrate their user data to another CSP.

In general, an enterprise can require access to all data produced by their employees. The enterprise can require employees to enable their admin access to their secret key. This would also allow the admin to reset their secret keys. Since employees only use their own personal password to encrypt their secret key (see Sect. 4.1), the personal password is not revealed to the employer. This is important in case the employee uses this password also for other purposes.

Another approach of using *PortableCloud* in an enterprise is to allow multiple clients to access a single portable cloud account. This is easily possible since the key store supports multiple access passwords. However, since all employees have the same access rights, all data can be accessed by all employees. Thus, this solution is only suitable for small enterprises.

Data Sharing with External Contacts. *PortableCloud* only allows sharing data with contacts that have a *PortableCloud* account at the same or a different CSP. It is often desirable to share data with external contacts, those who do not have a *PortableCloud* account. A common solution for this is public links. By sharing public links with external contacts, *e.g.,* via email, external contacts can gain access to shared data. For example, in Dropbox[6], files can be shared with external contacts using a public hyperlink. Overleaf[7] even allows external contacts to edit documents shared through a public hyperlink.

In general, a public link contains an access token that ensures that only contacts who know the link are able to access the shared data. By choosing a sufficiently large random access token, it becomes very difficult for an adversary

---

[6] dropbox.com.

[7] www.overleaf.com.

to gain access to the shared data. Thus, public links provide fairly good access control.

A disadvantage of public links is that, they are not easy to remember due to the embedded access token. External contacts have to manage their public links manually. Another problem with public links occurs if the shared data is encrypted. In this case, also the decryption key has to be shared with external contacts. For example, to access encrypted data through a web application, the external contact would need to provide the decryption key to the web application. Note that, in the case of a web application, the entity who provides the web application has to be trusted for not leaking the decryption key from the web application to an adversary. Another solution is to drop the privacy requirement for the data that is shared through public links and reveal the decryption key to the CSP. However, this is usually not desirable.

Data Update Notifications. When having access to the shared data, it becomes interesting to know when the data has been updated at the CSP. This is because polling for data updates can be expensive especially when monitoring a huge amount of data. For example, if a user has access to multiple files from different contacts, frequently checking for data updates becomes expensive for the user as well as for the involved CSPs. A more efficient way to monitor data updates is a publish-subscribe model [21,22]. Here, a subscriber can register with a publisher and the publisher notifies the subscriber if updates are available.

The simplest approach for a user to notify contacts about data updates is to send a *data update command* to all contacts who have access to the updated data. This approach does not leak information about who has access to the data. However, contacts do not get notified immediately when other contacts modify data because the contact who changed the data may not know who else has access to the data and thus is not able to send data update commands to other sharing contacts.

A different approach is to let the server notify contacts when data has been updated. For example, a contact registers with the user's CSP to indicate that she is interesting in updates on certain data sets. When data has been changed the CSP automatically sends data update commands to the registered contacts. A problem with this approach is that the CSP can gain information on who has access to which data. Protection of this information requires some other privacy-preserving techniques [23].

7   Related Work

Although migrating a system to the cloud is a challenging task, migration also brings scalability while offering flexible pricing options [24]. Migrating a local service to the cloud can reduce the cost to run and maintain servers but can also increase the dependency on external third parties and a potential deterioration of the service quality due to less control over the system [17].

For enterprises, it is not easy to decide if the migration from their IT system to the cloud is really beneficial. Cloud Genius assists users in finding an optimal CSP that provides IaaS, *i.e.,* it finds the IaaS that is able to run a certain VM image at better service conditions [25]. The problem of vendor lock-in can be addressed by using unified programming APIs and domain-specific languages to model application components and cloud requirements [5]. In a so-called meta-cloud, an agent continuously checks for alternative CSPs with better conditions for the specified requirements [5].

One way to share data is to use a distributed peer-to-peer data sharing system, such as PeerDB [26]. However, the data in PeerDB is not encrypted. Moreover, for sharing data, both peers are expected to be online.

Various security and privacy issues in cloud computing have been identified [27]. When transferring the data from/to the cloud, confidentiality and integrity must be ensured. When sharing data with other parties, there must be mechanisms to control access rights. To ensure privacy when storing data in the cloud, the usual way is to encrypt data. However, users may not have enough expertise to manage their keys. The data integrity can usually not be verified on the cloud storage without transferring the data to a local machine. When deleting data in the cloud, the user usually cannot ensure that no data copies remain at the CSP. One way of dealing with privacy issues is to keep users anonymous while storing the user's data in cleartext in the cloud [28]. K2C allows users to share encrypted data with other users but users have to manage their encryption keys in a local key store [29]. A more convenient approach is to store the encryption keys in an encrypted key store in the cloud [9]. Even when data is encrypted, it is possible to perform a search query on the encrypted data while respecting multi-user access policies [8].

The cloud storage system DepSky [30] stores encrypted and signed data at multiple CSPs. DepSky uses a secret sharing scheme [31], which means that shares of the secret key are distributed to different CSPs. While DepSky allows users to replicate data at different CSPs, it does not offer any contact management.

There are various popular cloud sharing systems available. The cloud software ownCloud[8] allows users to setup a personal cloud server. However, while own-Cloud enables public data access, private data can only be shared securely with users of the same server and not with users of other ownCloud servers. ownCloud only supports server side encryption, which requires trusting the server that hosts the ownCloud instance. Data sharing platforms, such Boxcryptor[9], support the client side encryption. However, these services neither support migration to another CSP nor do they allow private data sharing with users of other CSPs.

Mona allows users to share data with contacts and revoke access if necessary [32]. While the identity of a contact is concealed from the CSP, the user knows about the provenance of the data. To define an access policy, a simple Role-Based Access Control (RBAC) mechanism can be used. Here, roles can

---

[8] https://owncloud.com/.
[9] https://www.boxcryptor.com.

be granted and revoked if necessary [33]. Moreover, hierarchical attribute-based encryption can be used to control and revoke data access [34].

When establishing a contact, the public keys of both parties have to be exchanged. This key exchange is vulnerable to man-in-the-middle attacks. SafeSlinger enables an easy and secure exchange of public keys between contacts as long as there a secure channel between them, *i.e.*, they can exchange a simple word phrase in person or via other channels [35].

8 Conclusions and Future Work

In this chapter, we addressed the problem of vendor lock-in, which makes it difficult for cloud users to migrate to an alternative CSP because the data cannot easily be transferred to a new CSP and data shared with contacts at the old CSP may become inaccessible after the migration. To fill the gap, we presented *PortableCloud*, a framework that makes it possible to migrate a data sharing system to a new CSP. In *PortableCloud*, users can share data with contacts located at the same or at different CSPs. *PortableCloud* provides mechanisms to store the data in an encrypted manner.

We discussed the cost of migrating a portable cloud and various aspects, necessary for designing *PortableCloud*. We described a migration agent that assists users in automatically finding a suitable CSP that could satisfy their needs.

As future work, we plan to complete the implementation of *PortableCloud*. Furthermore, investigating accountability aspects of portable clouds would be an interesting research direction.

References

1. Capitani, D., di Vimercati, S., Foresti, S., Jajodia, S., Paraboschi, S., Pelosi, G., Samarati, P.: Preserving confidentiality of security policies in data outsourcing. In: WPES 2008, pp. 75–84 (2008)
2. Joint, A., Baker, E., Eccles, E.: Hey, you, get off of that cloud? Comput. Law Secur. Rev. **25**, 270–274 (2009)
3. Armbrust, M., Fox, A., Griffith, R., Joseph, A.D., Katz, R., Konwinski, A., Lee, G., Patterson, D., Rabkin, A., Stoica, I., Zaharia, M.: A view of cloud computing. Commun. ACM **53**, 50–58 (2010)
4. De Chaves, S., Uriarte, R., Westphall, C.: Toward an architecture for monitoring private clouds. Commun. Mag. **49**, 130–137 (2011). IEEE
5. Satzger, B., Hummer, W., Inzinger, C., Leitner, P., Dustdar, S.: Winds of change: from vendor lock-in to the meta cloud. IEEE Internet Comput. **17**, 69–73 (2013)
6. Hacigümüş, H., Iyer, B., Mehrotra, S.: Ensuring the integrity of encrypted databases in the database-as-a-service model. In: Data and Applications Security XVII, vol. 142, pp. 61–74 (2004)
7. Asghar, M.R., Ion, M., Russello, G., Crispo, B.: Securing data provenance in the cloud. In: Open Problems in Network Security. LNCS, vol. 7039, pp. 145–160 (2012)

8. Asghar, M.R., Russello, G., Crispo, B., Ion, M.: Supporting complex queries and access policies for multi-user encrypted databases. In: CCSW 2013, pp. 77–88 (2013)
9. Ferretti, L., Colajanni, M., Marchetti, M.: Distributed, concurrent, and independent access to encrypted cloud databases. Parallel Distrib. Syst. **25**, 437–446 (2014)
10. Josefsson, S.: PKCS#5: password-based key derivation function 2 (PBKDF2) test vectors. Technical report (2011)
11. Asghar, M.R.: Privacy Preserving Enforcement of Sensitive Policies in Outsourced and Distributed Environments. Ph.D. thesis, University of Trento (2013)
12. Garfinkel, S.: PGP: pretty good privacy (1995)
13. Burr, W.E., Nazario, N.A., Polk, W.T.: A proposed federal PKI using X.509 v3 certificates. NIST (1996)
14. Sandhu, R., Samarati, P.: Access control: principle and practice. Commun. Mag. **32**, 40–48 (1994). IEEE
15. Zhao, G., Rong, C., Li, J., Zhang, F., Tang, Y.: Trusted data sharing over untrusted cloud storage providers. In: Cloud Computing Technology and Science (CloudCom), pp. 97–103 (2010)
16. Ward, C., Aravamudan, N., Bhattacharya, K., Cheng, K., Filepp, R., Kearney, R., Peterson, B., Shwartz, L., Young, C.: Workload migration into clouds challenges, experiences, opportunities. In: Cloud Computing (CLOUD), pp. 164–171 (2010)
17. Khajeh-Hosseini, A., Sommerville, I., Bogaerts, J., Teregowda, P.: Decision support tools for cloud migration in the enterprise. In: Cloud Computing (CLOUD), pp. 541–548 (2011)
18. Stefanov, E., van Dijk, M., Shi, E., Fletcher, C., Ren, L., Yu, X., Devadas, S.: Path ORAM: an extremely simple oblivious ram protocol. In: CCS 2013, pp. 299–310 (2013)
19. Goldreich, O., Ostrovsky, R.: Software protection and simulation on oblivious RAMs. J. ACM **43**, 431–473 (1996)
20. Gong, X., Kiyavash, N., Borisov, N.: Fingerprinting websites using remote traffic analysis. In: Proceedings of the 17th ACM Conference on Computer and Communications Security, CCS 2010, pp. 684–686 (2010)
21. Cabrera, L.F., Jones, M.B., Theimer, M.: Herald: achieving a global event notification service. In: 2001 Proceedings of the Eighth Workshop on Hot Topics in Operating Systems, pp. 87–92. IEEE (2001)
22. Cooper, B.F., Ramakrishnan, R., Srivastava, U., Silberstein, A., Bohannon, P., Jacobsen, H.A., Puz, N., Weaver, D., Yerneni, R.: PNUTS: Yahoo!'s hosted data serving platform. Proc. VLDB Endow. **1**, 1277–1288 (2008)
23. Pal, P., Lauer, G., Khoury, J., Hoff, N., Loyall, J.: P3S: a privacy preserving publish-subscribe middleware. In: Narasimhan, P., Triantafillou, P. (eds.) Middleware 2012. LNCS, vol. 7662, pp. 476–495. Springer, Heidelberg (2012). doi:10.1007/978-3-642-35170-9_24
24. Zhao, J.F., Zhou, J.T.: Strategies and methods for cloud migration. Int. J. Autom. Comput. **11**, 143–152 (2014)
25. Menzel, M., Ranjan, R.: CloudGenius: decision support for web server cloud migration. In: WWW 2012, pp. 979–988 (2012)
26. Ng, W.S., Ooi, B.C., Tan, K.L., Zhou, A.: PeerDB: a P2P-based system for distributed data sharing. In: Data Engineering, pp. 633–644 (2003)
27. Takabi, H., Joshi, J.B., Ahn, G.J.: Security and privacy challenges in cloud computing environments. Secur. Priv. **8**, 24–31 (2010)

28. Khan, S., Hamlen, K.: AnonymousCloud: a data ownership privacy provider framework in cloud computing. In: Trust, Security and Privacy in Computing and Communications (TrustCom), pp. 170–176 (2012)
29. Zarandioon, S., Yao, D., Ganapathy, V.: K2C: cryptographic cloud storage with lazy revocation and anonymous access. In: Security and Privacy in Communication Networks, vol. 96, pp. 59–76 (2012)
30. Bessani, A., Correia, M., Quaresma, B., André, F., Sousa, P.: DepSky: Dependable and Secure Storage in a Cloud-of-clouds. In: EuroSys 2011, pp. 31–46 (2011)
31. Butoi, A., Tomai, N.: Secret sharing scheme for data confidentiality preserving in a public-private hybrid cloud storage approach. In: UCC 2014, pp. 992–997 (2014)
32. Liu, X., Zhang, Y., Wang, B., Yan, J.: Mona: secure multi-owner data sharing for dynamic groups in the cloud. Parallel Distrib. Syst. **24**, 1182–1191 (2013)
33. Sandhu, R.S., Coyne, E.J., Feinstein, H.L., Youman, C.E.: Role-based access control models. Computer **29**, 38–47 (1996)
34. Wang, G., Liu, Q., Wu, J., Guo, M.: Hierarchical attribute-based encryption and scalable user revocation for sharing data in cloud servers. Comput. Secur. **30**, 320–331 (2011)
35. Farb, M., Lin, Y.H., Kim, T.H.J., McCune, J., Perrig, A.: SafeSlinger: Easy-to-use and secure public-key exchange. In: MobiCom 2013, pp. 417–428 (2013)
36. Asghar, M.R., Ion, M., Russello, G., Crispo, B.: ESPOON: enforcing encrypted security policies in outsourced environments. In: The Sixth International Conference on Availability, Reliability and Security, pp. 99–108. IEEE Computer Society (2011)
37. Asghar, M.R., Russello, G., Crispo, B.: E-GRANT: enforcing encrypted dynamic security constraints in the cloud. In: Future Internet of Things and Cloud (FiCloud), pp. 135–144 (2015). Special Track on Security, Privacy and Trust
38. Muhammad, R.A., Mihaela, I., Giovanni, R., Bruno, C.: ESPOON$_{ERBAC}$: enforcing security policies in outsourced environments. Comput. Secur. (COSE) **35**, 2–24 (2013). Elsevier
39. Raymond, J-F.:Traffic analysis: protocols, attacks, design issues, and open problems. In: Designing Privacy Enhancing Technologies, pp. 10–29. Springer (2001)

Cost Analysis Comparing HPC Public Versus Private Cloud Computing

Patrick Dreher$^{(\boxtimes)}$, Deepak Nair, Eric Sills, and Mladen Vouk

Department of Computer Science, P.O. Box 890, Raleigh, NC 27606, USA
{padreher,dnair,edsills,vouk}@ncsu.edu

Abstract. The past several years have seen a rapid increase in the number and type of public cloud computing hardware configurations and pricing options offered to customers. In addition public cloud providers have also expanded the number and type of storage options and established incremental price points for storage and network transmission of outbound data from the cloud facility. This has greatly complicated the analysis to determine the most economical option for moving general purpose applications to the cloud. This paper investigates whether this economic analysis for moving general purpose applications to the public cloud can be extended to more computationally intensive HPC type computations. Using an HPC baseline hardware configuration for comparison, the total cost of operations for several HPC private and public cloud providers are analyzed. The analysis shows under what operational conditions the public cloud option may be a more cost effective alternative for HPC type applications.

Keywords: High performance cloud computing · Economic analysis · Public cloud · Private cloud

1 Introduction

Over the past ten years cloud computing technology has capitalized on advances in computing hardware, storage and network technologies. These advances have helped produce new cloud architectures and software environments capable of supporting a variety of cloud computing "pay-as-you-go" service model options. Early economic analysis of the public cloud price points for general purpose applications showed potential savings to move certain types of applications from the private IT to public cloud providers. These early cost comparisons generated interest from industry, government and academia in applying these "pay-as-you-go" service model within their organizations. If such a business paradigm could be implemented, it would allow an organization to convert costly capital line items to less expensive operating costs on their balance sheet while simultaneously continuing to meet some of their users' computational needs and requirements. Chen et al. [1] investigated the overall performance of private clouds for regular or small scale commercial applications but which are not necessarily

© Springer International Publishing AG 2017
M. Helfert et al. (Eds.): CLOSER 2016, CCIS 740, pp. 294–316, 2017.
DOI: 10.1007/978-3-319-62594-2_15

computationally intensive and Walker [2] has done detailed calculations of the total cost of a CPU hour.

In recent years public cloud providers greatly expanded the number and type of public cloud computing hardware configurations along with the base and incremental pricing options for computation, storage and network transmission of outbound data from their cloud computing facilities. This expanded set of choices and options has prompted several new studies and analyses of the public versus private cloud question, taking into account these expanded public cloud hardware platforms and pricing options for enterprise/business based cloud implementations. Many of these new studies have expanded on the original basis for cloud implementation of applications characterized by extremely low system utilization, highly dynamic user demand requirements and critical response times. With these new advances in cloud hardware architectures and software environments the application of cloud systems to areas of scientific and other high performance computing (HPC) type applications needs to be re-analyzed.

Several high performance computing studies compared the difference in performance for an HPC job run on a private HPC cluster and different public cloud provider platforms such as Amazon EC2 [3–5] or Microsoft Azure [6]. Other studies have suggested ways to improve the usage of cloud computing features for the benefit of HPC applications [7] and schedulers [8]. There has also been related work [9,10] specifically focused on how the Infrastructure as a Service (IaaS) cloud option performed when handling scientific workflow simulation environments. While the IaaS study was not specifically confined to analyzing computational performance for HPC clouds, the analysis of scientific workflow simulation environments for Infrastructure as a Services has a similar type of tradeoff criteria between the observed performance and incurred monetary cost for analyzing HPC cloud computing options.

This paper focuses on the cost analysis for HPC cloud computing implementations. A test case cost comparison with a given HPC hardware baseline configuration is established to determine under what conditions will a public cloud rather than a private cloud be more cost effective for computations, storage and network data transfers for HPC type applications. The paper is organized as follows. Section 2 describes the design of a common HPC baseline configuration for a cloud computing system that will serve as a common platform when comparing HPC options offered by various public and private cloud providers. Section 3 summarizes the various pricing models offered by this selected set of cloud providers. Section 4 provides a detailed analysis for the various computation pricing options among the selected private and public cloud providers. Section 5 reviews the detail analysis of the storage options and Sect. 6 summarizes the network transfer pricing options among these private and public cloud providers. Finally, Sect. 7 offers some comments, observations and a summary as to how these costs impact decisions as to when and where it may be advantageous to utilize public and private cloud computing platforms for HPC type applications.

2   The HPC Baseline Configuration

Cloud computing systems have successfully been adopted to service some types of applications characterized by extremely low system utilization, highly dynamic user demand requirements and critical response times. However, these characteristics do not always describe the properties of scientific and other high performance computing (HPC) applications. This paper is focused on examining cost comparisons among several public and private cloud computing providers using a typical HPC type baseline hardware configuration and under what conditions do the cloud computing costs favor a public or private cloud option.

As a starting point, it is important to define this baseline hardware configuration so that the analysis can be standardized when comparing the performance levels across different cloud providers. The baseline configuration must be sufficiently robust to handle the more computationally intense high performance computing demands and requirements and must be capable of effectively delivering similar expected service levels and productivity when compared against these same applications run on a supercomputer.

The detailed technical aspects of utilizing cloud computing for HPC applications has been covered elsewhere [11,12] and will not be the primary focus here. It is assumed that both the public and private cloud vendors have fully operational cloud systems that can deliver access to HPC level resources and services throughout the entire time interval selected for comparison.

Table 1 lists the types of performance characteristics and capabilities that one would expect from a cloud computing provider offering hardware platforms suitable for HPC and related computationally intensive applications. It was decided to use the parameters listed in this table as the baseline standard when comparing hardware characteristics among the systems supported by the various cloud vendors. These characteristics reflect the specifications of the Intel Xeon E5 2650 v3 processor. This baseline hardware system consists of two CPU sockets on the motherboard each connected to a chip with 10 cores. The baseline total physical

Table 1. Processor baseline specifications.

Number of cores	10
Number of threads	20
Operating frequency	2.3 GHz
Max turbo frequency	3.0 GHz
L2 cache	10×256 KB
L3 cache	25 MB
Manufacturing tech	22 nm
64-bit support	Yes
Hyper-threading support	Yes
Virtualization technology support	Yes

core count for these systems will be 20 with an option to enable hyperthreading so that each physical processor can virtualize two cores. Finally, these needs to be some assumption for local storage and network connectivity. For this baseline comparison it will be assumed that there is a total RAM of 160 GB available and a 10 GB network connection to provide support for I/O.

3 Cloud Provider Pricing Models

The baseline configuration sets a standard for comparison across all vendors and cloud providers studied in this report. It should be noted that some of the public cloud vendors offer other choices and selections for accessing HPC cloud instances that do not align with their baseline configuration proposed for these comparisons. However, these other choices proved to be more expensive than the option of scaling the per unit hardware and then analyzing and comparing the cost effectiveness for each vendor's cloud instance.

3.1 Public Cloud Options

There are numerous cloud vendors today offering various service options and hardware configurations. The public cloud vendors selected for comparison in this analysis include Amazon [13], Google [14], and Microsoft Azure [15].

Amazon AWS. The Amazon Elastic Compute Cloud (Amazon EC2) is a commercial cloud computing vendor offering a wide selection of different types of options and configurations. These instances are spread across various categories including General Purpose, Memory Optimized, Storage Optimized, and Clusters with options for GPUs with previous or current generation processors.

For the purpose of analyzing the different vendor choices, the M4.10xLarge configuration was selected because it best matches the common baseline configuration. The instance has 40 Virtual Cores with 160 GB of RAM with 10 GB interconnect[1]. It has a support for Elastic Block Storage (EBS) and so the I/O for the EBS will be dedicated and will not interfere with the connection to the Internet.

Among the various public cloud vendors Amazon offers some of the most complex pricing structures for cloud computing within the commercial cloud computing marketplace. Pricing structure offerings include:

– <u>On Demand</u> - Reservations initiated by the hour whenever required and can be terminated when the work is done. These reservations will be charged based on the total number of hours used
– <u>Reserved</u> - Reservations for up to a year with monthly payments required for the entire duration of the reservation.

---

[1] The 'M' in M4.10xLarge denotes the General Purpose intention of the instance and '4' indicates that the processors are current generation processors. Each virtual CPU is a hyperthread of an Intel Xeon core.

- Fully Prepaid Reservation - Payment at the outset for the entire duration independent of the level of monthly usage for that prepaid cloud instance. If the use of the reservation is terminated within the prepaid duration, the unused resources can be sold by the user on Amazon marketplace to other bidders.

  One Year - One year reservation with full amount paid at the outset
  Three Years - Three year reservation with full amount paid at the outset
- Partially Pre-paid Reservation - Half of the full amount paid at the outset with hourly rate based on a discounted amount and remaining amount due paid in monthly installments.

  One Year - One year reservation with partial amount paid at the outset
  Three Year - Three year reservation with partial amount paid at the outset
- Spot Pricing - Bid price submitted by users on an open market basis for unallocated resources with no guarantee of the computation time user will be able to secure through a spot reservation. The user with the highest bid will be awarded the reservation on the unused cloud resource.[2]

The On-Demand option provides the most flexibility for scheduling access to cloud resources but that convenience and flexibility also makes it the most expensive option. At the present time the M4.10xLarge has a default quota of 5 instances when booked On-Demand and 20 instances when Reserved. The quota can be increased by contacting the customer service department. Additional servers in the region may be obtained depending on availability at the time of the request.

Users who know at the outset that they will need full utilization of cloud computing resources over the course of a year can save money by electing to purchase a fully pre-paid one year reservation versus the on-demand option. If after purchasing the fully pre-paid one year reservation option, the user determines the entire one year block of reserved computation time will not be needed, the excess time can be sold in the Amazon public cloud marketplace. The seller can set a one-time fee that for the instance and can also pro rate the price depending on the remaining time available if the instance goes unsold for some time. Conversely, if the user determines that more computing capacity will be needed in that one year time window, more computing capacity can be added dynamically to a reservation by either buying from Amazon or from other users who are selling their extra computation time in the marketplace.

The 1 year and 3 year reserved prepaid options show cumulative cost jumps at the appropriate yearly anniversary dates while the 1 year and 3 year partial prepaid options show smaller incremental jumps at the appropriate yearly anniversary dates and upward sloping cumulative costs over time. These various pricing plans provide the users with options to manage their operational costs

---

[2] The spot pricing is the cheapest way one can secure access to cloud resources. However if another user initiates a reservation instance in the midst of the spot reservation, the spot reservation gets preempted for the reserved instance and the usage will not be charged to the spot reservation user.

and cash flows but also illustrate the growing complexity in properly choosing the most cost effective pricing option for their cloud computing services requirements.

Amazon also supports an additional option for High Performance Computing jobs by providing compute Intensive Virtual Machines guaranteeing low latency in intra-network communication among virtual machines, but are more expensive and have less RAM per core than the baseline configuration[3].

Google Cloud. Google Cloud Engine provides users with services for creating and running virtual machines to utilize the compute power of Google Infrastructure. The types of instances they offer are standard, high memory and high CPU instances. These options have various sizes ranging from 1 virtual core with 1 GB RAM to 32 virtual core machines with 208 GB of RAM.

Google also allows users to create custom machines. A custom machine is a configuration allows both the number of virtual CPUs and the total RAM for the instance to be specified by the user. The cost for the instance will be the sum of the costs for the individual CPUs and RAM based on the rate per CPU and rate per GB of RAM specified by Google. The user can specify 1 vCPU or any even number of vCPUs from 2 to 32 with a maximum of 6.5 GB of RAM per vCPU.

For the purpose of the comparison in this paper an instance which confirmed with the baseline configuration was required. There is no single instance allowed by Google which provided the required configuration. Hence two custom instances were used with 20 vCPUs and 80 GB of RAM for each instance. The custom instance constructed consisted of 40 vCPUs[4] and 160 GB of RAM. The custom instance option can help in reducing the total cost of the instances considering that if a smaller instance is required which is not provided by Google pre configured, the required instance can be created using the custom instance

Google prices the level of usage for the instance to the minute for which access was provided, regardless of whether or not the resource was actually used during that reservation time. Google does not provide HPC centric cluster machines, and their compute intensive machines are very low on RAM with only 0.98 GB of RAM per virtual core.

Microsoft Azure. Microsoft, along with a host of cloud services, allows Windows and Linux Virtual Machines to be created and managed on Microsoft Azure Cloud. Azure has many types of instances like General Purpose, Compute Optimized, Memory Optimized and Intranet work Optimized with Infiniband support. For the purpose of comparison to the baseline configuration, the instance

---

[3] For instance the most powerful Compute Intensive instance provided by Amazon is C4.8xLarge which has 32 virtual cores and 60 GB of RAM.

[4] A virtual CPU is equivalent to a single hyperthread on a 2.6 GHz Intel Xeon E5 (Sandy Bridge), 2.5 GHz Intel Xeon E5 v2(Ivy Bridge), or 2.3 GHz Intel Xeon E5 v3 (Haswell) depending on the processor which makes up the instance.

of Compute and Memory Optimized D15 v2 was chosen because it provided 20 cores and 140 GB of RAM. Thus two instances of the same configuration will have 40 cores and 280 GB of RAM with the latest processor. The instance also provides a 1000 GB of standard storage attached to the node. In Azure each virtual core is equivalent to one hyperthread in the physical core of the machine. Microsoft rounds the amount of time the instance is used to minutes.

Microsoft supports multiple payment options for Azure services (Pay-as-you-go, Prepaid, Microsoft Resellers, and Enterprise Agreements). The pay-as-you-go option allows the user to create on demand reservations with costs calculated on a per minute usage basis. The prepaid reservation is for 12 months and the full amount must be paid regardless of usage pattern during that time period. Microsoft Resellers provide Open License Keys which can be bought and used to access the Azure infrastructure for reserving an instance. Enterprise Agreements involve users making commitments for large amounts of compute minutes in return of a discounted rate.

Microsoft also supports compute intensive virtual machines which have infiniband support and ones without infiniband support which are specific for HPC [16]. They also have closer proximity and hence are more tightly coupled than regular virtual machines which helps HPC applications. However they are also more expensive than the HPC baseline configuration outlined in this paper.

3.2   Private Cloud Option

The hardware and operational pricing for a hypothetical private cloud computing configuration were based on the costs for a proven and tested production level private cloud architecture called the Virtual Computing Laboratory (VCL) [17–20]. For the purpose of building a baseline configuration for fair comparison with the respective public cloud options, an Intel Xeon E5 2650 v3 blade was chosen. This motherboard has two CPU sockets each connected to a chip with 10 cores. The chips have hyperthreading enabled and so the each physical core can virtualize two cores. This provides a total of 40 virtual cores, similar to the public cloud hardware options. The blade comes with a RAM of 128 GB and by adding a 32 GB RAM stick, the total RAM becomes 160 GB and is equivalent to the public cloud hardware configuration. There is also a fast intra-network connection between the cores similar to the public cloud configuration.

4   Cost Analysis of HPC Cloud Computation Options

When reserving an HPC instance there are two major types of reservations from which to choose. All public and private cloud providers offer some form of the guaranteed reservation option. Some cloud providers also offer options to competitively bid for any unused cloud resources in an open market system. Both of these choices offer both advantages and drawbacks in terms of the users' needs, requirements and time constraints for accessing HPC cloud computation resources.

4.1   HPC Cloud Guaranteed Reservation Options

The public cloud providers examined here are Amazon, Google and Azure. The Virtual Computing Laboratory is the private cloud option. For the guaranteed reservation option the user's job will run to completion, although the time required to complete may vary depending on the prioritization level, with higher prioritization options costing more money. With each provider, a cloud option was selected that best matched the HPC baseline cloud configuration discussed in Sect. 2. In order to make a quantitative comparison from among the many choices offered by HPC public cloud providers that can satisfy our example HPC baseline configuration requirements, it is essential to gather and enumerate in the next subsections the detailed pricing structures and options offered by each HPC public cloud provider.

Amazon AWS. For the compute instance which is selected (M4.10xLarge), the configuration matches the Baseline Configuration and so only one of the instances needs to be reserved.

- The hourly rate for the On-Demand reservation of this instance is $2.394.
- For the one year reservation option, the effective discounted hourly rate is $1.645.
    If the partial amount is prepaid at the outset, the total effective hourly rate becomes $1.406. The prepaid amount of $6158.28 is initially costed and the hourly rate of $0.703 is charged for all reservations made throughout the year.
- For the Fully paid reservation, the full amount of $12071.28 is costed immediately at the time of the first reservation.
- The 3 year reservations.
    For a partial prepayment plan, $12483 is initially costed and the remaining amount is paid in monthly payments of hourly rate $0.475. The effective hourly rate becomes $0.95.
    The effective hourly rate for the 3 year full prepaid reservation is $0.893 with a total amount of $23468.04 is costed immediately at the time of the first reservation.

Figure 1 shows a graph of cost versus time for these various Amazon pricing plans for cloud computing instances. This figure graphically illustrates the complexities in picking the correct pricing option for the users' time dependent computational requirements.

Google Cloud. For the custom instance that was chosen, multiplying the quantity of virtual cores and RAMs with their respective rates provided by Google and adding them together will give the hourly rate of the custom instance. Google provides built-in incentives in their pricing model. The more the amount of usage in a month, the more the discount is applied to the rate. For our example the full rate for the month with minimum utilization was calculated to be $2.1456/h,

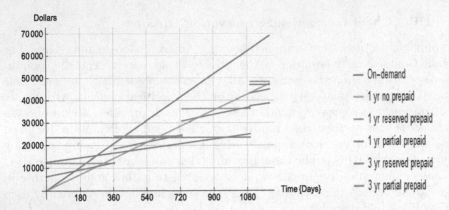

Fig. 1. Cumulative costs versus time for various Amazon cloud computing pricing plans.

Table 2. Google rates for vCPU and GB memory.

Item	$/h full usage	Quantity	$/h full usage
vCPU	0.02444	40	0.9776
Memory (GB)	0.00328	160	0.5248

while for sustained utilization of 100%, the effective monthly rate went down to $1.5024/h (Table 2).

The itemized list below summarizes the pricing structure.

- In a month the usage of 25% of total time is at the full rate of the instance which is $2.1456/h.
- The second 25% is at a discount of 20% in the full rate of the instance which turns out to be $1.7164/h.
- The third 25% is at a rate of $1.2874 which is a discount of 40% of the full rate of the instance.
- And the last 25% at 60% discount of the full rate of the instance which is $0.8582/h.
- If an instance is used for the entirety of a month, an effective discount hourly rate would be $1.5012/h which gives an effective discount of 30% on the actual rate by the end of the month.

Microsoft Azure. Microsoft's Azure provides an instance D15 v2 which has the closest resemblance to the baseline configuration. The D series of virtual machines have an additional feature of Solid State Drives attached to the machine directly which gives high speed Disk IO. The storage of the SSD is limited. The v2 series is the advanced version of the D series with more powerful, latest generation Intel processors coupled with Intel Turbo Boost Technology which can increase the speed of the processors. The hourly rate of one instance is $1.853. Since two instances of D15 v2 are considered for matching the baseline configuration, the hourly rate becomes twice the rate of one instance which is $3.706. This instance comes with 140 GB of RAM and 1000 GB of Solid State Drive support.

The high rates charged by Azure reflect the addition of Solid State Drives and isolation to dedicated hardware for a customer. Both of these additions help increase the speed of the data throughput because of data proximity and high IO and enhance the processing of HPC and compute intensive jobs.

Private Cloud Option. Using historical costs in the private VCL cloud system the following hardware was selected and costs assigned to the baseline configuration defined in this paper. An Intel Xeon E5 2650 v3 blade cost ($5,500) has two CPU sockets on the motherboard each connected to a chip with 10 cores which brings the total physical core count to 20. The chips are hyperthreading enabled so the each physical core can virtualize two cores. Thus the virtual core count is 40 cores. The blade comes with a RAM of 128 GB. Adding the cost of a 32 GB RAM stick ($200) makes the total RAM to be 160 GB which confirms to the baseline configuration. The blades will require a rack cost ($2,000) for its installation. Considering that the blade will be added to an existing and working data center, the entire cost of the rack need not be borne. The total cost of the rack will be divided by the number of blades the rack can house. In this instance the rack held 84 blades, giving a proportionate cost of $23.81 for each blade.

The blade, RAM and the rack costs will be capital costs which will be paid one time and will be amortized over three years. The rack space rent ($4,380) will be a recurring fee, a fraction proportionate to the one blade ($52.14) needs to be borne by the user. The blades will also require power supply which also would be divided by the total number of blades housed by the data center. Each rack will require power circuits to deliver the power. That needs to be bought or paid for in proportionate manner. Each rack could use two 60 A circuit (rent $3,480 each) or four 30 A circuits (rent $1,980 each). In case of the example in consideration, four 30 A circuits (total rent $7,920) are being used which

Table 3. Private cloud baseline costs.

Equipment	Price	Price per blade
Capital costs Intel Xeon E5 2650 v3 blade	$5,500	-
32 GB RAM	$200	-
Rack	$2,000	$23.81
Switch	$20,000	$238.1
SFP	$200	$200
Operational costs rack space rent	$4,380	$52.14
Four 30 A circuits	$7,920	$94.29
Developer	$100,000	$100

ultimately cost more than two 60 A circuit (total rent $6,960). So for one blade's power usage a proportionate fee (rent - $94.29) needs to be paid by the user. The costs to outfit a private cloud instance are summarized in Table 3.

4.2   Comparison of Cloud Computation Reserved Options

Figure 2 shows a graph of the accumulated cost versus time the for on-demand cloud computing access option offered by Amazon, Google, Azure and a private cloud configuration using the HPC baseline hardware configuration referenced in this paper. This sample figure illustrates that for short periods of time the cost for each public cloud option is less than the cost for the private cloud option. However, if there is demand for HPC type cloud computations for extended periods of time, then at some point the cumulative cost for each public cloud intersects and exceeds the cumulative costs for the private cloud. These calculations show that private cloud costs for sustained cloud computing usage are more cost effective over the longer term in a private cloud implementation than in the public cloud sector. However, the cost of operations is only one component in determining the economic effectiveness when comparing public versus private clouds for HPC applications. This will be discussed in more detail in Sect. 7 summarizing HPC cloud computing planning based on cost analysis.

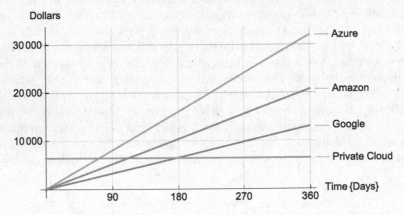

Fig. 2. Azure, Amazon, Google, and private cloud on-demand cumulative costs versus time.

4.3   HPC Cloud Spot Market Options

In 2009 Amazon was the first cloud vendor to introduce the concept of a "spot market" into the mix of options for purchasing cloud services [21]. The basic idea is to allow users to bid on unused CPU cycles under the constraint that their cloud instance would run as long as their bid price exceeded the current spot price. Users can place a bid for the highest amount they wish to pay for usage of their cloud facility instances. Whenever the Amazon current price for a Spot Instance (SI) is equal to or less than the user's bid, access to those cloud computing facilities is provided to the user. As long as the user's bid price is higher than the instance price, the user gets access to the instance at the market rate of that instance. However, if the spot price increases and/or other users bid prices that are higher than the user's bid price, the instance that was allocated to the user gets preempted and the user does not pay for their partial usage of the instance.

This type of purchasing option has attracted attention within the cloud computing community. Several projects have experimented with this spot market option working from the assumption that the spot market operates on the basis that spot prices are set through a uniform price, sealed-bid, market-driven auction [22,23]. The price for the instance increases or decreases in time depending on the supply and demand for each type of instance.

The argument put forward is that by analyzing the spot market for the historical trends, an informed bid can be placed that has a high probability of giving the user the maximum amount of time for that spot instance without having the instance preempted by higher bids. Those interested in using the spot market for general and commercial computing have studied these historical trends. Mazzucco and Dumas [24] analyzed the spot market and made observations on how to achieve maximum availability on the spot instances and Mattess, Vecchiola and Buyya [25] have published an analysis of the economic benefits of the practice of leasing these spot instances. Although these techniques have improved the success rate for accessing the spot market, there are occasions when the spot price goes even above the reservation price for that instance and preemption of the instance is unavoidable. For these type of situations other techniques such as checkpointing and migration have been suggested as alternate strategies [26].

However, not everyone in the cloud community subscribes to this view of how the cloud spot market operates. There have been studies that suggest that the algorithm for the fluctuation of the spot price of each instance type in this market is entirely controlled by Amazon and that the the spot price is set according to a constantly changing reserve price that discounts actual client bids [27]. In contract Bonacquisto et al. [28] argued that the Amazon spot market was more aligned with the method of resource maximization through allocations and instead proposed a model for a procurement auction market as an alternative to maximize the utilization rate for a provider's data center resources.

Independent of these contrasting ideas, this work has extended the concept of the spot market to HPC instances and has investigated whether spot market strategies and methods can benefit HPC applications. A first step in determining whether there are other strategies that may help maximize the economic benefits for accessing these cloud computing resources is to gain access to the SI bid prices for each region and availability zone. The M4.10xLarge General purpose baseline configuration was used as the basis for this spot market study. Amazon currently provides historical spot price data for this configuration in both the U.S. East and West regions on a three month rolling basis. To obtain this historical data a python code was written using the Boto Python Library [29,30] to connect with the AWS API [31] and extract the M4.10xLarge General purpose spot pricing data.

The hourly spot prices for the M4.10xLarge General purpose cloud instance from 7th May, 2016 to 18th July, 2016 were downloaded for the US East Region, availability zones 1B and 1C, and for the US West Region, availability zones 1B, 1C, 2A and 2B. The spot price data within each availability zone was verified to be free of misprints or corrupt data. Autocorrelation calculations in each availability zone determined that the daily bid prices were uncorrelated. Using the set of hourly prices spanning a 24 h period, the full correlation matrix for all of the availability zones was calculated for each day between 7th May, 2016 to 18th July, 2016. It was observed that the correlations among availability zones seemed to vary over time.

Among all of the different combinations of regions and availability zones there were certain pairs of availability zones that showed either strongly correlated or strongly anti-correlated correlation matrix values as a function of time during that period. As shown in Fig. 3 it was observed that the spot pricing correlation matrix values for the W1B/W1C and the W2A/W2B availability zones were highly correlated for large portions of time between the May 7th through July 18th time frame. In contrast, Fig. 4 shows periods of strongly anti-correlated spot pricing for the W1B/W2B, W1C/W2A and W1C/W2B availability zones.

These anti-correlated spot pricing patterns between different pairwise combinations of availability zones were unexpected. It should be noted that these anti-correlations do not occur at regular or periodic intervals and so they cannot be anticipated in advance. As a result, in order to potentially take advantage of these favorable anti-correlations in spot pricing positions in the marketplace, the spot price market must be regularly and closely monitored. Regardless of whether or not the spot market operates as a true auction market or if the spot price is controlled by Amazon based on a changing reserve price approach, if these hourly changes and daily spot pricing correlations are carefully monitored, it may be possible to minimize spot pricing costs by migrating applications needing HPC level baseline hardware platforms between different availability zones

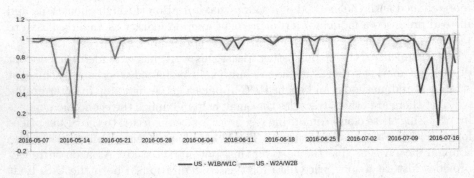

Fig. 3. Correlation versus time showing positive correlations for specific pairwise Amazon availability zones.

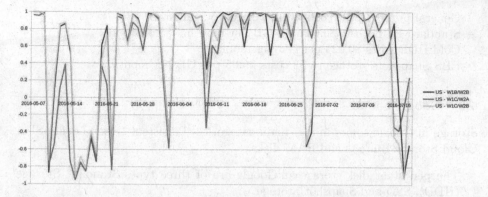

Fig. 4. Correlation versus time showing intervals of negative correlation for specific pairwise Amazon availability zones.

that are highly anti-correlated during a given time frame. The success of such a program will highly depend on the type of HPC calculations being performed. If the application requires that large quantities of data be pre-staged in a particular region before the application is launched, then it is not practical to attempt to switch availability zones between different regions. However, if the HPC type application is compute bound with little to no need for access to large data sets for the computations, it may be practical to consider launching spot price bids between zones that are highly anti-correlated at a particular time frame in order to get the best probability for uninterrupted running of applications using the spot market option.

5   Cost Analysis of the Cloud Storage Options

The cloud computing storage option can be attached to the compute node itself or located at a remote physical location. The storage options range from slow and cheap archival storage up through options for more expensive solid state drives.

5.1   Amazon AWS

The data associated with EC2 compute nodes must have capabilities to be stored and accessed as needed. The Amazon instances come with attached storage per instance or with Elastic Block Storage (EBS) support. The I/O from the EBS to EC2 has dedicated throughput between 500 to 4000 Mbps. This dedicated throughput minimizes contention between Amazon EBS I/O and other Network I/O from the EC2. The listing below shows the various types of storage including SSDs and HDDs which are provisioned at different rates pro-rated to the hour of its provisioning.

- General purpose SSD costs \$0.10 per GB per month
- Standard HDD costs \$0.045 per GB per month.
- Cold HDD costs \$0.025 per GB per month
- EBS Snapshots storage to S3 costs \$0.05 per GB per month.

5.2   Google Cloud

Storage in Google comes in four main categories- Persistent Disks, Local SSDs, Cloud Storage Buckets and RAM disks.

- The persistent disk storage at Google are of three types Standard Storage (HDD), SSD and Snapshot Storage.
- The Standard Storage of HDD costs \$0.04 per GB per month.
- The SSD costs \$0.17 per GB per month
- The Snapshot Storage costs \$0.026 per GB per month because it is the slowest of them all.
- The maximum storage per instance is 64 TB with data replication and encryption.

Local SSD are physically attached to the machines that are used and hence are the fastest to access. The cost is \$0.218 per GB per month and has a limit of 3 TB. The data is not replicated but it is encrypted. The Cloud Storage Buckets are the slowest and cost anywhere from \$0.01 to \$0.026 per GB per month and provide replication and encryption. The RAM Disks are the extensions for the RAM for in memory storage of files. They are the most expensive types of storage costing between \$3.37 to \$3.71 per GB per month with a limit of 208 GB per instance.

5.3   Microsoft Azure

Azure provides standard disks for storage in multiple forms of data redundancy that include Local Redundancy Storage (LRS), Zone Redundancy Storage (ZRS), Geo Redundancy Storage (GRS) and Read access Geo Redundancy Storage (RA GRS). Azure storage option have at least 3 replications of data. LRS stores all the 3 replicas of the data in the same region as the storage account location and is the least expensive. Zone Redundancy Storage replicates the data within 2 or 3 facilities within the same region or multiple regions and it is available only for block data. Geo Redundancy replicates data 5 times into facilities which are very far away from each other with a total of 6 copies of the data. GRS costs more than LRS because of maximum redundancy. The characteristics for each data redundancy are listed below and summarized in Table 4 below.

5.4   Private Cloud

Private cloud storage options cover wide ranges of both price and performance. A capacity storage system from a top-tier vendor is used as the storage baseline.

Table 4. Microsoft Azure LRS and GRS rates per GB per month.

Storage	LRS	GRS
1 TB	$0.05	$0.095
50 TB	$0.05	$0.08
500 TB	$0.05	$0.07
1000 TB	$0.05	$0.065
5000 TB	$0.045	$0.06

This type of storage system is typically used for HPC storage where there is requirement to maximize storage capacity. The baseline storage system has 240 6 TB near-line SAS disk drives with redundant controllers supporting various attachment methods including iSCSI, fibre channel, and NFS. Raw capacity is about 1.4 PB and usable capacity is about 1.1 PB. The list price for the storage system is $1,350,000 including three years of maintenance. Discounts vary by organization and circumstance of the purchase but generally would be expected to be in range of 15–60%. Annualized capital expense over the three year warranty period is between $0.030 per GB/month and $0.014 per GB/month. Operating expense for the baseline system is about $8.60/TB/year or about $0.001 per GB/month. Overall baseline storage expense is between $0.015 and $0.031 per GB/month.

5.5   Comparison of the Cloud Storage Options

The types of storage options vary by the provider and each vendor has different pricing algorithms. The data show that Google offers the most inexpensive regular storage options such as hard drives, solid state drives and snapshot storage drives. Although Microsoft provides multi region replication for data recovery as one of the storage options, it may not be as useful for HPC jobs.

6   Cost Analysis of Cloud Network Transfer

Network transfer of data can be a potentially expensive when using public clouds. Cloud providers do not charge for network ingress of data but data extraction may incur substantial costs from public cloud providers.

6.1   Amazon AWS

Amazon provides free network transfer into its EC2 compute nodes or storage nodes. It also provides free data transfer between EC2 nodes in the same region. However Amazon charges for data transfer into another AWS region and for removal of data from their cloud system into the internet. The incremental costs for data transfer are summarized below.

- The first 1 GB of data transfer from the compute node to the internet is free each month.
- Above that, the first 10 TB will cost $0.9 per GB.
- The next 40 TB of data transferred out will cost $0.085 per GB.
- The next 100 TB of data transfer in a month will cost $0.07 per GB.
- Data transfer between 150 TB and 500 TB will cost $0.05 per GB per month.
- If there is more than 500 TB of data transferred per month from the EC2 compute node to the Internet, the customer service should be called to get a better deal on the data.

6.2 Google Cloud

The data transfer into the compute engine is free. The data transfer from Google to external Internet locations is charged in tiers listed below.

- The data transfer up to 1 TB from the compute engine to the internet has a cost of $0.12 per GB.
- The next 9 TB of data transfer has a cost of $0.11 per GB.
- All the data transfer more 10 TB has a cost of $0.08 per GB.

6.3 Microsoft Azure

Azure also does not charge for inbound data transfer to the compute nodes. Azure does have an outbound data transfer rate after the initial transfer of 5 GB of data from the compute instance to the Internet with the incremental costs for data transfer summarized below.

- For the data transfer of 5 GB to 10 TB of data, the rate is $0.087 per GB per month.
- For the transfer of next 40 TB of data the rate is $0.083 per GB per month.
- For the next 100 TB the rate is $0.07 per GB per month.
- For the next 350 TB of data from 150 to 500 TB, the rate is $0.05 per GB per month.

6.4 Private Cloud

For the network transfer of data among the blades and to the Internet, a network switch ($20,000) must be included. Assuming 84 blades per rack, the apportioned cost per blade is approximately $238 In addition, each port of the switch will require an SFP (cost - $200) for fast communicating with the blade over fiber optic cables. The SFP and the switch will be capital costs and incur a one time fee. The private cloud will also require personnel to operate and maintain the cluster and network. Salaries for personnel can vary and are apportioned to users proportionate to the number of blades the developer handles (salary - $100–$250 per blade and network connection).

6.5   Comparison of the Cloud Network Transfer Options

Table 5 gives a tabulated summary of network transfer incremental pricing for each of the public cloud providers as a function of the total size of the data to be transferred[5].

Table 5. Network transfer rate costs.

Total data transfer	Amazon AWS	Google cloud	Microsoft Azure
1 GB	Free	$0.12	Free
5 GB	$0.09	$0.12	Free
1 TB	$0.09	$0.12	$0.087
10 TB	$0.09	$0.11	$0.087
50 TB	$0.085	$0.08	$0.083
150 TB	$0.07	$0.08	$0.07
500 TB	$0.05	$0.08	$0.05

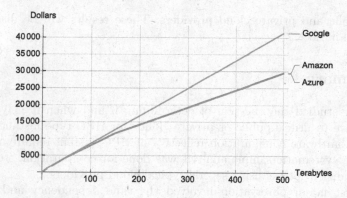

Fig. 5. Costs Charged by Cloud Providers to Transfer Data to an external location.

These data transfer costs are illustrated in Fig. 5 showing a graph of the cumulative network transfer costs for Amazon, Google, and Azure versus the amount of data transferred (in TB). To further illustrate the impact of data transfer on the cost comparisons of public versus private cloud options, a three-dimensional plot was constructed with axes of cost in dollars, time in days and data transferred in terabytes. Figure 6 shows planes in the figure representing the cumulative on-demand computation costs for a private cloud provider and an Amazon on-demand computation instance. The additional plane on the graph shows the Amazon add-on costs for transfer of data up to 50 TB over and above the Amazon computation on demand instance. These calculations show the sensitivity of adding data transfer requirements when performing a costs analysis

---

[5] The Private Cloud does not have a fee for the data transfer over the network.

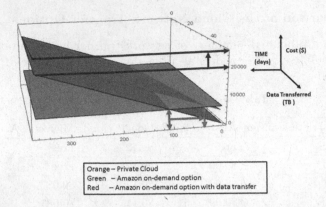

Fig. 6. Three dimensional plot showing cumulative costs for an on-demand private cloud, an Amazon on-demand public cloud and the additional add-on Amazon costs for transferring up to 50 TB of data over and above the Amazon computation on demand cost.

among public and private cloud providers. These results will be discussed in more detail in Sect. 7.

7   Summary

This paper studied the question of determining if and when it may be more cost effective to utilize public or private clouds for HPC type instances. Using a baseline hardware configuration reflective of HPC system requirements as a test case, a systematic sample analysis was done for computation, storage and network transfer options using several typical cloud providers.

The first major observation involved the time dependency and compute requirements of the HPC application in determining whether it would be more cost effective to use a public or private cloud. As discussed in Subsect. 4.2 and illustrated in shown in Fig. 2, the public cloud option is shown to be more cost effective than the private cloud for HPC type applications of short duration or intermittent need for cloud computation resources. However, as the need for extended or prolonged access to computation resources grows, at some point in time the cumulative costs for HPC applications are more cost effective being assigned to a private cloud provider.

The second major observation showed the cost sensitivity when large quantities of data must be transferred from public cloud providers to external Internet locations. As shown earlier in Subsect. 6.5 the three dimensional plot of cumulative cost, time and data transferred (Fig. 6) has strong implications for choosing a cost effective HPC cloud computing option. This plot compared the on-demand cumulative computation costs for an Amazon public cloud option and a typical private cloud HPC option. Previously discussed in Subsect. 4.2, Fig. 2 showed that the intersection point in time for a crossover of cumulative costs between an on-demand Amazon baseline instance and a private cloud instance is approximately 110 days. The arrows in Fig. 6 pointing to the time axis illustrate how the additional costs of data transfer will shorten the crossover point for determining the cost effective choice between public and private cloud computing options from approximately 110 days to approximately 40 days. The arrows in the same figure pointing to the cost axis show an additional cost of approximately $5,000 that will be incurred if the Amazon HPC baseline cloud instance user must also pay for data transfer.

The third major observation involved the introduction of the ideas of Spot Market Instances into the mix of cloud computing reservations. Subsection 4.3 studied the hourly spot prices for the M4.10xLarge sample baseline configuration in two Amazon regions and five availability zones. The results of the correlation matrix calculations showed that during the time period when the spot prices were tracked, certain pairwise correlations and anti-correlations were detected among various availability zones. This indicated that any spot price bid submitted to one zone may be highly correlated to the price in the other zone. However, there were periods of time when pairs of availability zones showed strong anti-correlated behavior of their spot instances. In effect, the spot price in one zone would be substantively different than the spot price in the other zone.

This observation has potentially interesting consequences for users dependent on the spot market for accessing the most cost effective HPC level computational resources. The strong anti-correlation signal would indicate when and where to switch availability zones and bid prices to assure the best access to the cloud resources at the most cost effective rate. By closely tracking the spot prices among the different availability zones and calculating the correlation matrix for a given hardware configuration, it might be possible to shift the workload from the availability zone with the higher spot price to the other availability zone with the lower spot price[6]. In effect, the user would be monitoring the spot market for inefficiencies in the pricing structure in order to maximize access to the public cloud resources in the most cost effective manner.

The cost analysis here shows that it is important to determine the full duration needed for a cloud computing system at the outset. If the cost analysis is focused on just the immediate short term or incremental need, then the public cloud is the option of choice. However, if the overall total duration for use of that cloud for the project ultimately extends over the longer term, then repeatedly using the public cloud for each incremental calculation can end up costing the user substantially more money than if the decision was made at the outset to commit the entire project to a private cloud.

In summary, to answer the question when it becomes more cost effective to use a private HPC cloud versus public HPC cloud requires a detailed analysis of computation, storage and network transfer costs associated with the specific HPC type hardware configuration. The evaluation is sensitive as to whether the

---

[6] This potential economic strategy may work if the user's job only involves computations and is not dependent on staging large quantities of data in a particular region and zone before submitting a bid price for access.

type of HPC application needs continuous CPU resources for long periods of time, the quantity of storage that application will require in the cloud, and the amount of data that must be ultimately be transferred from the cloud computing vendor to an external location. The sample process described here for the HPC baseline configuration would need to be repeated for the specific HPC hardware configuration platform being considered as well as the public and private cloud computing vendors being considered. It is also recognized that cost is not the only consideration when determining the most economically effective public or private cloud computing configuration for an HPC application. The throughput performance of an HPC application must also be factored into the overall analysis to determine the most economical option for an HPC application in the cloud. The authors are analyzing HPC level benchmark performance of each vendor's baseline configuration and the results will be reported in a future publication.

Acknowledgments. This work is supported in part through NSF grant 0910767, 1318564, 1330553, the U.S. Army Research Office (ARO) grant W911NF-08-1-0105 managed by the NCSU Science of Security Initiative and the Science of Security Lablet, by the IBM Share University Research and Fellowships program funding, and the Argonne Leadership Computing Facility at Argonne National Laboratory, which is supported by the Office of Science of the U.S. Department of Energy under contract DE-AC02-06CH11357. One of us (Patrick Dreher) gratefully acknowledges support with an IBM Faculty award.

References

1. Chen, Y., Sion, R.: To cloud or not to cloud? musings on costs and viability. In: Proceedings of the 2nd ACM Symposium on Cloud Computing, pp. 29:1–29:7 (2011)
2. Walker, E.: The real cost of a CPU hour. IEEE Comput. **42**, 3541 (2009)
3. Zhai, Y., Liu, M., Zhai, J.: Cloud versus in-house cluster: evaluating amazon cluster compute instances for running MPI applications. In: State of the Practice Reports, pp. 11:1–11:10 (2011)
4. Gupta, A., Milojicic, D.: Evaluation of HPC applications on cloud. In: Fifth Open Cirrus Summit, pp. 22–26 (2011)
5. Ostermann, S., Iosup, A., Yigitbasi, N., Prodan, R., Fahringer, T., Epema, D.: A performance analysis of EC2 cloud computing services for scientific computing. In: Avresky, D.R., Diaz, M., Bode, A., Ciciani, B., Dekel, E. (eds.) CloudComp 2009. LNICSSTE, vol. 34, pp. 115–131. Springer, Heidelberg (2010). doi:10.1007/978-3-642-12636-9_9
6. Ding, F., Mey, D., Wienke, S., Zhang, R., Li, L.: A study on today's cloud environments for HPC applications. In: Helfert, M., Desprez, F., Ferguson, D., Leymann, F. (eds.) CLOSER 2013. CCIS, vol. 453, pp. 114–127. Springer, Cham (2014). doi:10.1007/978-3-319-11561-0_8
7. Brandt, J., Gentile, A., Mayo, J., Pebay, P., Roe, D., Thompson, D., Wong, M.: Resource monitoring and management with OVIS to enable HPC in cloud computing environments. In: IEEE International Symposium Parallel Distributed Processing, IPDPS 2009, pp. 1–8 (2009)

8. Gupta, A., Kal, L.V.: Towards efficient mapping, scheduling, and execution of HPC applications on platforms in cloud. In: Parallel and Distributed Processing Symposium Workshops Ph.D. Forum, pp. 2294–2297 (2013)
9. Gómez Sáez, S., Andrikopoulos, V., Hahn, M., Karastoyanova, D., Leymann, F., Skouradaki, M., Vukojevic-Haupt, K.: Performance and cost trade-off in IaaS environments: a scientific workflow simulation environment case study. In: Helfert, M., Méndez Muñoz, V., Ferguson, D. (eds.) CLOSER 2015. CCIS, vol. 581, pp. 153–170. Springer, Cham (2016). doi:10.1007/978-3-319-29582-4_9
10. Saez, S., Andrikopoulos, V., Hahn, M., Karastoyanova, D., Leymann, F., Skouradaki, M., Vukojevic-Haupt, K.: Performance and cost evaluation for the migration of a scientific workflow infrastructure to the cloud. In: Proceedings of the 5th International Conference on Cloud Computing and Service Science, CLOSER 2015, p. 110. SciTePress (2015)
11. Coghlan, S., Yelick, K., Draney, B., Canon, R.S: The Magellan report on cloud computing. In: Office of Advanced Scientific Computing Research (ASCR), US Department of Energy (2011). http://science.energy.gov/~/media/ascr/pdf/programdocuments/docsMagellan_Final_Report.pdf
12. Vouk, M., Sills, E., Dreher, P.: Integration of high-performance computing into cloud computing services. In: Handbook of Cloud Computing, pp. 255–276 (2010). Chap. 11
13. Amazon High Performance Computing (2016). https://aws.amazon.com/hpc/
14. Google Compute Engine (2016). https://cloud.google.com/compute/
15. Microsoft Azure (2016). https://azure.microsoft.com/en-us/
16. Microsoft Big Compute: HPC & Batch (2016). https://azure.microsoft.com/en-us/solutions/big-compute/
17. Vouk, M.: Cloud computing issues, research and implementations. J. Comput. Inf. Technol. 16(4), 235–246 (2008)
18. Dreher, P., Vouk, M., Sills, E., Averitt, S.: Evidence for a cost effective cloud computing implementation based upon the NC state virtual computing laboratory model. In: Advances in Parallel Computing, High Speed and Large Scale Scientific Computing, vol. 18, pp. 236–250 (2009)
19. Schaffer, H.E., Averitt, S.F., Hoit, M.I., Peeler, A., Sills, E.D., Vouk, M.A.: NCSUs virtual computing laboratory: a cloud computing solution. In: IEEE Computer, pp. 94–97 (2009)
20. Apache VCL (2016). https://vcl.apache.org/
21. Amazon EC2 Spot Instances. http://aws.amazon.com/ec2/spot-instances/
22. Zhang, Q., Gurses, E., Boutaba, R., and Xiao, J., Dynamic resource allocation for spot markets in clouds. In: Workshop on Hot Topics in Management of Internet, Cloud, and Enterprise Networks and Services Hot-ICE (2011)
23. Chen, J., Wang, C., Zhou, B.B., Sun, L., Lee, Y.C., Zomaya, A.Y.: Tradeoffs between profit and customer satisfaction for service provisioning in the cloud. In: HPDC (2011)
24. Mazzucco, M., Dumas, M.: Achieving performance and availability guarantees with spot instances. In: IEEE International Conference on High Performance Computing and Communications, pp. 296–303 (2011)
25. Mattess, M., Vecchiola, C., Buyya, R.: Managing peak loads by leasing cloud infrastructure services from a spot market. In: IEEE International Conference on High Performance Computing and Communications, pp. 180–188 (2010)
26. Yi, S., Andrzejak, A., Kondo, D.: Monetary cost-aware checkpointing and migration on amazon cloud spot instances. IEEE Trans. Serv. Comput. 5(4), 512–524 (2012)

27. Agmon Ben-Yehuda, O., Ben-Yehuda, M., Schuster, A., Tsafrir, D.: Deconstructing amazon ec2 spot instance pricing. In: 2011 IEEE Third International Conference on Cloud Computing Technology and Science (CloudCom), pp. 304–311 (2011)
28. Bonacquisto, P., Di Modica, G., Petralia, G., Tomarchio, O.: Dynamic pricing in cloud markets: evaluation of procurement auctions. In: CLOSER 2014. CCIS, vol. 512, pp. 31–46 (2015)
29. GitHub Repository for Boto Python Library (2016). https://github.com/boto/boto
30. Amazon SDK for Python to access Amazon public data (2016). https://aws.amazon.com/sdk-for-python/
31. Bhatia, K.: The data science of AWS Spot Pricing (2015). https://medium.com/cloud-uprising/the-data-science-of-aws-spot-pricing-8bed655caed2#.f9w14i4iq

MING: Model- and View-Based Deployment and Adaptation of Cloud Datacenters

Ta'id Holmes[(⊠)]

Infrastructure Cloud, Deutsche Telekom Technik GmbH, Darmstadt, Germany
t.holmes@telekom.de

Abstract. For the configuration of a datacenter from bare metal up to
the level of infrastructure as a service (IaaS) solutions, currently, there
is neither a standard nor a common datamodel that is understood across
deployment automation tools. Following a model- and view-based approach, MING (明) aims at holistically describing cloud datacenters. Establishing a respective metamodel, it supports different stakeholders with
tailored views and permits utilization of arbitrary deployment tools for
providing the basic cloud service model. In addition to initial deployments, it targets (model-based) adaptation of datacenters for covering
operational use cases such as extending a cloud with additional resources
and for providing for software upgrades and patches of the deployed solutions.

Keywords: Adaptation · Cloud · Code generation · Configuration ·
Datacenter · Deployment · DSL · IaaS · MBE · Metamodel · OpenStack ·
SDN

1 Introduction

The cloud computing paradigm continues to change the way end-users consume
services and communicate (cf. [31]). At the same time service providers adapt
(*cloudify*) software services (cf. [2]) for profiting from the benefits of cloud computing (cf. [18]). Among them are the possibility to meet changing workloads
through elastic scaling, to make efficient use of resources, e.g., through multite-
nancy, an on-demand usage and pricing model, as well as defined models and
roles in regard to infrastructure as a service (IaaS), platform as a service (PaaS),
and software as a service (SaaS) (cf. [33]).

Thus, cloud computing penetrates all levels from datacenters (DCs) to end-
users. On an infrastructure level cloud computing offers (virtualized) hardware
resources and network capabilities as IaaS. This impacts the design and deployment of DCs.

For this reason the planning, implementation, and operation of DCS has
changed. Above all, cloud DCs distinguish themselves from traditional DCs by

---

This is an extended version of a former contribution [23] incorporating post-
deployment aspects.

© Springer International Publishing AG 2017
M. Helfert et al. (Eds.): CLOSER 2016, CCIS 740, pp. 317–338, 2017.
DOI: 10.1007/978-3-319-62594-2_16

having an IaaS solution deployed. The IaaS solution establishes the correspondent service model while managing the DC resources. These resources comprise computational power from central processing units (CPUs), random-access memory (RAM), storage in the form of objects stores or block devices from hard drive disks (HDDs) and solid-state drives (SSDs), and network interfaces. All of these need to be registered and managed by the IaaS solution.

OpenStack [42] is a popular IaaS solution with a big community and support across industries. For easing the installation and deployment of OpenStack various automated deployment technologies and tools exist. Often they are provided from an operating system (OS) vendor as a kind of value proposition.

Software-defined networking (SDN) (cf. [32]) and storage (cf. [56]) solutions may be deployed in addition to OpenStack. Providing a scalable networking architecture beyond other benefits, the former eases the planning of DCs. The later offers storage as a service (e.g., in form of a block device or object storage) towards OpenStack.

For realizing the installation of bare machines and the deployment of SDN, storage, and IaaS solutions, currently, various information needs to be aggregated in configurations by experts who are familiar with the technologies. Often the information relates to different aspects and is scattered and tangled and needs to be kept consistent. Changes in the design or characteristics of a DC may impact the configuration fundamentally: e.g., number of availability zones in a DC, dedication of a certain node to some aggregate[1], or networking.

Ideally, it would be possible to describe the various aspects of a cloud DC (e.g., the hardware and the networking) and its deployment (i.e., OSs and the services) conceptually in a domain-specific language (DSL) that is tailored towards the respective experts so that from such information as contained in the respective views the automated DC deployment can take place. While several deployment tools exist (that in fact can all be made use of following the model-based approach) there is no technology agnostic datamodel for specifying a DC deployment that is understood by tools.

Therefore, MING (明), a model-based approach, is proposed: it permits the model- and view-based description of DCs and respective IaaS deployments by means of a platform-independent model (PIM). This way, separation of concerns (SoC) is realized supporting different stakeholders. Also, integration with existing tools is realized using code generation. Finally, for supporting operational aspects MING aims at supporting post-deployment adaptation scenarios.

The remainder is structured as follows: The following section further introduces the context by explaining tasks when deploying and maintaining a cloud DC. Prior to presenting the MING approach in Sect. 5, Sect. 3 relates to the state of the art and Sect. 4 presents some background on model-based engineering (MBE).

---

[1] In order to profit from Enhanced Platform Awareness (EPA) host aggregates can be defined in OpenStack. Resources can be allocated within such aggregates for hosting services that want to make use of features such as hugepages, Non-Uniform Memory Access (NUMA), or CPU pinning for achieving high throughput.

Next, Sect. 6 presents some details of the current prototype. Section 7 discusses on the benefits, risks, and limitations and Sect. 8 concludes.

2 Automated Datacenter Deployment

The planning, setup, implementation, and operation of a DC comprises multiple activities involving hard- and software. Prior to focusing on the latter, i.e., the automated software installation and deployment, this section first looks at the structure of DCs.

2.1 Structure of Datacenters

Nowadays, a DC follows a Leaf-Spine topology (cf. [1]) and is connected to the core network through datacenter routers (DCRs). For this, each top of the rack router (ToR) is fully meshed to the spine layer. Establishing predictability in latency as well as redundancy for achieving high availability, any spine router is thus connected to every DCR as well as ToR. Similarly, a dedicated operations, administration, and management (OAM) network comprises bottom of the rack routers (BoRs).

Figure 1 depicts a simple metamodel for DCs (that is also part of the MING metamodel, cf. Fig. 2). A DATACENTER comprises one or several AVAILABILITYZONES. These may be fire compartments that are separated from each other. Each AVAILABILITYZONE contains RACKS for mounting equipment such as routers and servers (NODE). A server comprises network interface controllers (NICs), CPU, RAM, and storage in form of HDDs and/or SSDs. Each NIC has a unique media access control (MAC) address. For networking (cf. Layer 3 of the Open Systems Interconnection (OSI) reference model [58]) a NIC will be configured at some stage with an Internet Protocol [10] (IP) address, e.g., using Dynamic Host Configuration Protocol [17] (DHCP). Within a NETWORK (i.e., IP and NETMASK) a NIC has a particular DEVICEID (e.g., the last bit(s) of the address).

INSTALLATION nodes are used for bootstrapping the DC deployment. Generally it is possible to group servers into different categories: STORAGE nodes contain a large amount of storage capacity while COMPUTE nodes have high computational power. NETWORK nodes comprise fiber optical NICs for high bit rates. Finally, MANAGEMENT nodes are dedicated for hosting IaaS services (see also Sect. 2.2).

Intelligent Platform Management Interface [25] (IPMI) may provide an administrative access to the servers through a dedicated network. Besides, all servers may be part of multiple physical or virtual (VLANID) networks. For example, storage nodes may have a backend network for replication in addition to a frontend network for the data.

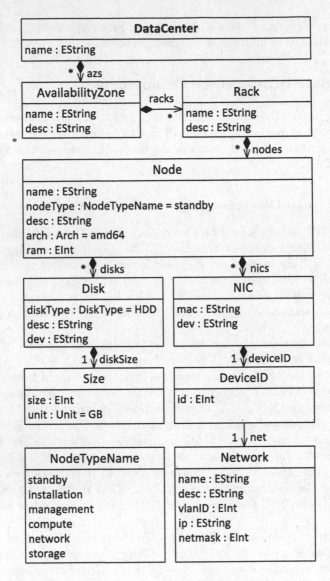

Fig. 1. A datacenter metamodel.

2.2   Software Installation

Given a DC with a completed physical setup including networking and cabling, deployment of an SDN solution, and configuration of the network devices installation of the bare machines can take place (see also Sect. 3.1). That is, on each node a base OS is installed over the net, e.g., using IPMI and Preboot Execution Environment [24] (PXE) together with a DHCP server. Yet, some information needs to be collected beforehand and placed at the DHCP server such as the

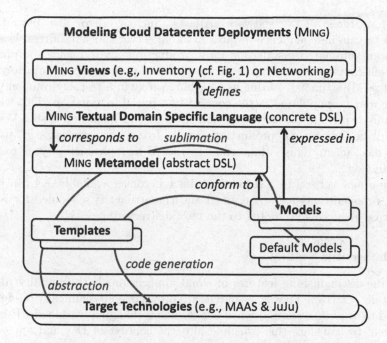

Fig. 2. Approach overview: the MING framework.

MAC addresses of the NICs. This information may be discovered automatically. For the software installation a local mirror can be made use of (see also Sect. 2.3).

Next, IaaS services can be installed (see also Sect. 3.2). STORAGE servers may deploy Ceph [56], a distributed object store. MANAGEMENT nodes host services such as OpenStack Identity Service (Keystone), OpenStack Compute (Nova) cloud controller, OpenStack Orchestration (Heat), and OpenStack Dashboard (Horizon). COMPUTE nodes run Nova. Finally, OpenStack Networking (Neutron) is used for NETWORK nodes and integrates with a deployed SDN solution through a respective plugin.

2.3 Software Projects, Artifacts, and Version Control

For such a described deployment, several software projects and subprojects have to be considered. In this context, we need to distinguish between up- and downstream projects. Upstream projects hold their own repositories where (community) development takes place. Downstream projects take releases delivered by upstream projects and customize these, e.g., for a respective OS or a container engine. Thus, they produce software artifacts such as Debian or RedHat Package Manager (RPM) packages or Docker [16], [4] images.

For an IaaS provider it is crucial to know what versions of the respective projects are (to be) deployed and to control the software update and upgrade. Addressing this requirement, a local mirror may be deployed on installation nodes that stores all the software artifacts. Such a mirror also permits the provider to gain independence towards external services and resources. For this, prior to an automated installation for a production environment, the provider holds a self-contained set of data (software images, packages, configuration management (CM) artifacts). Acting as a caching proxy in a test environment, such a mirror may be populated in the course of an initial installation. This way, all the required software packages can be captured and identified easily. Versions that shall be foreseen for production can be frozen in a review, e.g., using a tagging mechanism. This is important in regard to reproducibility as required for production.

Once a new version of a software artifact becomes available, e.g., in public mirrors, an operations team is notified and the artifact is scheduled for testing and in case of success promoted to the next delivery stage.

2.4 Datacenter Scaling

One of the distinguishing features of cloud applications is their ability to scale horizontally (cf. [18]). That is, resources are provisioned dynamically on-demand (referred to as *elastic scaling*) when experiencing changing workloads. It is only consequent to translate this principle into the context of Dcs and apply it to physical resources as well.

One of the implications is that an IaaS deployment may start small and grow bigger over time. As a result, a DC is only partially deployed with racks, initially. The more and more applications are onboarded over time additional racks are ordered and integrated into the DC and the IaaS deployment. This iterative deployment of DCs is interesting from an economic point of view. As mentioned, a Leaf-Spine topology and SDN eases the scaling within a DC. In the context of OpenStack the Tuskar project of TripleO offers an approach how to address the scaling of IaaS deployments (see also Sect. 3.2).

Generally, new resources such as compute or storage nodes shall be integrated into the cloud, e.g., OpenStack or Ceph, posterior to an initial deployment. Also it shall be possible to seamlessly replace faulty hardware.

Having briefly introduced the deployment of cloud DCs, the following section discusses different existing deployment tools for realizing the installation.

3 State of the Art

Prior to presenting the MING approach let us first have a look at the state of the art. At the end of this section, a current shortcoming of the existing deployment technologies is summarized for positioning the contribution of MING. As outlined in the previous section there are at least two distinct phases in the deployment of DCs: bare machine and IaaS installation with some of the tools focusing on the former and others on the latter.

3.1   Bare Machine Installation

The following selection of tools and projects specializes on installing a base OS on bare machines (cf. [11] for the evaluation of some frameworks).

Cobbler [13] is a lightweight build and provisioning system for the deployment of physical and virtual machines. Objects and variables are used for configuring the provisioning. These are then applied in templates, e.g., for generating preseed files. This way, i.e., through templates, Cobbler also integrates with Kickstart [44]. Generally, integration with existing tools and CM systems is encouraged. In addition to a command-line interface (CLI) there is a also a web user interface (UI). Cobbler is currently used by Compass and Fuel (see Sect. 3.2).

Fully Automatic Installation [19, 28, 29] – with a particular focus on unattended automated installations – builds, as Cobbler, on top of technologies such as PXE, DHCP, and Trivial File Transfer Protocol [48] (TFTP). Originally focusing on Debian-based [47] distributions, FAI has been adopted for CentOS [50]. It realizes profiles in addition to a class concept that can help to describe complex setups.

OpenStack Bare Metal Provisioning [39] **(Ironic)** is used for the provisioning of physical machines within OpenStack. Thus, in contrast to the other tools of this category, it is not a self-contained system. Its functionality is used by TripleO and will also be relied on by Fuel.

Metal as a service [8] **(MAAS)** is used for the provisioning of OSs in combination with JuJu [7] and Charms (see Sect. 3.2). Similar to Cobbler and FAI, a MAAS server acts as a DHCP server for the provisioning of machines. Hosts (physical or virtual) can be put under control of a MAAS server. Configuration such as MAC to IP mapping can be done in a YAML Ain't Markup Language (YAML) file (see also Sect. 6).

3.2   OpenStack Installation

The automated provisioning, configuration, and installation of services is addressed by CM systems such as Ansible [14], Chef [12], JuJu, or Puppet [43]. Thus, after each node has been installed with an OS, the installation and configuration of IaaS services can be realized using CM. For the deployment of OpenStack there is a variety of existing tools:

Compass [37] supports different CM systems through a plugin architecture. By establishing abstraction layers, it also decouples resource discovery and bare metal installation. Besides, it facilitates operations support system (OSS) integration through a northbound interface (NBI).

Crowbar [15] is a project that relies on the Chef CM system for the deployment of applications such as OpenStack or Hadoop [49]. In contrast to other solutions of this category it does not presume but also realizes bare metal installation and comes with a web UI.

Fuel [38] offers a web UI frontend for the deployment of OpenStack in addition to a CLI. Cobbler is currently used under the hood, yet, migration to OpenStack Bare Metal Provisioning (Ironic) is intended. Puppet is used for CM. Some features comprise the automated discovery of nodes and the possibility to perform pre-deployment checks.

JuJu is an orchestration technology that is also used for MAAS. JuJu bundles describe the orchestration of applications. Charms, classified by Wettinger et al. [57] as environment-centric artifacts, deploy the actual OpenStack services of JuJu bundles. Thus, the deployment of OpenStack is specified in a JuJu YAML file referencing different charms that are related to respective upstream projects.

Packstack [41] provides Puppet modules for OpenStack projects. Using Puppet for CM, the various OpenStack services can be deployed. Thus, some frontend deployment tools such as RPM Distribution of OpenStack [45] (RDO) (see below) make use of Packstack. Currently, distributions based on RPM are supported.

RDO is a web-based deployment tool based on Foreman [27], a Ruby on Rails [21] application and frontend for the CM with Puppet. Therefore Packstack is used.

TripleO [40] is an exception to the CM-based solutions. Instead of relying on such, TripleO aims at realizing the functionality using OpenStack's own cloud features for facilitating installation, management, and operation. For this, a deployment cloud (a.k.a. undercloud) needs to be setup first. Using Ironic workload cloud(s) (a.k.a. overcloud(s)) are deployed. The deployment and configuration of nodes is realized using Heat. For this golden images need to be prepared. These consist of a base OS with elements on top (resembles FAI class and profile concept). During provisioning a node will configure itself using the parameters from a Heat Orchestration Template (HOT) that constitutes the deployment plan. Finally, an overcloud can be scaled using the Tuskar subproject.

3.3   Positioning and Contribution of Ming

Currently, there is neither a standard nor a common datamodel for the configuration of an OpenStack deployment in a cloud DC. As a result, none of the projects exposes its configuration in a form that can be used by other projects. This however would be interesting in order to evaluate different frameworks and avoid tool dependencies. TripleO – aiming at avoiding any third party dependency for the deployment of OpenStack – is a particular case. It is using HOT for realizing the CM. This way, it decouples the configuration from the automated deployment. It may be argued that HOT is an established format that other tools could implement. Yet, this is not feasible, as it dictates orchestration through Heat. Not only Heat but also JuJu is an orchestration technology. In both cases, therefore, configuration needs to be expressed in a particular syntax and way by experts leading to the problems mentioned such as scattering and tangling.

MING in contrast truly decouples configuration from automated deployment technologies. It declarativly permits the view-based modeling of DCs and their deployments, facilitating SoC. As a result, stakeholders that are not familiar with the used deployment technologies and/or other concerns of the deployment are supported as well. Similar in spirit with Cobbler MING integrates with arbitrary tools and frameworks (as also envisioned by Compass) through code generation. Last but not least, it establishes a deployment tool-agnostic metamodel that serves as a datamodel for DC deployments.

4 Background on Model-Based Engineering

This work follows an MBE approach as outlined in the next section. Thus, this section briefly gives some background information regarding MBE. MBE is a paradigm of software engineering that establishes models as first-class entities (cf. [36]). Metamodels are used for formally describing concepts on a distinct level of abstraction (cf. [5]). Models, that need to conform to such metamodels, can be validated and are usually used in model transformations. Finally, model transformations (cf. [34]) map models such as in a model-to-model (M2M) transformation or a code generation (model-to-text (M2T) transformation). In case of forward engineering, resulting models generally conform to metamodels with a lower level of abstraction. This is for example the case when a PIM is mapped to a platform-specific model (PSM) or when code is generated.

In this work, MBE is used for establishing a metamodel that describes the IaaS deployment of a DC from bare metal. As the metamodel is agnostic towards deployment tools is can be classified as a PIM. Conforming models are transformed through code generation to target technologies.

4.1 Domain-Specific Languages

While metamodels are abstract in their nature, one or more (concrete) DSLs can define each a particular syntax for a metamodel (also called *abstract DSL*) that is tailored towards stakeholders of a respective domain (cf. [35,55]). This way, it is possible to express and represent models using a graphical or textual syntax as defined by the language. Models that are expressed in a DSL can be transformed and represented in another DSL. Similarly to general purpose programming languages and compilers, models expressed in higher-level DSLs can be translated to lower-level language artifacts. In fact, models expressed in DSLs can also be mapped to general purpose programming languages.

4.2 View Models

The principle of SoC is proven to be a successful approach to manage complexity. In MBE, it is realized using view models. That is, different concerns are separated into distinct views. This enables different stakeholders to relate to concerns which are relevant to them more easily. Within the domain of software architecture, for example, view models have been established and standardized (cf. [26]).

In this work, a DSL that is bound to the MING metamodel defines various views that describe different aspects of the deployment such as inventory, networking, and IaaS configuration options.

5 Approach: Abstracting from Technologies

The approach aims at a tool agnostic, declarative specification of configuration for realizing an IaaS deployment in a cloud DC from bare machines. For this, configurations from existing provisioning and deployment technologies need to be sublimated using reverse MBE. That is, models are established through abstraction. Given a valid deployment plan DC specific values and repetitive code need to be identified in a first step.

5.1 Establishing Models Through Abstraction

For capturing respective information in models, a conceptual metamodel is derived and templates are created in a next step. Finally, integration with the target technologies is realized using code generation. That is, the same code is generated using the MBE approach. As a result, a conceptual modeling layer is established with sublimated configuration in form of models.

Figure 2 depicts the MING framework. As an interface for populating, expressing, and representing models, a textual DSL is bound to the MING metamodel (i.e., abstract DSL). In order to support SoC, distinct views permit the expression of different aspects. One view, e.g., covers the DC related inventory information as depicted in the metamodel shown in Fig. 1. Other views capture networking, node assignments (e.g., to an aggregate), OpenStack specific configuration, and credentials.

For supporting convention over configuration, defaults can be expressed in models too, that are applied to models in case of missing configuration. Finally, model transformation processes the (resulting) models and generates code using templates.

5.2   Model-Based Adaptation

The above described code generation is suitable for initial deployments. As the approach is generally limited to such, post-deployment adaptation scenarios are not supported *per se*, however. Such adaptation scenarios comprise the addition of new DC resources to the cloud, e.g., in the form of new racks or the reassignment of nodes to some aggregates as pointed out in Sect. 2.4. The latter may become necessary if free resources run low within such an aggregate, e.g., as required for services that make use of EPA features. In such a case (unused) – e.g., compute – nodes may be reassigned to a corresponding host aggregate. This implies evacuation and a complete redeployment of the node, i.e., migration of remaining resources such as server instances, a wipe-out of existing disks, a new OS installation, deployment of appropriate services, and, finally, integration into the target aggregate.

Without support of operational aspects the models – apart from describing and realizing initial deployments – would be of no further use. Yet, because of their conceptual level of abstraction and the availability of multiple views, they are also interesting for later stages of the DC lifecycle.

Ideally, it would be possible to change such models and automate a respective adaptation. In either case it would be beneficial to describe the impact for the various possible changes and simulate an adaptation. This way, cloud engineers – while changing the models – could interactively receive information on the respective impact and further guidance. Table 1 lists and describes some possible model changes based on the inventory concepts of the MING metamodel as shown in Fig. 1. Such changes can be stored in a diff-model, i.e., a model that comprises model differences between two models conforming to the same metamodel.

In case of resources it is possible to base an execution engine on processing such a diff-model (cf. [22]). That is, for every change a model transformation is performed for generating respective adaptation actions related to API calls (e.g., the provisioning of a resource).

In this regard it seems promising to adopt such an adaptation approach that permits for an incremental deployment. Yet, beyond resources the MING metamodel also comprises other concerns such as networking or configuration options. In contrast to resources that can be added or removed – just as described in each difference of a diff-model with the kind of change and the respective model

Table 1. Examples of model changes related to post-deployment adaptation scenarios.

Model Element	Kind	Description
AvailabilityZone.racks	Addition	A rack to be integrated into a DC's availability zone
Rack.nodes	Addition	(new) node(s) to be integrated into the cloud
Rack.nodes	Deletion	(faulty) node(s) to be removed from the cloud
Node.nodeType	Modification	Reassignment of a node to another category
Node.disks	Addition	(new) disk(s) to be added to a node
Node.disks	Deletion	(faulty) disk(s) to be removed from a node

element and as supported by existing API calls – the change impact and resulting adaptation actions cannot be determined without further domain knowledge. In some cases a change implies a trivial update of a configuration option; in other cases it is not possible at all to apply the described change to an existing IaaS deployment without affecting tenants during production.

Some adaptation scenarios relate to the patch management of an OS as well as SDN, storage, and IaaS solutions. That is, updates such as security patches and upgrades in case of new software releases shall be applied in existing deployments. The version control of respective services accounts for a corresponding view. That is, distinct models hold respective information such as releases and versions.

6 The Ming Prototype

For the implementation of the MING prototype, the Eclipse Modeling Framework [51] was chosen as a modeling foundation. Eclipse Xtext [53] (Xtext) served for defining the DSL and its views and for obtaining a respective DSL editor. EMF Compare [52] is used for calculating a diff-model. Finally, model transformations were implemented in Eclipse Xtend [54] (Xtend).

In the following, the engineering process is described that was executed when realizing the MING prototype. Also some excerpts from code generation templates are depicted. Next, examples in form of DSL views demonstrate how the MING prototype is used for specifying different deployment configuration options. Finally, post-deployment scenarios are discussed.

6.1 Initial Engineering

Prior to adopting the MBE approach for basing the automation on models a MAAS configuration and OpenStack JuJu bundle were engineered and tested. These files served as the target code and constituted a starting point for the reverse engineering. For this, values that are specific to a DC deployment were identified first. Next, loop statements and rules were introduced for generating repetitive code and for improving its quality. In this process, the target code was transformed to a M2T code generation template. Also, a metamodel was established. For this, a grammar of a DSL was defined with the intention to act as an interface for stakeholders for expressing and representing various aspects related to the deployment of DCs. For supporting SoC, different concepts were separated into distinct views. Next, the configuration options from the target code were sublimated into models. That is, these values were stored in models that conform to the metamodel as used by the M2T transformation. Finally, the original code was produced from the models using code generation. From this point on, it became possible to base the overall deployment on models. Relating to certain views, it also became possible to reuse models easing the deployment of multiple DCs.

Figure 3 depicts an excerpt from the M2T transformation for generating a MAAS configuration. Three FOR loops iterate over all DC's availability zones,

```
nodes:
  «FOR az : dc.azs»
  «FOR rack : az.racks»
  «FOR node : rack.nodes»
  - name: «model.deployment.name»-«node.name»
    «IF node.nodeType == NodeTypeName.CN»
    tags: «getComputeAggregate(zones, node, az)»
    «ELSEIF node.nodeType == NodeTypeName.SN»
    tags: storage-«getCephPool(zones, node).name»
    «ELSEIF node.nodeType == NodeTypeName.MN»
    tags: api
    «ELSEIF node.nodeType == NodeTypeName.NN»
    tags: gateway-«getGatewayZone(zones, node).name»
    «ELSE»
    tags: standby
    «ENDIF»
    architecture: «node.arch»/generic
    mac_addresses:
    «FOR nic : getNICs(node)»
    - «nic.mac»
    «ENDFOR»
    power:
      type: ipmi
      address: «getIPMI(node)»
      user: «model.credentialsIPMI.username»
      pass: «model.credentialsIPMI.password»
      driver: LAN_2_0
    «enrichWithIPs(node)»
    «FOR nic : getNICs(node).filter[it.ip4 != null]»
    sticky_ip_address:
      mac_address: «nic.mac»
      requested_address: «nic.ip4»
    «ENDFOR»
  «ENDFOR»
  «ENDFOR»
  «ENDFOR»
```

Fig. 3. Code Generation for Metal as a Service (MAAS) nodes with Xtend.

racks, and finally nodes. As a result an entry is generated for every node containing all of its MAC addresses and assigned IP addresses. The latter are specified in a sticky_ip_address section. Code generation assures the consistency between the MAC addresses.

Please note that a separate template that supports a different target technology can process the same models. That is, while the models describe the overall DC deployment, they are agnostic to actual deployment automation technologies.

6.2   Continuous Improvements

Yet, following an agile approach, the engineering of the target code was subject to an iterative process. For this reason also the templates, the metamodel, and the models had to be revised repeatedly. For comprising increments such as new features or bugfixes, the YAML files were improved and tested by cloud engineers. When a new stable version of target code became available, a new reverse engineering cycle was started for capturing the respective expert knowledge in model transformations. For this, the differences in the target code were analyzed and the template aligned accordingly. If necessary, new concepts were added to

```
phase2:
  inherits: phase1
  services:
    «FOR cluster : zones.cephClusters»
    ceph-«cluster.name»:
      charm: cs:«deployment.dist.getName()»/ceph
      num_units: «cluster.nodes.size»
      options:
        osd-devices: «deployment.ceph.osdDevices»
        osd-reformat: '«IF deployment.ceph.osdReformat»True«ELSE»False«ENDIF»'
        osd-format: '«deployment.ceph.osdFormat.literal»'
        ceph-public-network: «getNetwork(networks, 'cephFE' + cluster.name)»
        ceph-cluster-network: «getNetwork(networks, 'cephBE' + cluster.name)»
      to:
        «FOR i : 0..cluster.nodes.size-1»
        - '«deployment.name»-dc-storage-«cluster.name»=«i»'
        «ENDFOR»
    «ENDFOR»
```

Fig. 4. Code generation of Ceph clusters in a JuJu bundle with Xtend.

the metamodel and the DSL views. This two-phase procedure with a handover of target code for reverse MBE has the advantage that domain experts can continue to work as usual and do not have to be involved closely into MBE activities which they may not be familiar with.

As an alternative to the reverse engineering, increments were sometimes directly realized in the metamodel and the templates. This way, complex changes can sometimes be addressed in a more efficient manner. Generally, however, this requires a close collaboration between MBE and domain experts or either expert knowledge in the domain on part of an MBE expert or a fair understanding of a templating language such as Xtend on part of the domain experts. A change that was easier to realize because of resulting repetitive code was the assignment of nodes to some Ceph clusters or host aggregates. Features such as modeling support for more than one Ceph cluster in a DC were directly realized using loop statements within the transformation template (see Fig. 4).

6.3   Example Configuration and Version Control

For configuring OpenStack various variables exist for the different services. Using the DSL editor a user profits from code completion, syntax highlighting, and validation. Figure 5 shows an example configuration view for the deployment of a DC from bare machines with a base OS distribution (Ubuntu) and storage (Ceph) and IaaS (OpenStack) solutions with respective services (e.g., Neutron and Nova).

Specific versions of the distribution and the various services are configured in a separate view as shown in Fig. 6(a). For example, a certain OS image is referenced for the bare machine installation there. That is, the underscored name is a reference to a definition of the image with metadata such as an Uniform Resource Locator [20], [3] (URL) and checksums.

In addition to the upstream projects, specific artifacts from downstream projects that target specific operating systems or deployment technologies are

```
IaaS Deployment SongThrush @ MingDC9

OperatingSystem Ubuntu

Ceph
        osdFormat btrfs
        osdReformat

OpenStack
    Heat
            workers 8
            hiddenTags "generated"

    Neutron
            externalBridge "br-ext"
            overlayNetType "gre vxlan"
            l3_ha
            l3_agents 2-4
            mtu 9000

    Nova
            liveMigration
            mtu 9000

    Percona
            maxConnections 12345
```

Fig. 5. Configuration of an IaaS deployment – DSL view.

recorded in another view. For the Ceph and OpenStack services, Fig. 6(b) shows versions of Debian packages as found in the Ubuntu Cloud Archive [9] that relate to respective upstream projects and versions. The DSL (editor) facilitates consistency between the versions of the up- and downstream projects through validation and scoping while offering a selection of existing matching packages in code completion.

Although these views exist, it is not compulsory to specify certain versions. That is, it is possible to use default versions in a deployment. Yet, at a certain point in time it is important to fix and freeze the set of versions of the various projects. Ensuring the roll-out of defined software artifacts, this permits for reproducibility of deployments in particular. In addition, it facilitates patch management in a production environment.

6.4   Code Generation

Together with the other views (e.g., an instance of the metamodel as shown in Fig. 1) and the default models all required information is complete for model transformation to take place. The current prototype supports generation of a MAAS configuration and a OpenStack JuJu bundle in form of YAML files. For the sublimated configuration using default options and versions, the ratio between the size of MING models and YAML code for MAAS and JuJu yielded 27%. That is, the models in MING are nearly four times more compact than the corresponding code as generated by the templates.

```
Upstream Project Versions
SongThrush @ MingDC9

Ubuntu 16.04.1 LTS (Xenial Xerus)

Ceph v10.2.0 (Jewel)

OpenStack 2016-04 (Mitaka)
      Ceilometer "6.0.0"
      Cinder "8.0.0"
      Glance "12.0.0"
      Heat "6.0.0"
      Horizon "9.0.1"
      Keystone "9.0.0"
      Neutron "8.1.2"
      Nova "13.0.0"
```

(a) Versions of IaaS Upstream Projects

```
Package Versions
SongThrush @ MingDC9

Ceph "10.2.0-0ubuntu0.16.04.2"

OpenStack
      Ceilometer "1:6.0.0-0ubuntu1"
      Cinder "2:8.0.0-0ubuntu1"
      Glance "2:12.0.0-0ubuntu2"
      Heat "1:6.0.0-0ubuntu1.1"
      Horizon "2:9.0.1-0ubuntu2"
      Keystone "2:9.0.0-0ubuntu1"
      Neutron "2:8.1.2-0ubuntu1"
      Nova "2:13.0.0-0ubuntu5"
```

(b) OS Packages of IaaS Solutions

Fig. 6. Version control – DSL views.

6.5   Post-Deployment Adaptation Scenarios

As described, the MING approach provides a framework for describing the various aspects of DC deployments. Relying on existing deployment tools, this permits automation from a modeling level. Beyond such deployment automation MING can help to facilitate operational scenarios as well. For this, models relating to a deployment are modified by DSL users and are analyzed in MING in a first step. This requires that the models are put under version control, e.g., using Git [30], and are up to date, i.e., truthfully reflect the DC deployment.

Based on the same MING metamodel, model differences between two model versions are calculated using EMF Compare. In case the types of changes are supported for automation, a respective adaptation process can be initiated. Besides, MING can help experts to describe and learn about the change impact and best practices also including other types of modifications. That is, documentation can be written for different types of modifications that will be looked up and presented to the experts. This way, MING can be used to interactively try out modifications, learn or document knowledge related to the changes, and, if available, trigger an adaptation process for enforcing the changes.

Scaling. One of the adaptation scenarios in a production environment relates to the scaling of a DC as described in Sect. 2.4. For this new racks with new nodes are added to an inventory model of the DC. There, nodes can be assigned a nodeType. Besides, nodes can be dedicated to some aggregate.

Figure 7(a) shows an example for an inventory DSL view of the MingDC9. A new rack (r9) is added to one of the availability zones (az1). It contains a storage node (n8) with a couple of NICs and disks. The DSL view in Fig. 7(b)

```
DataCenter MingDC9

AvailabilityZone "az1"

Rack "r9"

Node "n8" type storage
      NIC network ipmi id 23
      NIC "01:02:03:04:05:06" network oam id 45
      NIC "07:08:09:0a:0b:0c"
      NIC "0d:0e:0f:10:11:12"
      HDD 2 TB
      HDD 2 TB
      HDD 2 TB
      HDD 2 TB
      HDD 2 TB
      SSD 512 GB
```

```
Aggregates for MingDC9

CephCluster "public"
"MingDC9.az1.r1.n3"
"MingDC9.az2.r2.n4"

Aggregate "EPA"
"MingDC9.az1.r1.n5"
"MingDC9.az2.r2.n6"
```

(a) Inventory   (b) Node Assignments

Fig. 7. Inventory and node assignments (excerpts) – DSL views.

illustrates an example with an excerpt referencing other nodes from the inventory model. It defines a Ceph cluster (public) and an host aggregate (EPA) with two nodes respectively (n3, n4 and n5, n6). These nodes are located in two racks (r1 and r2) in different availability zones (az1 and az2). Part of the inventory, node n8 can now be added to the Ceph cluster public. That is, a reference ("MingDC9.az1.r9.n8") can be added to the list. After having finished the modifications, the DSL user may now start an adaptation analysis. For this, the model differences (i.e., a new rack within the inventory and a new node in the Ceph cluster) are identified based on the previous version of the MING views. Next, the user receives previously documented information that describes the adaptation process that has been automated and can be triggered. For enforcing the adaptation, the new node is installed with an OS (see also Figs. 5 and 6(a)), a Ceph service (see Fig. 6(b)), and integrated into the Ceph cluster.

Patch Management. Another use case is support for security patches and software updates as pointed out in Sect. 2.3. For this, the responsible DSL views are related to the version control of software projects and artifacts (see Fig. 6). In the DSL editor new available versions (e.g., Charms and/or (related) OS packages) can be highlighted to the user for selection. For enforcing the changes, an adaptation process may possibly rely on the (re)generation and deployment of a JuJu bundle if supported by the respective Charms.

7   Discussion

The model-based approach enables the utilization of diverse technologies. In this work, MAAS and JuJu constituted target technologies. For supporting other deployment tools, the process described in the Sects. 6.1 and 6.2 can be used. Given availability of respective templates, this enables evaluation of different deployment tools. That is, from the same models target code is generated for various deployment tools using respective M2T transformations. This in turn fosters a common datamodel for establishing a standard for configuring an IaaS deployment from bare machines. With the separation of the models in distinct views further benefits can be leveraged:

Discovery of nodes and their components automatically yields a certain view. Credentials as stored in another view are generated initially if not set for a certain deployment. In both cases parts of the overall configuration are provided and the DSL user only needs to focus on the other aspects of a deployment.

For a given DC it is possible to specify different deployments. That is, while the physical setup (i.e., availability zones, racks, and nodes) as captured in one view does not change, views related to the deployment such as the assignment of nodes or the deployment and configuration of OpenStack services may be different. Yet, only a part of the overall configuration is adapted and existing views can be reused avoiding software clones. This way, deployments can be tested and the differences between them can be described precisely by comparing two models.

In case a different DC shall be deployed similarly, it is possible to reuse views such as the configuration of OpenStack services. This eases the deployment of multiple DCs with a tested configuration.

The possibility to specify default configuration options in models permits custom user-defined defaults. The fact that these are then applied on the views has two major advantages: It simplifies the models by moving default configuration options out of the views making the files more compact. Also, it simplifies code generation by only processing the resulting model and makes it independent from any (user-defined) defaults.

Regardless of the benefits of this model-based approach, it is always possible to continue work with the output. Thus, MING does not introduce any dependency in regard to the underlying automation.

Not all fine-grained configuration options of the bare machine provisioning or IaaS solutions are reflected in the MING metamodel. Thus, in case these shall be lifted to the modeling layer, the metamodel and the views need to be extended. In case multiple technologies are supported using code generation, certain features of one technology may not be supported by alternatives. For example, the deployment of IaaS services may be realized using Kernel-based Virtual-Machine (KVM), Linux Containers (LXC), or Docker. In case such a configuration option is specified but not supported by a technology a fallback may be realized during code generation. For early feedback this can be addressed through validators in the DSL. That is, when selecting an option that is not supported by some technologies a warning is displayed in the DSL editor.

As pointed out in Sect. 5.2 adaptation actions can generally be derived from inventory changes. The change impact of some other model differences needs to be examined case by case. For this, respective actions can be identified, documented, and implemented if possible. Although arduous, such an endeavor certainly is worthwhile as it realizes operational support for different scenarios as described from a technology agnostic modeling level. In this regard it should be pointed out that – even in production – not all possible model changes need to be covered necessarily. In the course of an agile development, different scenarios could be formulated as user stories and be part of a prioritized (Scrum [46]) backlog.

There is a need to keep the models up to date. That is, changes that are performed in a datacenter (hard- and software) also need to be reflected in the models. Ideally, the models would be in sync with reality; furthermore, they would be causally connected. That is, model changes would imply appropriate adaptations. For this, MING could be extended with a dedicated models@runtime layer (cf. [6]).

In addition to deployment and adaptation the model-based framework could cover monitoring aspects as well. This would further strengthen MING from an operational point of view. Relating to, e.g., host aggregates that run low on resources (see also Sect. 5.2) an adaptation process could automatically be triggered for balancing out resources.

8   Conclusion

Gaining independence from existing deployment technologies MING establishes a metamodel that serves as a technology agnostic datamodel for the configuration of IaaS deployments in DCs from bare metal. The model-based approach integrates with available deployment tools and realizes SoC through view models backed by a textual DSL. For facilitating operational aspects posterior to initial deployments, MING analyses changes on a modeling level. It supports experts to both document and learn about different types of adaptations and their change impact and can, if available, trigger an appropriate process.

Acknowledgments. The author would like to thank the members of the extended Infrastructure Cloud team, i.e., Alexandros Tsirepas, Andreas Flick, Axel Clauberg, Basil Ahmed, Bernard Tsai, Daniel Brower, George Wu, Herbert Damker, Karsten Reincke, Ken Jung, Matthias Britsch, Michael Linke, Michael Machado, Normen Kowalewski, Patrick Münch, Rainer Schatzmayr, Robert Schwegler, Seth Chen, Stefan Schraub, Steve Liu, Thomas Hillen, Thomas Oswald, Tobias Brausen, and Tomislav Sukser for their dedicated endeavors making this work possible, valuable feedback, and helpful comments.

References

1. Alizadeh, M., Edsall, T.: On the data path performance of leaf-spine datacenter fabrics. In: IEEE 21st Annual Symposium on High-Performance Interconnects, HOTI 2013, Santa Clara, CA, USA, 21–23 August 2013, pp. 71–74. IEEE Computer Society (2013)
2. Andrikopoulos, V., Binz, T., Leymann, F., Strauch, S.: How to adapt applications for the cloud environment. Computing **95**, 493–535 (2013)
3. Berners-Lee, T., Masinter, L., McCahill, M.: Uniform Resource Locators (URL), December 1994. http://ietf.org/rfc/rfc1738.txt. Accessed Sept 2016
4. Bernstein, D.: Containers and cloud: from LXC to Docker to Kubernetes. IEEE Cloud Comput. **1**(3), 81–84 (2014)
5. Bézivin, J.: On the unification power of models. Soft. Syst. Model. **4**(2), 171–188 (2005)
6. Blair, G.S., Bencomo, N., France, R.B.: Models@ run.time. Computer **42**(10), 22–27 (2009). IEEE
7. Canonical, Ltd.: JuJu. http://jujucharms.com. Accessed Sept 2016
8. Canonical, Ltd.: MAAS: Metal as a Service. http://maas.io. Accessed Sept 2016
9. Canonical, Ltd.: Ubuntu Cloud Archive. https://wiki.ubuntu.com/OpenStack/CloudArchive. Accessed Sept 2016
10. Cerf, V.G., Khan, R.E.: A protocol for packet network intercommunication. IEEE Trans. Commun. **22**, 637–648 (1974)
11. Chandrasekar, A., Gibson, G.: A comparative study of baremetal provisioning frameworks. Technical report CMU-PDL-14-109, Parallel Data Lab, Carnegie Mellon University, December 2014. http://pdl.cmu.edu/PDL-FTP/associated/CMU-PDL-14-109_abs.shtml. Accessed Sept 2016
12. Chef Software, Inc.: Chef. http://getchef.com. Accessed Sept 2016
13. DeHaan, M.: Cobbler. http://cobbler.github.io. Accessed Sept 2016
14. DeHaan, M.: Ansible (2012). http://ansible.com. Accessed Sept 2016
15. Dell. Inc.: Crowbar. http://crowbar.github.io. Accessed Sept 2016
16. Docker, Inc.: Docker (2013). http://docker.com. Accessed Sept 2016
17. Droms, R.: Dynamic Host Configuration Protocol. RFC 2131, The Internet Engineering Task Force, March 1997. http://ietf.org/rfc/rfc2131.txt. Accessed Sept 2016
18. Fehling, C., Leymann, F., Retter, R., Schupeck, W., Arbitter, P.: Cloud Computing Patterns - Fundamentals to Design, Build, and Manage Cloud Applications. Springer, Wien (2014)
19. Gärtner, M., Lange, T., Rühmkorf, J.: The fully automatic installation of a Linux cluster. Technical report 379, Computer Science Department, University of Cologne, December 1999. http://e-archive.informatik.uni-koeln.de/id/eprint/379. Accessed Sept 2016
20. Hansen, T., Hardie, T., Masinter, L.: Guidelines and Registration Procedures for New URI Scheme, February 2006. http://ietf.org/rfc/rfc4395.txt. Accessed Sept 2016
21. Hansson, D.H.: Ruby on Rails (2005). http://rubyonrails.org. Accessed Sept 2016
22. Holmes, T.: Facilitating migration of cloud infrastructure services: a model-based approach. In: Paige, R.F., Cabot, J., Brambilla, M., Hill, J.H. (eds.) Proceedings of the 3rd International Workshop on Model-Driven Engineering on and for the Cloud co-located with the 18th International Conference on Model Driven Engineering Languages and Systems, MoDELS 2015, Ottawa, Canada, 29 September 2015. CEUR Workshop Proceedings, vol. 1563, pp. 7–12. CEUR-WS.org (2015)

23. Holmes, T.: Sublimated configuration of infrastructure as a service deployments – Ming: a model- and view-based approach for cloud datacenters. In: Cardoso, J., Ferguson, D., Muñoz, V.M., Helfert, M. (eds.) 6th International Conference on Cloud Computing and Services Science. vol. 2, pp. 308–313. SciTePress (2016)
24. Intel Corporation: Preboot Execution Environment (PXE) Specification Version 2.1, September 1999. http://download.intel.com/design/archives/wfm/downloads/pxespec.pdf. Accessed Sept 2016
25. Intel Corporation, Hewlett-Packard Company, N.E.C., Corporation, Dell Inc.: Intelligent Platform Management Interface Specification v2.0 rev. 1.1, October 2013. https://www-ssl.intel.com/content/www/us/en/servers/ipmi/ipmi-second-gen-interface-spec-v2-rev1-1.html. Accessed Sept 2016
26. International Organization for Standardization: ISO/IEC 42010:2011 Systems and software engineering - Architecture description, December 2011. http://iso.org/iso/catalogue_detail.htm?csnumber=50508. Accessed Sept 2016
27. Kelly, P., Levy, O.: Foreman (2009). http://theforeman.org. Accessed Sept 2016
28. Lange, T.: Fully Automatic Installation (2000). http://fai-project.org. Accessed Sept 2016
29. Lange, T.: 10 Jahre FAI Projekt. Technical report 603, Computer Science Department, University of Cologne, July 2010. http://e-archive.informatik.uni-koeln.de/id/eprint/603. Accessed Sept 2016
30. Torvalds, L.: Git, April 2005. http://git-scm.com. Accessed Sept 2016
31. Maggiani, R.: Cloud computing is changing how we communicate. In: International Professional Communication Conference, pp. 1–4 (2009)
32. McKeown, N., Anderson, T., Balakrishnan, H., Parulkar, G.M., Peterson, L.L., Rexford, J., Shenker, S., Turner, J.S.: Openflow: enabling innovation in campus networks. Comput. Commun. Rev. **38**(2), 69–74 (2008)
33. Mell, P.M., Grance, T.: The NIST definition of cloud computing. Technical report, SP 800-145, National Institute of Standards & Technology (2011)
34. Mens, T., Gorp, P.V.: A taxonomy of model transformation. Electr. Notes Theor. Comput. Sci. **152**, 125–142 (2006)
35. Mernik, M., Heering, J., Sloane, A.M.: When and how to develop domain-specific languages. ACM Comput. Surv. **37**(4), 316–344 (2005)
36. Miller, J., Mukerji, J.: MDA Guide Version 1.0.1. http://omg.org/cgi-bin/doc?omg/03-06-01. Accessed Sept 2016
37. OpenStack Foundation: Compass. http://syscompass.org. Accessed Sept 2016
38. OpenStack Foundation: Fuel. http://wiki.openstack.org/Fuel. Accessed Sept 2016
39. OpenStack Foundation: OpenStack Bare Metal Provisioning (Ironic). http://wiki.openstack.org/Ironic. Accessed Sept 2016
40. Openstack Foundation: OpenStack on OpenStack (TripleO). http://wiki.openstack.org/TripleO. Accessed Sept 2016
41. OpenStack Foundation: Packstack. http://wiki.openstack.org/Packstack. Accessed Sept 2016
42. OpenStack Foundation: OpenStack, July 2010. http://openstack.org. Accessed Sept 2016
43. Puppet Labs, L.L.C.: Puppet. http://puppetlabs.com. Accessed Sept 2016
44. Red Hat, Inc.: Kickstart (2011). http://github.com/rhinstaller/pykickstart. Accessed Sept 2016
45. Red Hat, Inc.: RPM Distribution of OpenStack (RDO) (2013). http://rdoproject.org. Accessed Sept 2016
46. Schwaber, K., Beedle, M.: Agile Software Development with Scrum, 1st edn. Prentice Hall PTR, Upper Saddle River (2001)

47. Software in the Public Interest, Inc.: Debian (1993). http://debian.org. Accessed Sept 2016
48. Sollins, K.R.: The TFTP Protocol (Revision 2). RFC 1350, The Internet Engineering Task Force, July 1992. http://ietf.org/rfc/rfc1350.txt. Accessed Sept 2016
49. The Apache Software Foundation: Hadoop (2011). http://hadoop.apache.org. Accessed Sept 2016
50. The CentOS Project: CentOS (2004). http://centos.org. Accessed Sept 2016
51. The Eclipse Foundation: Eclipse Modeling Framework Project (EMF) (2002). Accessed Sept 2016. http://eclipse.org/modeling/emf
52. The Eclipse Foundation: EMF Compare, October 2006. http://wiki.eclipse.org/EMF_Compare. Accessed Sept 2016
53. The Eclipse Foundation: Xtext (2006). Accessed Sept 2016. http://eclipse.org/Xtext
54. The Eclipse Foundation: Xtend, June 2011. http://eclipse.org/xtend. Accessed Sept 2016
55. Völter, M., Benz, S., Dietrich, C., Engelmann, B., Helander, M., Kats, L.C.L., Visser, E., Wachsmuth, G.: DSL Engineering - Designing, Implementing and Using Domain-Specific Languages. dslbook.org (2013)
56. Weil, S.A., Brandt, S.A., Miller, E.L., Long, D.D.E., Maltzahn, C.: Ceph: a scalable, high-performance distributed file system. In: Bershad, B.N., Mogul, J.C. (eds.) 7th Symposium on Operating Systems Design and Implementation, pp. 307–320. USENIX Association (2006)
57. Wettinger, J., Breitenbücher, U., Leymann, F.: Standards-based DevOps automation and integration using TOSCA. In: 7th IEEE/ACM International Conference on Utility and Cloud Computing, pp. 59–68. IEEE (2014)
58. Zimmermann, H.: OSI reference model - the ISO model of architecture for open systems interconnection. IEEE Trans. Commun. **28**(4), 425–432 (1980)

A Generic Framework for Representing Context-Aware Security Policies in the Cloud

Simeon Veloudis[1], Iraklis Paraskakis[1(✉)], Yiannis Verginadis[2], Ioannis Patiniotakis[2], and Gregoris Mentzas[2]

[1] South East European Research Centre (SEERC), International Faculty, CITY College, The University of Sheffield, 24 Proxenou Koromila Street, 54622 Thessaloniki, Greece
{sveloudis,iparaskakis}@seerc.org
[2] Institute of Communications and Computer Systems, National Technical University of Athens, Athens, Greece
{jverg,ipatini,gmentzas}@mail.ntua.gr

Abstract. Enterprises are increasingly embracing cloud computing in order to reduce costs and increase agility in their everyday business operations. Nevertheless, due mainly to confidentiality, privacy and integrity concerns, many organisations are reluctant to migrate their sensitive data to the cloud. In order to alleviate these security concerns, this chapter proposes the *PaaSword framework*: a generic PaaS solution that provides capabilities for guiding developers through the process of defining appropriate *policies* for protecting their sensitive data. More specifically, this chapter outlines the construction of an extensible and declarative formalism for representing policy-related knowledge, one which disentangles the definition of a policy from the code employed for enforcing it. It also outlines the construction of a suitable Context-aware Security Model, a framework of concepts and properties in terms of which the policy-related knowledge is expressed.

Keywords: Context-aware security · Ontologies · Linked USDL · Policies · Access control · Data privacy · Security-by-design · Governance of policies

1 Introduction

There is generally consensus among analysts that cloud computing is being adopted by enterprises at an ever-increasing pace [1]. The main force that drives this trend is the new economy-based paradigm that cloud computing introduces [2] which enables significant cost savings, whilst accelerating the deployment and utilisation of new applications [3]. Nevertheless, the increasing adoption of cloud computing transforms the enterprise IT environment into a matrix of interwoven infrastructure, platform and application services that are delivered remotely, over the Internet, by diverse service providers [4]. These services may span not only different technologies and geographies, but also entirely different domains of ownership and control. This creates a set of unprecedented security vulnerabilities stemming mainly from the fact that corporate data reside in externally-controlled servers or untrusted cloud providers. Exploiting these vulnerabilities may result in data confidentiality and integrity breaches [5]. Evidently,

M. Helfert et al. (Eds.): CLOSER 2016, CCIS 740, pp. 339–359, 2017.
DOI: 10.1007/978-3-319-62594-2_17

the benefits offered by the cloud computing paradigm cannot fully materialise without addressing these new security challenges [6].

A promising approach to alleviating the security concerns associated with cloud computing is to assist application developers in defining effective security controls for safeguarding the sensitive data accessed through the cloud applications that they develop. To this end, in [6] we proposed the *PaaSword framework*, a generic PaaS solution that provides capabilities for guiding developers through the process of defining appropriate *policies* for protecting sensitive data. More specifically, three are the main kinds of policy that the PaaSword framework aims at supporting: (i) *Data encryption policies*, which determine the strength of the cryptographic protection of a sensitive object; (ii) *data fragmentation and distribution policies*, which determine the manner in which sensitive data objects are fragmented and distributed to different physical servers for privacy reasons; (iii) *access control policies*, which determine when to grant, or deny, access to sensitive data.

In order to *effectively* guide developers in defining security policies, the PaaSword framework bears two seminal characteristics. Firstly, it hinges upon an adequate scheme that takes into account the inherently dynamic and heterogeneous nature of cloud environments. Secondly, it captures the *knowledge* that lurks behind such a scheme (e.g. actions, subjects, locations, environmental attributes, etc.) using a generic and extensible formalism, one which can be tailored to the particular needs of different cloud applications. The first characteristic calls for the incorporation of the notion of *context* in policies, i.e. the consideration of dynamically-changing contextual attributes that may characterise data accesses. It therefore involves the development of a re-usable and generic *Context-aware Security Model* which goes beyond the traditional context-insensitive security (e.g. DAC, MAC, RBAC [7]). The second characteristic calls for the adoption of a declarative approach to modelling policy-related knowledge, one which is orthogonal to the code of any particular cloud application and which can be easily adapted to suit the needs of any such application.

The aim of this chapter is twofold. On the one hand, it outlines the construction of the Context-aware Security Model. On the other hand, it outlines the construction of an extensible and declarative formalism for representing policy-related knowledge, one which disentangles the definition of a policy from the code employed for enforcing it, bringing about the following advantages: (i) it allows the policy-related knowledge to be extended and instantiated to suit the needs of a particular application, independently of the code employed by the application; (ii) it forms an adequate basis for reasoning generically about the correctness and consistency of the security policies, hence about the effectiveness of the security controls that these policies give rise to.

The rest of this chapter is organised as follows. In Sect. 2, we elaborate on a *context-aware security model* that will be used as the underlying vocabulary for describing the three kinds of policy that the PaaSword framework supports. In Sect. 3, we introduce a *policy model* that allows for the semantic description of the policies. In Sect. 4, we present a case study that demonstrates the use of the Context-aware Security Model, as well as of the policy model. Finally, Sect. 5 briefly discusses relevant work and Sect. 6 concludes the chapter by presenting the next steps for the implementation and evaluation of the proposed approach.

2   Context-Aware Security Model

In this section, we present a Context-aware Security Model, which can be used by the developers in order to annotate database Entities, Data Access Objects (DAO) or any other web endpoints that give access to sensitive data managed by cloud applications. This context model conceptualises the aspects, which must be considered during the selection of a data-access policy. These aspects may be any kind of information which is machine-parsable [8]; indicatively they may include the user's IP address and location, the type of device that s/he is using in order to interact with the application as well as his/her position in the company. These aspects can be interpreted in different ways during the security policy enforcement. In particular, the Context-aware Security Model can set the basis for determining which data is accessible under which circumstances.

2.1   Context-Aware Security Meta-model

In Fig. 1, we present a meta-model that captures the main facets of the `Context-aware Security Model` along with their associations. Specifically, this model comprises of two different kinds of facets that may give rise to:

- Dynamic security controls – These controls grant or deny access to sensitive data on the basis of dynamically evolving contextual attributes, which are associated with the entity requesting the access. The relevant model facets are:
  - `Security Context Element`
  - `Permission`
  - `Context Pattern`
- Static security controls – These controls are independent of any dynamically evolving contextual attributes. They mainly correspond to the distribution and cryptographic protection features that certain data artefacts must have and affect the bootstrapping phase of a cloud application. The relevant model facet is the:
  - `Data Distribution and Encryption Element (DDE)`

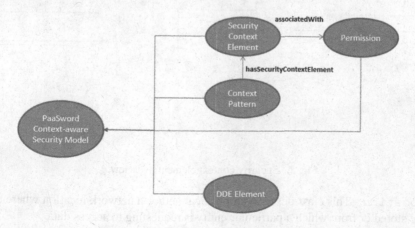

Fig. 1. Context-aware security meta-model.

According to this meta-model, instances of these aforementioned facets formulate the `Context-aware Security Model`. Furthermore, `Context Pattern Elements` are directly associated to `Security Context Elements` (through the `hasSecurityContextElement` property) in order to be defined, while the latter can be associated with certain `Permission Elements`. Due to space limitations, we discuss only the context model facets that are relevant to access control.

2.2   Context Model Facets

This section provides an elaboration of the initial set of facets that have been included in the part of the model that gives rise to dynamic security controls. We note that all these model facets are focused on the aspects relevant to access control for cloud services.

Security Context Element
The `Security Context Element` refers to the following five top-level concepts (see Fig. 2):

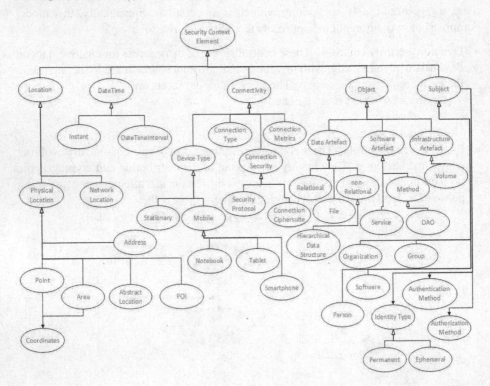

Fig. 2.   Security context element overview.

- `Location` - This class describes a physical and/or a network location where data are stored or from which a particular entity is requesting to access data.

- `DateTime` - This class describes the specific chronological point expressed as either instant or interval that characterises an access request (extends `owl-time:TemporalEntity`).
- `Connectivity` - This class captures the information related to the connection used by the subject for accessing sensitive data (see Fig. 2).
- `Object` - This class refers to any kind of artefacts that should be protected based on their sensitivity levels. These artefacts may refer to (non-) relational data, files, software artefacts that manage sensitive data or even infrastructure artefacts used.
- `Subject` - An instance of this class represents the agent seeking access to a particular data artefact. This can be an organization, a person, a group or a service (extends `foaf:Agent`, `goodrelations:BusinessEntity`, `goodrelations: ProductOrService`).

In Fig. 3, we provide further details regarding the `Connectivity` top level concept that include subclasses, imported or extended external classes, data and object properties. The identifier `pcm` (stands for PaaS Control Model) recognises the namespace underlying the classes and properties of the proposed vocabulary. Due to space limitations the details of all the top level concepts are not explained in this chapter but they are available in the following URL: http://imu.ntua.gr/software/context-aware-security-model.

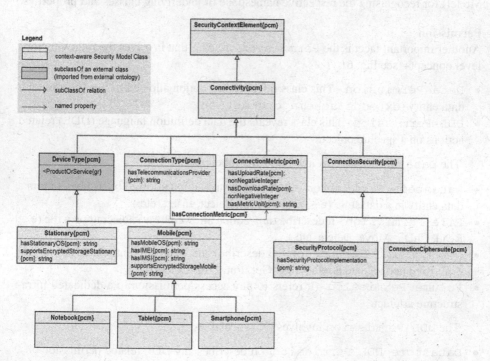

Fig. 3. UML class diagram for the connectivity context element.

Context Pattern

The next facet of this model is the `Context Pattern model` that includes the following top-level concepts:

- `Location pattern` - It refers to recurring motives of data accesses that are recognised with respect to the `Location` context element.
- `DateTime pattern` - It refers to recurring motives of data accesses that are recognised with respect to the `DateTime` context element.
- `Connectivity pattern` - It refers to recurring motives of data accesses that are recognised with respect to the `Connectivity` context element.
- `Object pattern` - It refers to recurring motives of data accesses that are recognised with respect to the `Object` context element.
- `Permission pattern` - It refers to recurring motives of data accesses that are recognised with respect to the `Permission` element.
- `Access Sequence Pattern` - It refers to data accesses that are recognised by any preceding access actions made by a particular subject (extends `Kaos:AccessAction`).

For the above vocabulary, we use the identifier `pcpm` (stands for PaaS Context Pattern Model) for recognising the respective namespace of underlying classes and properties.

Permission

Another important facet is the `Permission model` that involves the following top-level concepts (see Fig. 6):

- `Data Permission` - This class refers to any action allowed by a subject upon a data entity (extends `schema.org:Action`)
- `DDL Permission` - This class reveals the data definition language (DDL) related actions on a specific object.

The `Data Permission` involves four subclasses:

- `Datastore Permission` – It describes any action allowed by a subject upon a data entity in a datastore (e.g. Search, List, Select, Insert, etc.)
- `File Permission` - It describes any action allowed by a Subject upon a file (e.g. Read, ChDir, Move, Delete, etc.)
- `WebEndpoint Permission` – It describes any web endpoint related action that is allowed upon a data artefact (e.g. Get, Put, Post, Delete).
- `Volume Permission` - It refers to any access permission to a dedicated infrastructure artefact.

The `DDL Permission` involves two subclasses:

- `Datastore DDL Permission` – It describes any DDL related permission on a datastore (e.g. Create, Alter, Drop).
- `File System Structure Permission` - It describes any DDL related permission on a file (e.g. CreateDir, RenameDir, CopyDir, DeepCopyDir, ChOwner, etc.).

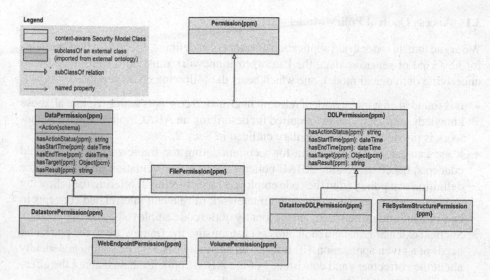

Fig. 4. UML class diagram for the `Permission` context element.

For the above vocabulary, we use the identifier ppm (stands for PaaS PaaS Permission Model) for recognising the respective namespace of underlying classes and properties.

In Sect. 3, we demonstrate the way that these contextual elements that give rise to dynamic security controls, can set the basis for developing a policy model for paas-enabled access control.

3 Policy Model for PaaS-Enabled Access Control

As already mentioned in Sect. 1, three are the main types of security policy that the PaaSword framework aims at supporting: (i) *Data encryption* policies. These determine the strength of the cryptographic protection that each sensitive object enjoys for confidentiality reasons. They give rise to security controls enforceable during bootstrapping of a cloud application. (ii) *Data fragmentation and distribution* policies. These determine the manner in which sensitive data objects must be fragmented and distributed to different physical servers for privacy reasons. They too give rise to security controls enforceable during application bootstrapping. (iii) *Access control* policies. These are essentially ABAC policies that determine when to grant, or deny, access to sensitive data on the basis of dynamically-evolving contextual attributes associated with the entity requesting the access. Context awareness is deemed of utmost importance for leveraging the security of cloud-based applications which by definition operate in dynamic and heterogeneous environments. Access control policies give rise to security controls dynamically enforceable during application execution time. Due to space limitations, in this chapter we only consider access control policies.

3.1   Access Control Policy Model

We argue that, in order to aid application developers in defining effective ABAC policies for any kind of sensitive data, the PaaSword framework must be underpinned by an underlying ontological model, one which bears the following characteristics:

- It is founded on a framework of relevant interrelated concepts which capture all those knowledge artefacts that are required for describing an ABAC policy. Such a framework is provided by the vocabulary outlined in Sect. 2.
- It uses an extensible formalism for accommodating the framework of interrelated concepts, hence expressing ABAC policies. Such a representation disentangles the definition of a policy from the code employed for enforcing it, offering the following seminal advantages: (i) It allows the framework of relevant interrelated concepts to be extended and instantiated, independently of the code employed by the application. Such an extension/instantiation aims at customising the framework to the particular needs of a given application. (ii) It forms an adequate basis for reasoning generically about the correctness and consistency of the ABAC policies, hence about the effectiveness of the security controls that these policies give rise to.

ABAC Policy Rules

Following an approach inspired by the XACML standard [9], an ABAC policy comprises one or more rules. A rule is the most elementary structural element and the basic building block of policies. A generic template for ABAC rules is provided in Table 1.

Table 1.   Generic ABAC rule template.

[actor] with [context expression] has [authorisation] for [action] on [controlled object]

The template defines a generic structure, in terms of relevant attributes, to which all ABAC rules in the PaaSword framework adhere. It comprises several attributes which are further elaborated below.

- *actor* identifies the subject who may request access to perform an operation on a sensitive object; it draws its values from the `pcm:Subject` class of the Security Context Element model defined in Sect. 2.2.
- *context expression* is a Boolean expression which identifies the environmental conditions that must hold in order to permit, or deny, the performance of an operation on a sensitive object. Context expressions are further elaborated in the following subsection.
- *authorisation* determines the type of authorisation (positive i.e. 'permit', or negative i.e. 'deny') that is granted.
- action identifies the operation that may, or may not, be performed on the protected sensitive object; it draws its values from the `ppm:Permission` class of the Security Context Element model defined in Sect. 2.2.

- *controlled object* identifies the sensitive object on which access is requested; it draws its values from the `pcm:Object` class of the Security Context Element model defined in Sect. 2.

In our ontological model, an ABAC rule takes the form of an instance of the class `pac:ABACRule` (see Fig. 5) of our framework. A number of object properties are attached to this class which are intended to capture the aforementioned attributes. As depicted in Fig. 5, these associate the `pac:ABACRule` class with an appropriate framework of relevant classes from the vocabulary of Sect. 2.2 which adequately capture the attributes of the ABAC rule template. The identifier `pac` (stands for PaaS Access Control) recognises the namespace underlying the classes and properties of the proposed ontological model.

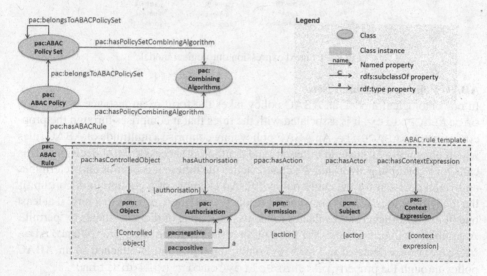

Fig. 5. ABAC ontological model.

Context Expressions

A context expression takes the form of an instance of the class `pac:ContextEx-`
`pression` of our framework (see Fig. 5). It specifies a number of constraints on the values of one or more instances drawn from the vocabularies `pcpm:ContextPat-`
`tern` and `pcm:SecurityContextElement` defined in Sect. 2.1. The class `pac:ContextExpression` is associated with these vocabularies through the object properties `pac:hasPatternParameter` and `pac:hasParameter` respectively depicted in Fig. 6. As we would expect, a context expression may combine two or more constraints using logical connectives (conjunction, disjunction, exclusive disjunction, negation). In order to capture such combinations of constraints, the `pac:Contex-`
`tExpression` class encompasses a subclass for each logical connective. A context expression may be defined recursively, in terms of one or more other context expressions. This is captured by associating the `pac:ContextExpression` class with

itself through the properties `pac:hasParameter` and `pac:hasPattern Parameter` (see Fig. 6).

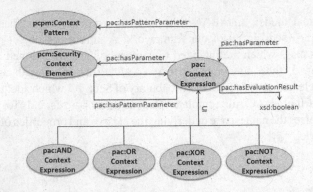

Fig. 6. Context expression ontological model.

ABAC Policies and Policy Sets

In our ontological model, an ABAC policy takes the form of an instance of the class `pac:ABACPolicy`. It is associated with the rules that it comprises through the property `pac:hasABACRule`. An ABAC policy may comprise a multitude of ABAC rules which potentially evaluate to different (and conflicting) access control decisions. This calls for a combining algorithm which reconciles the different decisions and determines an overall decision for the entire policy [9]. An example of a combining algorithm is the 'deny-overrides' algorithm, whereby a policy evaluation resolves to 'deny' if at least one of its constituent rules evaluates to 'deny', or if none of them evaluates to 'permit'. A combining algorithm takes the form of an instance of the class `pac:CombiningAlgorithms` depicted in Fig. 5. A combining algorithm is attached to an ABAC policy through the property `pac:hasPolicyCombiningAlgorithm`.

Following an approach inspired by the XACML standard [9], access control policies are grouped into *policy sets*. In our ontological model, a policy set takes the form of an instance of the class `pac:ABACPolicySet` (see Fig. 5). A policy is associated with its enclosing policy set through the property `pac:belongsToABACPolicySet`. A policy set may exhibit a hierarchical structure and comprise one or more other ABAC policy sets. This recursive inclusion is captured by rendering the `pac:belongsToA-BACPolicySet` property applicable to ABAC policy sets too (see Fig. 5). ABAC policy sets are also associated with combining algorithms. As in the case of policies, these reconcile the potentially different access control decisions to which the policies comprising a policy set may evaluate.

It is to be noted here that analogous policy models have been devised for the rest of the policy types outlined at the beginning of Sect. 3.

3.2   Access Control Policies in Linked USDL

Section 3.1 outlined a model for the generic representation of ABAC policies. This section demonstrates how this model can be incorporated into the ontological framework provided by Linked USDL [10], and in particular, into USDL-SEC – Linked USDL's security profile (USDL stands from Unified Service Description Language). By capitalising on USDL-SEC, our approach avoids the use of bespoke, non-standards-based, ontologies for the representation of ABAC policies (see Sect. 5.2 for a relevant outline of such ontologies). Instead, it is based on a diffused ontological framework which has recently attracted considerable research interest.

In addition, the adoption of Linked USDL brings about the following advantages [11]: (i) Linked USDL relies on existing widely-used RDF(S) vocabularies (such as GoodRelations, FOAF and SKOS), whilst it can be easily extended through linking to further existing, or new, RDF(S) ontologies. In this respect, it promotes knowledge sharing whilst it increases the interoperability, reusability and generality of our framework. (ii) By offering a number of different profiles, Linked USDL provides a holistic and generic solution able to adequately capture a wide range of business details. This is important for our work as it allows us to capture the business aspects of the security policies encountered within our framework. (iii) Linked USDL is designed to be easily extensible through linking to further existing, or new, RDF(S) ontologies. This is particularly important for our model as it facilitates seamless integration with the Context-aware security model of Sect. 2. (iv) It provides ample support for modelling, comparing, and trading services and service bundles. It also provides support for specifying, tracking, and reasoning about the involvement of entities in service delivery chains. This is important for our work for it allows comparisons to be drawn between different policy models that may potentially be offered through our framework.

The USDL-SEC Vocabulary

USDL-SEC identifies five top-level concepts: *Security Profile, Security Goal, Security Mechanism, Security Technology* and *Security Realization Type* (see Fig. 7). The *Security Profile* is a root concept that encompasses the different security profiles to which a cloud service, or application, may adhere. Each security profile is associated with one or more security goals. This gives rise to the *Security Goal* concept which encompasses a number of sub-concepts each representing a distinct security goal. A complete list of all security goals provided by USDL-SEC is depicted in Fig. 7. Each security goal is associated with one or more security mechanisms through which it is implemented. This gives rise to the *Security Mechanism* concept which encompasses a number of sub-concepts each representing a particular kind of security mechanism – a complete list of all the security mechanism kinds provided by USDL-SEC is depicted in Fig. 7. Each security mechanism is associated with one or more security technologies through which it is realised, giving rise to the *Security Technology* concept. In addition, a security mechanism is associated with a particular layer of the ISO/OSI protocol stack at which it is realised (e.g. the network or the application layer). This gives rise to the concept *Security Realization Type* that specifies such a layer.

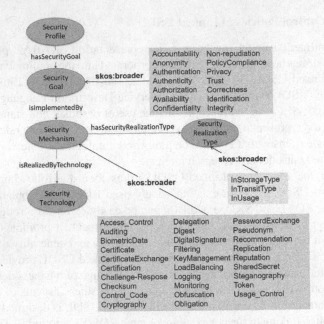

Fig. 7. USDL-SEC.

The concepts and their associations identified above are formalised in terms of classes of the ontology, and each concept association takes the form of an object property[1]. In fact, four object properties are introduced: hasSecurityGoal which associates a security profile with its corresponding security goal; isImplementedBy which associates a security goal with the mechanism that achieves it; isRealized-ByTechnology which associates a security mechanism with the technology that implements it; hasSecurityRealizationType which associates a security mechanism with the ISO/OSI layer at which it is realised. In addition, the fact that a concept forms a sub-concept of another concept is captured through the SKOS broader property.

The above framework of classes and properties lays the foundations for constructing a set of ontological templates suitable for the semantic representation of the three kinds of security policy that the PaaSword project supports. The following subsection, provides an account of how this framework is reified in order to give rise to an ontological template for the representation of ABAC policies. Analogous accounts apply to the other two kinds of policy, namely Data Encryption policies and Data Fragmentation and Distribution policies.

Incorporating ABAC Policies into USDL-SEC

At the highest level of abstraction, the ABAC policy model forms, essentially, a particular security profile to which a cloud application may adhere. In this respect it is

---

[1]   All USDL-SEC classes and properties are prefixed with the usdl-sec namespace. To avoid notational clutter, this namespace is omitted here.

modelled as an instance of USDL-SEC's `SecurityProfile` class, namely `pac:PaaSAccessControlProfile`. A security profile is associated, through the object property `hasSecurityGoal`, with one or more security goals from the USDL-SEC class `SecurityGoal`. In the case of ABAC policies, the security goal is *authorisation*. This is modelled in Fig. 8 by associating the instance `pac:PaaSAccess-ControlProfile` with an instance, say `pac:AccessControlGoal`, of the `Authorization` class through the property `hasSecurityGoal`. The `Authorization` class forms a sub-concept of `SecurityGoal`.

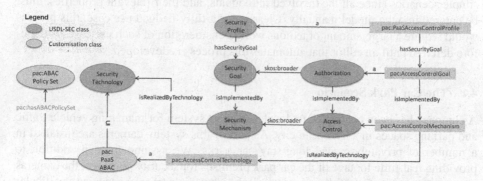

Fig. 8. USDL-SEC customisation (only classes and properties used in this chapter are depicted).

The authorisation goal is achieved by means of a suitable access control mechanism. USDL-SEC provides a layer of abstraction, namely the concept `SecurityMechanism`, for the specification of such a mechanism. In particular, it provides the class `AccessControl`, a sub-concept of `SecurityMechanism`, an instance of which, say `pac:AccessControlMechanism`, represents the access control mechanism offered by the PaaSword framework. This instance is associated with the `pac:AccessControlGoal` instance through the property `isImplementedBy`.

The access control mechanism represented by the instance `pac:AccessControlMechanism` is realised by means of some underlying concrete security technology. USDL-SEC provides a layer of abstraction, namely the concept `SecurityTechnology`, for the specification of such a technology. In our model, the access control mechanism is realised by the access control technology provided by the PaaSword framework. This is modelled by introducing the `pac:PaaSABAC` subclass (see Fig. 8), along with the instance `pac:AccessControlTechnology` which represents this access control technology. This instance is associated with the access control mechanism through the property `isRealizedByTechnology` (see Fig. 8). The `pac:PaaSABAC` subclass is associated, through the property `pac:hasABAC-PoliceSet`, with the class `pac:ABACPolicySet` (the top concept of the ABAC policy model of Fig. 5). This essentially captures the fact that the access control mechanism is realised through the policies encompassed in one or more ABAC policy sets.

It is to be noted here that the policy models devised for the rest of the policy types outlined at the beginning of Sect. 3 are incorporated into USDL-SEC in an analogous manner.

4   Case Study

The purpose of this section is to demonstrate how the Context-aware Security model of Sect. 2 and the policy model of Sect. 3 can be further reified in order to give rise to concrete security policies. In particular, we present a number of policies – and their constituent rules – that form part of a fictitious, albeit realistic, scenario. This scenario is deliberately kept simple and it is by no means intended to form a fully-fledged use case. It is to be noted here that we are currently obliged to confine ourselves to such simple scenarios since all the involved individuals, and their relevant properties, must be entered into the model manually. Clearly, for a fully-fledged use case, this process would require a large amount of tedious work; consideration of such use cases is therefore deferred until an editor that automates this process is developed.

4.1   The Car Park Scenario

A company has been contracted to develop a smart system for managing vehicle traffic and parking spaces in a European city. As part of this system, cameras are installed in a number of privately-owned long-stay car parks. We assume that, in addition to providing real-time footage of the car park premises for security purposes, the cameras also capture and store the following data: (i) Which vehicle (if any) occupies a particular parking space; vehicles are identified by their registration plate numbers. (ii) The date a vehicle entered the car park; we assume that in long-stay car parks the exact time of entry, or exit, of a car is not of interest as costs are typically charged on a per-day basis. (iii) The date a vehicle exited the park. We assume that these data are stored in the database table `ParkingPositions` depicted in Fig. 9.

SpaceId	RegistrationPlate	EntryDate	ExitDate	CostPerDay(€)
1001	NIP5146	17/11/2015	21/11/2015	7
1002	NZY8547	13/11/2015	24/11/2015	7
⋮	⋮	⋮	⋮	⋮

Fig. 9. The `ParkingPositions` table.

In order to compress costs and ensure storage elasticity, it has been decided to migrate the data contained in these tables to the cloud. To this end, SIEMENS' R&D department undertook the development of a cloud application, namely AppDB, through which car park administrators can obtain access to these data. The application is available for Windows, Mac OS and Android devices.

In order to alleviate security concerns, the application was developed with the aid of the framework presented in Sect. 3, which allowed the incorporation of a number of security controls during application design time. These controls implement a set of security policies which take the form of reifications of the ABAC policy model presented in Sect. 3 (see Fig. 5).

4.2   ABAC Policy

As part of the long-stay car park scenario we require that a user of the AppDB application, typically an employee of a car park should only be allowed read/write access to the table of Fig. 9 from specific locations and during specific hours. This requirement aims at preventing situations whereby the user of the AppDB accesses the table from a public place thus giving the opportunity to a third party to inadvertently, or on purpose, look at – or even alter – the data stored in the table.

This requirement gives rise to two access control rules, one allowing read access from specific locations and during specific times, and one allowing write access from the same locations and during the same times. These rules are represented by the individuals ex1:ABACRule_1 and ex1:ABACRule_2 respectively depicted in Fig. 10.

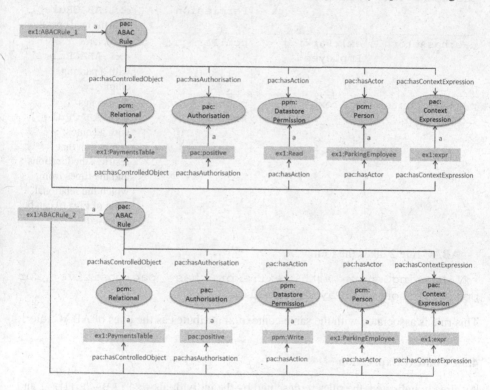

Fig. 10.   Instantiated ABAC rule template.

These rules are further elaborated below. ABAC rule 1 takes the form:

```
ex1:ParkingOwner  with  ex1:expr  has  pac:positive  for
ppm:Read on ex1:PaymentsTable
```

It reifies the generic rule template of Fig. 5. The rule is associated with a number of contextual attributes through the properties depicted in Fig. 10. These attributes are represented by the individuals illustrated in Fig. 10 and detailed in Table 2.

Table 2. ABAC rules contextual attributes.

Object property	Individual	Instance of	Description
pbe:has Controlled Object	ex1:PaymentsTable	pcm: Relational	Associates ex1:ABACRule_1 with the relational table of Fig. 9 that it protects
pbe:has Authorisation	pac:positive	pac: Authorisation	Associates ex1:ABACRule_1 with the positive (permit) authorisation
pac:hasAction	ppm:Read	ppm:Datastore Permission	Associates ex1:ABACRule_1 with the read operation
pac:hasActor	ex1:Parking Employee	pcm:Person	Associates ex1:ABACRule_1 with a parking employee.
pac:has Context Expression	ex1:expr	pac:Context Expression	Associates ex1:ABACRule_1 with a context expression that restricts the locations and the times from which the relational table of Fig. 10 can be accessed

ABAC rule 2 takes the form:

```
ex1:ParkingOwner  with  ex1:expr  has  pac:positive  for
ppm:Write on ex1:PaymentsTable
```

This rule is associated with the same contextual attributes as the ones of ABAC rule 1.

4.3 Context Expression

As already indicated, the rules represented by the individuals ex1:ABACRule_1 and ex1:ABACRule_2 involve a context expression which restricts the locations from which, and the times during which, the relational table of Fig. 9 can be accessed. This context expression is represented by the individual ex1:expr depicted in Fig. 10. This individual is an instance of the class pac:ANDContextExpression (see Fig. 6) and therefore the parameters that it involves are logically conjuncted. These parameters are associated with ex1:expr through the object property pac:hasParameter. They are modelled in Fig. 11 by the individuals ex1:EmployeeWorkingHours and ex1:expr1. The former individual is an instance of the class pcm:DateTimeInterval introduced by the Context-aware Security model of Sect. 2. It is restricted to the required working hours (e.g. say 09:00 to 17:00) through the properties `pcm:hasBeginning`, `pcm:hasEnd` and `pcm:hasTimeZone`. The former two properties associate `ex1:EmployeeWorkingHours` with the `xsd:dateTime` values T09:00:00 and T17:00:00 respectively. The latter property associates `ex1:EmployeeWorkingHours` with the `xsd:string` value +02:00.

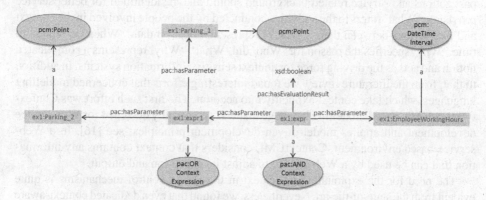

Fig. 11. Context expression for `ex1:ABACRule_1` and `ex1:ABACRule_2`.

The latter individual, namely `ex1:expr1`, is yet another context expression modelled as an instance of the class `pac:ORContextExpression`. This means that its two parameters, namely `ex1:Parking_1` and `ex1:Parking_2` (see Fig. 11), are logically disjuncted. These parameters are instances of the class `pcm:Point` from the Context-aware Security model presented in Sect. 2. They represent the two parking buildings from the long-stay car park scenario.

5 Related Work

The research outlined in this section is divided into two main categories: (i) works that are related to the Context-aware Security of Sect. 2; works that pertain to the Policy Model of Sect. 3.

5.1 Research Related to the Context-Aware Security Model

In the literature, there is a plethora of context models. For example, [12, 13] review models of context that range from key-value models, to mark-up schemes, graphical models, object-oriented models, logic-based models and ontology-based models. An interesting context model is the one proposed in [14], which was initially developed for mobile devices and later extended for use in service-based applications in [15]. Another example is the one in [16] who developed an ontological model of the W4H classification for context. The W4H ontology provides a set of general classes, properties, and relations exploiting the five semantic dimensions: identity (who), location (where), time (when), activity (what) and device profiles (how). Furthermore, authors exploited the concepts of the W4H ontology by including domain-independent common context concepts from existing work; e.g. FOAF, vCard, the OWL-Time Ontology, etc. The five dimensions of context have been also pointed out earlier by Abowd and Mynatt [17] who stated that context should include the 'five W': Who, What, Where, When, and Why. For example, by 'Who', they mean that it is not enough to identify a person as a customer; the person's past actions and service related background should also be identified for better service provision. 'What' refers to the activities conducted by the people involved in the context and interactions between them. 'Where' represents location data. 'When' is related to time. 'Why' specifies the reason for 'Who' did 'What'. 'Why' represents a complicated notion and acts as the driving force for context sensitive information systems. In addition to that, from the literature review we found interesting efforts that concerned modelling languages, which take context explicitly into account. The first such effort was ContextUML a UML-based modelling language that was specifically designed for Web service development and applies model-driven development principles; see [16]. In a Web-service-based environment, ContextUML considers that context contains any information that can be used by a Web service to adjust its execution and output.

The need for the exploitation of context in the access control mechanisms is quite evident from the state-of-the-art. Nevertheless, we found that even dedicated context-aware extensions to traditional access control models (e.g. Role-based Access Control - RBAC) either do not cover all the contextual elements with a reusable security related context model or are proven hard to maintain in dynamic environments where users often change roles or are not known a priori [18]. On the other hand, pure ontological models (e.g. [16]), or even Attribute-based Access Control (ABAC) approaches (e.g. [19]) they do not seem to cover all the security requirements associated with the lifecycle of a cloud application (i.e. bootstrapping and run-time). Specifically, either they do not cover the full range of contextual elements that are associated with all the security aspects of sensitive data managed by cloud applications or they are based on heavy inferencing that is considered as inefficient for such dynamic environments [20].

5.2   Research Related to the Policy Model

Turning now to policies and policy-based applications, syntactic descriptions promote a declarative approach to policy expression, one which aims at replacing a trend whereby policies are encoded imperatively, as part of the same software that checks for their compliance. Several markup languages have been proposed for the declarative description of policies, some prominent examples being [9, 21–23]. These generally provide XML-based syntaxes for expressing policy rules and sets. Nevertheless, such syntactic descriptions fail to capture the knowledge lurking behind policies. In this respect, they are merely data models that lack any form of semantic agreement beyond the boundaries of the organisation that developed them. Any interoperability relies on the use of vocabularies that are shared among all parties involved in an interaction.

In order to overcome the aforementioned limitations, semantically-rich approaches to the specification of policies have been brought to the attention of the research community. These generally embrace Semantic Web representations for capturing what we term action-oriented policies, i.e. policies which control when a particular actor or subject can perform a specified action on, or through the use of, a particular resource. These approaches typically employ ontologies in order to assign meaning to actors, actions and resources. Several works in the area of semantic policy representation have been reported in the literature [24–26]. In [24], the authors presented KAoS – a general-purpose policy management framework which exhibits a three-layered architecture comprising:

- A human interface layer, which provides a graphical interface for policy specification in natural language.
- A policy management layer, which uses OWL [27] to encode and manage policy-related knowledge.
- A policy monitoring and enforcement layer, which automatically grounds OWL policies to a programmatic format suitable for policy-based monitoring and policy enforcement.

In [25] the authors proposed Rei – a policy specification language expressed in OWL-Lite [27]. It allows the declarative representation of a wide range of policies which control which actions can be performed, and which actions should be performed, by a specific entity. Furthermore, it defines a set of concepts (rights, prohibitions, obligations, and dispensations) for specifying and reasoning about access control rules. In this respect, it provides an abstraction which allows the specification of a desirable set of behaviours which are potentially understandable – hence enforceable – by a wide range of autonomous entities in open and dynamic environments.

In [26], the authors recognise that cloud computing, and in particular the concept of multi-tenancy, calls for policy-driven access control mechanisms. They propose an ontology-based framework to capture the common semantics and structure of different types of access control policies (e.g. XACML policies, firewall policies, etc.), and facilitate the process of detecting anomalies in these policies. Their ontology captures the underlying domain concepts involved, the policy structure and the policy attributes. Particular types of access control policies are obtained by appropriately instantiating the ontology.

6 Conclusions

We have presented suitable vocabularies of concepts and properties, namely the Security Context Element, the Context Pattern and the Permission, which adequately capture the knowledge lurking behind ABAC policies. We have also proposed a generic ontological model for the abstract representation of ABAC policies, which disentangles the definition of a policy from the actual code employed for enforcing it, bringing about the advantages outlined in Sect. 3.1. The model is underpinned by the Security Context Element vocabulary, and is incorporated into the ontological framework offered by USDL-SEC (Linked USDL's security profile). Such a model forms the basis of the PaaSword framework – essentially a security-by-design framework which aims at aiding cloud application developers in defining effective access control policies for any kind of sensitive data.

Any effective use of the ABAC policy model requires a mechanism through which it can be suitably customised in order to allow for the specification of concrete ABAC policies. Such a customisation amounts to an extension and/or instantiation of the abstract classes and properties presented in Sect. 3. It is the responsibility of such a mechanism to ensure that this extension/instantiation takes place according to a set of predefined governance policies. In the future, we intend to investigate the construction of a higher-level ontological framework that will generically accommodate these governance policies and thus pave the way for the construction of a generic customisation mechanism that can be easily adapted to the particular needs of the potential adopter of our framework.

Acknowledgements. The research leading to these results has received funding from the European Union's Horizon 2020 research and innovation programme under grant agreement No. 644814. The authors would like to thank the partners of the PaaSword project (www.paasword.eu) for their valuable advice and comments.

References

1. Cisco: Cloud: What an Enterprise Must Know, Cisco White Paper (2011)
2. Vaquero, L.M., Rodero-Merino, L., Caceres, J., Lindner, M.: A break in the clouds: towards a cloud definition. SIGCOMM Comput. Commun. Rev. **39**(1), 50–55 (2008)
3. Micro, T.: The Need for Cloud Computing Security. Trend Micro (2010)
4. NIST: Cloud Computing Reference Architecture, National Institute of Standards and Technology (2011)
5. CSA: The Notorious Nine. Cloud Computing Top Threats in 2013. Cloud Security Alliance (2013)
6. Verginadis, Y., Michalas, A., Gouvas, P., Schiefer, G., Hübsch, G., Paraskakis, I.: PaaSword: a holistic data privacy and security by design framework for cloud services. In: Proceedings of the 5th International Conference on Cloud Computing and Services Science, CLOSER 2015, 20–22 May, Lisbon, Portugal (2015)
7. Ferrari, E.: Access Control in Data Management Systems. Synthesis Lectures on Data Management, vol. 2, no. 1, pp. 1–117. Morgan & Claypool (2010)
8. Dey, A.K.: Understanding and using context. Pers. Ubiquit. Comput. J. **5**(1), 4–7 (2001)
9. OASIS: OASIS eXtensible Access Control Markup Language (XACML) (2013). http://docs.oasis-open.org/xacml/3.0/xacml-3.0-core-spec-os-en.html
10. Linked USDL (2014). http://linked-usdl.org/
11. Pedrinaci, C., Cardoso, J. Leidig, T.: Linked USDL: a vocabulary for web-scale service trading. In: 11th Extended Semantic Web Conference (ESWC) (2014)
12. Strang, T., Linnhoff-Popien, C.: A Context modeling survey. In: Workshop on Advanced Context Modelling, Reasoning and Management, UbiComp 2004 - The Sixth International Conference on Ubiquitous Computing, Nottingham, England (2004)
13. Bettini, C., Brdiczka, O., Henricksen, K., Indulska, J., Nicklas, D., Ranganathan, A., Riboni, D.: A survey of context modelling and reasoning techniques. Pervasive Mob. Comput., 161–180 (2010)

14. Miele, A., Quintarelli, E., Tanca, L.: A methodology for preference-based personalization of contextual data. In: ACM Proceedings of the 12th International Conference on Extending Database Technology: Advances in Database Technology, EDBT 2009, Saint-Petersburg, Russia, pp. 287–298 (2009)
15. Bucchiarone, A., Kazhamiakin, R., Cappiello, C., Nitto, E., Mazza, V.: A context-driven adaptation process for service-based applications. In: ACM Proceedings of the 2nd International Workshop on Principles of Engineering Service-Oriented Systems, PESOS 2010, Cape Town, South Africa, pp. 50–56 (2010)
16. Truong, H.-L., Manzoor, A., Dustdar, S.: On modeling, collecting and utilizing context information for disaster responses in pervasive environments. In: ACM Proceedings of the First International Workshop on Context-Aware Software Technology and Applications, CASTA 2009, Amsterdam, The Netherlands, pp. 25–28 (2009)
17. Abowd, G., Mynatt, E.: Charting past, present, and future research in ubiquitous computing. ACM Trans. Comput. Hum. Interact. (TOCHI), 29–58 (2000). Special issue on human-computer interaction in the new millennium
18. Heupel, M., Fischer, L., Bourimi, M., Kesdogan, D., Scerri, S., Hermann, F., Gimenez, R.: Context-aware, trust-based access control for the di.me userware. In: Proceedings of the 5th International Conference on New Technologies, Mobility and Security, NTMS 2012, Istanbul, Turkey, pp. 1–6. IEEE Computer Society (2012)
19. Jung, C., Eitel, A., Schwarz, R.: Cloud security with context-aware usage control policies. In: Proceedings of the INFORMATIK 2014 Conference, pp. 211–222 (2014)
20. Verginadis, Y., Mentzas, G., Veloudis, S., Paraskakis, I.: A survey on context security policies. In: Proceedings of the 1st International Workshop on Cloud Security and Data Privacy by Design, CloudSPD 2015, Co-located with the 8th IEEE/ACM International Conference on Utility and Cloud Computing, Limassol, Cyprus, 7–10 December (2015)
21. Specification of Deliberation RuleML 1.01 (2015). http://wiki.ruleml.org/index.php/Specification_of_Deliberation_RuleML_1.01
22. Security Assertions Markup Language (SAML) Version 2.0. Technical Overview (2008). https://www.oasis-open.org/committees/download.php/27819/sstc-saml-tech-overview-2.0-cd-02.pdf
23. WS-Trust 1.3 (2007). http://docs.oasis-open.org/ws-sx/ws-trust/200512/ws-trust-1.3-os.doc
24. Uszok, A., Bradshaw, J., Jeffers, R., Johnson, M., Tate, A., Dalton, J., Aitken, S.: KAoS policy management for semantic web services. IEEE Intel. Sys. **19**(4), 32–41 (2005)
25. Kagal, L., Finin, T., Joshi, A.: A policy language for a pervasive computing environment. In: 4th IEEE International Workshop on Policies for Distributed Systems and Networks, POLICY 2003 (2003)
26. Hu, H., Ahn, G.-J., Kulkarni, K.: Ontology-based policy anomaly management for autonomic computing. In: 7th International Conference on Collaborative Computing: Networking, Applications and Worksharing (CollaborateCom) (2011)
27. OWL Web Ontology Language Reference. W3C Recommendation (2004). http://www.w3.org/TR/owl-ref/

Towards Outsourced and Individual Service Consumption Clearing in Connected Mobility Solutions

Michael Strasser[1]([⊠]) and Sahin Albayrak[2]

[1] Bosch Software Innovations GmbH, Schöneberger Ufer 89-91,
10785 Berlin, Germany
michael.strasser@bosch-si.com
[2] DAI Laboratory, Technical University Berlin, Ernst-Reuther-Platz 7,
10587 Berlin, Germany
sahin.albayrak@dai-labor.de

Abstract. The work on hand targets the missing capabilities for data access and service clearing in connected mobility service solutions. Interviewed experts pointed out that current solutions of the mobility domain lack appropriate clearing capabilities as well as access on internally processed data. We approach both limitations and elaborate a message protocol and respective interfaces which enables participants of a service solution to outsource their clearing process individually. The design of the interfaces and the message protocol are discussed. Its feasibility is justified with the help of an additionally developed clearing service prototype. This prototype uses the interfaces in an automated fashion and calculates the costs for a set of transactions in accordance to a given contract. The presented interfaces and message protocol support service solution developers in their development of new marketplace capabilities demanded by participants. What kind of clearing might be applicable in an ecosystem of interconnected marketplaces is furthermore discussed. The work on hand closes the gap of limited data access and service clearing. The findings of this work can be used as a foundation for a standard regarding service clearing or any other use case that requires access on internal marketplace data.

1 Introduction

The number of service consumers and operators (providers) for mobility services and service provisioning solutions increases constantly. It is anticipated that this trend continues due to the growing mobility density. This implies that, due to the increasing number of citizens and commuters, the need for more mobility reasons raises but the infrastructure is limited due to geographical reasons. This subsequently leads to a higher number of service offerings, demands and transactions [1]. As a logical consequence, more business relationships will be established via service marketplaces. These marketplaces represent a Business-to-Business (B2B) environment in which participants can trade services.

© Springer International Publishing AG 2017
M. Helfert et al. (Eds.): CLOSER 2016, CCIS 740, pp. 360–381, 2017.
DOI: 10.1007/978-3-319-62594-2_18

The increasing demand for convenient mobility will increase the need for and usage of mobility services. That circumstance leads to more transactions which are processed by service operators and routed by mobility service marketplaces. The service consumption has to be charged respectively but an appropriate clearing approach is currently missing. It is likely that the growth of service consumptions results in a higher effort for service clearing (transaction billing). Therefore it is feasible to assume that service operators want to outsource their service clearing process to a third party. This party has to have comprehensive service clearing knowledge and approved and standardized processes.

In this work we demonstrate how a marketplace participants can outsource its service consumption clearing to an individual third party which is also registered with the marketplace. One of the major hurdles is that a clearing operator has to have access on participants data stored inside a marketplace. However, due to the assumption that such a marketplace is operated via the cloud, tenant relevant data might be isolated and strictly kept separate. Thus a clearing operator, even though its is an ordinary marketplace participants, has to have certain privileges to access other tenants' data. A service operator which offers a service, for instance for charging stations in the smart mobility context, contracts a clearing service of a clearing operators (called SCCO henceforth). Once they are business partners, the charging station operator can assign the SCCO to its services (individually) which is then responsible for clearing (charging) the charging station service consumptions. Therefore, a service marketplace has to provide mechanisms to ensure accurate data access on internally stored data via respective and secure interfaces. And more important, provide a mechanism to enable access on tenant data if a requester has a privileged role like the SCCO. Interviews with experts from the mobility domain have been conducted in [2]. The authors discussed the situation of mobility trading environments with experts who presented the current state of the art, outlined the actual limitations and future improvements of mobility service solutions and presented Critical Success Factors (CSF) which should be in place. The experts point out the need for individual service clearing (as a CSF) in service trading environments as this functionality is currently not available. Individual in this context means that a marketplace participant can choose to do the clearing either on its own or contract and instruct a professional SCCO for doing it on its behalf. As discussed in [3], a service infrastructure has to provide appropriate mechanisms in place that enable flexible billing. However, due to missing marketplace Application Programming Interfaces (API)s to access and obtain internally stored data, service clearing and other use-cases cannot be realized.

The contribution of this work is an analysis and implementation of a fully functional clearing mechanism for mobility services marketplaces to close the gap described in Sect. 2. We furthermore discuss the clearing concept of a Directory Agent (DA). This DA is the management entity of an ecosystem to which individual service marketplaces can connect to exchange data between each other [4,5]. The ecosystem supports the data exchange across marketplace boundaries which enables service consumption of foreign services. This request roaming requires a charging concept too. The work on hand is a special selection and contribution to [6]. It provides new technical insights about how to implement the outsourced clearing approach with a Business Process Management (BPM) and Business Rules Management (BRM) application. The rest of the paper is organized as follows. Section 2 presents the current state of the art in B2B service clearing in mobility service marketplaces. Section 3 presents and discusses interfaces which have been identified as necessary to enable communication and are therefore implemented within our marketplace solution. The evaluation of these interfaces is done in Sect. 4 where a rule based service-clearing algorithm is presented. How clearing might be done within an ecosystem, composing of an ecosystem manager and connected service marketplaces is briefly discussed in Sect. 5. The paper ends with a conclusion in Sect. 6 and an outlook in Sect. 7.

2   State of the Art and Problem Description

Research projects like Green eMotion, CROME, Olympus, Streetlife explore possibilities about how to provide services for the mobility domain. These services strive various mobility domains, like parking, routing, charging, sharing, public transport, etc. The trade of such services can be done via a service marketplace, also called *e-hub* [7], *service provisioning platform* [8], *Value Added Service Supplier* [9] *intermediary* [10], *Platform* [11] or *electronic marketplace, trading communities, trading exchanges* [12]. Research projects mostly try to achieve specificity identified use cases but clearing is often treated as secondary. On the other side, commercial companies like Hubject, Parku, Multicity or Smartlab also provide Information and Communication Technology (ICT) based solutions over which mobility service providers and consumers can trade services in a B2B fashion like goods [13]. An electronic marketplace matches sellers and buyers and encourage them for service trades by providing appropriate trading capabilities [12]. Current marketplaces for electric mobility services lack the possibility to enable sophisticated and comprehensive data access on internally processed data from outside. This limitation prevents future use cases and the realization of CSFs. Interviewed experts have emphasized that it is mandatory to access transaction data, contract data, participant information or service descriptions via APIs. By the current date, neither an appropriate message protocol nor respective interface standards are available to guide the development of respective interfaces regarding comprehensive communication with the outside world. Without comprehensive access, internal data is locked inside a marketplace and cannot be processed further by any participant except the marketplace itself.

Participants pay a membership fee to the marketplace (depends on a marketplaces' revenue model [14]) but also pay the service operators for using their services. An identified business case (indicated by interviewed experts and literature) is B2B service consumption clearing, also known as service charging. Clearing is considered as a CSF of electronic marketplaces [2,14,15]. The experts point out that service provisioning marketplaces have currently no elaborated clearing approach or completely lack the possibility to enable service clearing within their system. [16] point out that a business model for clearing does currently not exists and that "the billing process is more cost-intensive than the overall billing amount". However, clearing mechanisms have to be in place as service transactions are not free of charge [1,14,17,18]. Marketplaces for electric charging infrastructure like Green eMotion and Smartlab's e-clearing.net do B2B service clearing for their participants right within their systems. Their participants have no choice to appoint an individual clearing operator or do it on their own. CROME or Hubject do not support service clearing inside their platform and leave it to their participants which have to agree on a common clearing method. The limited service clearing mechanism is a disadvantage of today's mobility service platforms and marketplaces. Current solutions do not provide appropriate interfaces which enable access to internally stored data or do not even store data accordingly (interviewed experts). This prevents the possibility for individual service clearing where participants outsource their clearing process by appointing an individual SCCO. The state of the art in mobility transaction clearing in marketplaces for mobility service is in summary that either the marketplace itself seizes service clearing or participants have to do it somehow outside the marketplace. Marketplaces like Amazon or iTunes as well as eBay and other known marketplaces or trading platforms offer a few payment options to their participants from which the participants can choose. Individual service clearing via SCCOs is not available in these solutions.

3 Accessing Internal Data Stored Inside Marketplaces

A comprehensive set of marketplace interfaces would enable the accomplishment of various use cases, not only individual service clearing but also interconnectivity between service marketplaces [4,19], Service Level Agreement (SLA) and condition monitoring, compensation claims and others. The marketplace, which is developed in this work, makes use of a Service Oriented Architecture (SOA) and uses Simple Object Access Protocol (SOAP) for the communication. The interfaces are furthermore differentiated between public and private. Context, service or participant relevant data can only be accessed via private interfaces. This kind of data might be classified as confidential and only accessible for registered marketplace participants. General information about a marketplace and its participants, services, supported domains or operation area can be retrieved via public interfaces which do not require a marketplace membership. Interfaces to deal with Cloud Computing resource elasticity or similar concepts are not in the scope of this work. Due to security mechanisms, a API requester is validated in respect to its marketplace membership and role, using digital certificates as well as additional credentials. The communication is furthermore encrypted and end customer data only stored anonymously. In case that a marketplace applies multi-tenancy and keeps tenant data strictly isolated, then the private interfaces have to arrange data access on other tenants data if required. The interfaces are role-base driven rather than data driven. That enables re-using and re-purposing of interfaces while being able to limit the data access to roles.

We separate the web interfaces between public and private but each web service composes of various operations. Due to this split the interfaces can have different assigned service level agreements (SLA)s. Extensible markup language (XML) schema validation is used to validate the input data which ensures parameter and data type integrity which intends to prevent internal malfunction. Figure 1 presents a simplified overview about how marketplace data is accessed from external. Even though the security layer can also be applied for public interfaces, it is assumed that security checks are costly and thus should only be conducted when necessary. Marketplace interfaces facilitate the realization of various use cases. Performance and efficiency monitoring, Service Level Agreement (SLA) validation, interconnectivity or service clearing are just a few which rely on APIs which allow to access internally stored marketplaces data from outside. The elaborated marketplace has to provide interfaces and processes to access contract and transaction data from other tenants.

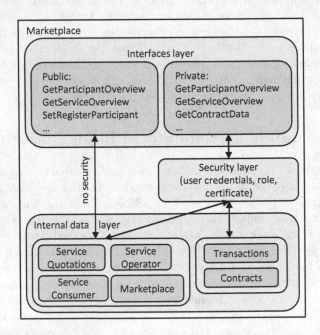

Fig. 1. Interfaces to access internal marketplace data (Source: [6]).

3.1   Service Consumption Data

The following use case exposes how transaction data incurs within a mobility service marketplace. A service operator (SO) has parking infrastructure. The SO is registered with a service marketplace (MP) and offers a service which enables to access his infrastructure. The service's functionalities are for example (i) to open a barrier arm, (ii) to guide a driver to an empty parking lot or

(iii) to determine the duration of the parking. A service consumer (SC), which is registered with the same MP, searches for a service with capabilities to access parking infrastructure. Therefore, the SC (i) contacts the SO, (ii) both close a digital contract about the service usage and settle a business relationship and (iii) establishes request forwarding within their systems. An end-customer (EC) of the SC cannot, at first, access the parking infrastructure of the SO because the SO does not know the EC. For the SO is the EC a stranger and thus forwards the request to the MP. The MP checks all of the SO's contracts and forwards the service request to all business partners of the SO. The SC retrieves the service request and recognizes the EC. The SC responses accordingly and the MP forwards the response to the SO. The SO opens its parking infrastructure for the EC because the SC has acted as guarantor. As soon the EC leaves the parking area, the SO sends a *park detail record*, which contains information about the parking process to the MP. The MP forwards the record to the SC. If a requests is roamed within the ecosystem via the DA as suggested in [4,19], then the clearing has to be done across marketplace boundaries. Furthermore is it necessary that the DA implements a clearing mechanism to charge those which have forwarded their requests within the ecosystem.

3.2 Individual Service Clearing in Service Marketplaces

In the current context, service clearing is done in respect to B2B service consumption among marketplace business partners (service operator and consumer). The outsource of the service clearing process to a third party is called *outsourced invoice* [20]. This approach is, according to the interviewed experts, not yet available by any mobility service solution. However, the experts emphasize that it is a fundamental functionality and according to [12], is it necessary that a trading platform provides mechanisms to arrange payment. Even though [21] describe in detail what kind of data has to be in place to clear an electric vehicle charging process via a service marketplaces, they miss the chance to implement and present a working solution. Futhermore, their presented data is too general to be used for a prototypical implementation. To outsource the service clearing process to a SCCO, a service operator must comply to the following process: A SCCO-A offers a clearing service via a marketplace (MP). Another service operator (SO-B), which offers parking infrastructure services, closes a contract with the SCCO-A. Then, the SO-B assigns the contracted clearing service of SCCO-A to its service XY. This entitles the SCCO-A to access those contracts and transactions which relate to the SO-B's XY service. The service description of XY should particularly emphasize that a third party (SCCO-A) has been entitled to accesses all accrued transactions and the contracts related to service XY. A service consumer (SC) who signs a contract with the SO-B for service XY agrees on the service's conditions and thus accepts that the SCCO-A accesses all clearing relevant data. This agreement is important because the transaction data represents a kind of an end-customer's mobility profile. Thus, the SC has to inform its end-customers about the third party access. The SCCO-A gathers the necessary data according to the defined clearing schedule and calculates the price which the SC has to pay to SO-B. Due to the individually, the SO-B can appoint another SCCO to its other services but also can do the clearing on its own without third party support. The role-heritage diagram in Fig. 2 introduces those roles used in the use-case diagram presented in Fig. 3. The role *Clearing* is inherited from the role *Service Provider*.

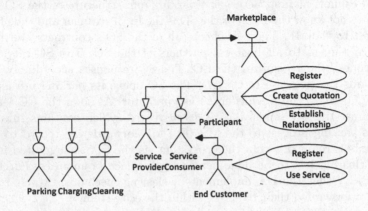

Fig. 2. Overview role inheritance. Own figure published in [6].

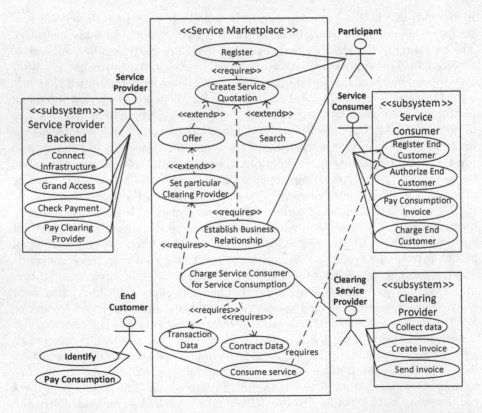

Fig. 3. Use-Case diagram for individual service clearing. Own figure published in [6].

The clearing role is, although a service operator (provider), somehow privileged as it has to have access on third party, tenant-isolated contract and transaction data. In contrast, a *parking* or *charging* operator do not have to access internal marketplace or tenant related data. The role *end customer* represents a Business-to-Customer (B2C) relationship with the *service consumer*. Solid arrows represent a membership while empty arrows represent an instance of a particular role. The generic use-case diagram illustrates the roles, systems and processes which are part of an individual and outsourced clearing process within a mobility service marketplace. The elaborated use-case diagram is neither an extension nor a modification of the e-roaming clearinghouse [16]. It presents a succeeding step in which e-roaming is actually charged according to previously defined conditions written in a contract. To achieve individual and outsourced service clearing in service marketplaces, various logical and functional challenges have to be approached. An overview of identified challenges is shown in the list below.

- Grant or reject access on internally stored data
- Authorize access at other participants (tenant) data
- Manage data access scope to prevent unauthorized access
- Define appropriate process to enable contract closing for clearing service
- Set up notification for service consumers about third party data access
- Manage data access in case a service's SCCO changes
- Manage data access to avoid multiple access by multiple SCCOs
- Control data access so that it cannot be charged multiple times
- Prevent that a SCCO outsources the clearing to another SCCO
- Evaluate the robustness, role-based approach and maintenance of the interfaces
- Check interface compatibility for other use cases like marketplace interconnectivity.

The list does not claim to be complete. Once our proposed solution is implemented into commercial solutions it is believed that further challenges will be identified. The challenges furthermore depend on the design of the service clearing as well as on the implemented contracting mechanisms as described in [5].

The authorization of a participant to access data of another participant is challenging. Cloud computing applications are most of the time multi-tenancy applications that strictly isolate tenant data and a marketplace should apply cloud computing principles [22] This separation can be achieved by storing data on different (virtual) hardware or middleware. However, this data separation via different locations makes it difficult for tenant A to access data of tenant B. Solutions have to incorporate hubs that bridges the tenants for certain use cases like service clearing. Another challenge is the access withdraw. In case a clearing contract expires, the assigned clearing provider has to be able to do the final clearing. Thus, although the data access is canceled, the final clearing requires access to the contracts and the transaction data until the day of the expiration. Moreover, a clearing provider is only allowed to access data of those service providers which have entitled them for doing so. The entitlement is not a permission to access everything but only a special portion. Clearing loops, which appear if a SCCO outsources its clearing process further to another SCCO, has to be detected and avoided. A SCCO is a privileged role which has a deeper access than other roles as it has to use, per default, foreign tenant data to create an appropriate invoice. The marketplace has to adjust its interfaces and internal processes accordingly. Legal regulations in respect to clearing are not part of the research's scope. The interviewed experts confirm the complexity of service clearing but unfortunately do not propose a suitable solution. No standardized protocol or manual exists on which marketplace APIs can be build on. The relevant domain protocols have capabilities to support the marketplace in its operation only. Marketplace participants are able to provide information to the platform but cannot retrieve necessary information about contracts and transactions or other data useful for future business cases. Access on data is only possible for data which has been provided by the participants before. No guidelines are available which emphasize the required steps to enable data access from outside and, even more important, no research was found which identified what interface functionalities are required by mobility marketplace participants. Therefore, this work provides insights into the required steps, tasks and data for doing so while discussing the need for individual and outsourced service clearing.

3.3   Elaboration of Service Clearing Interfaces

Table 1 presents the major interfaces for outsourced and individual service clearing in mobility service marketplaces. Those interfaces have been implemented with a BPM application. These interfaces are used by a SCCO to access clearing relevant data from the service marketplace. The clearing functionality has been realized with several light wise interfaces to achieve a better performance, to reduce maintenance work and to enable interface re-usability. Re-usable interfaces might support future use cases like marketplace interconnectivity.

It is pointed out that marketplace interfaces should not, if applicable, serve one particular use-case only. Use-case relevant interfaces process and return data

Table 1. Interfaces for service clearing in service solutions.

Interface name	Description
GetClearingRelatedContracts	Returns a list of contracts in which the requester is set as SCCO. This is required to Identify the service consumer to which the invoice will be send
GetContractData	Returns contract details like participant contact data and usage price plan. This is necessary to determine the prices for the transactions
GetTransactionData	Returns the transaction data that belongs to the to be charged contract

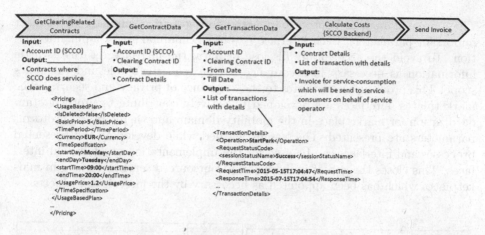

Fig. 4. Interface sequence for service clearing (Extended from source: [6]).

for a particular use case, thus they are probably not applicable for other use-cases. The developed marketplace and its interfaces are use-case independent and role-based driven. Thus participants and there roles are differentiated and the data returned respectively. The interfaces in Table 1 return data according to the provided certificate and identifiers. The first is used by a SCCO only while the remaining two can be used by all registered marketplace participants for informative reasons. A role validation is necessary to decide whether access is granted or rejected and which data has to be delivered to the requester. Due to the interfaces' use case independent implementation it is possible to use them individually or in combination. Individual usage would be, for example, to check transactions or to check when a specific contract expires. Combined usage is about using an interface's output for another interface's input. This is done for service clearing and presented in Fig. 4.

At first, a SCCO checks its contracts to identify its clients by using the *GetClearingRelatedContracts* interface. These contracts define the services for which the SCCO is entitled to do the service clearing. Once a SCCO knows which services are under its clearing responsibility it uses the *GetContractData* interface to retrieve the contracts which have been signed between its clients and their business partners. With those contracts, a SCCO knows the service consumption conditions (price plan) and other contact data like the address. The *GetTransactionData* interface is used to retrieve all transaction data that belongs to a specific contract. Once all data is gathered, a SCCO computes the costs of all retrieved transactions according to the contract and the specified price plan. The final cost calculation is implemented as a rule-based service-clearing service using a BRM application. After the final cost has been calculated the SCCO sends the invoice to its clients' business partners on behalf of its client. As previously mentioned, a service operator is able to entitle different clearing operators to do the service clearing for different services.

However, at one point in time each service can have at most one assigned SCCO. From a SCCO point of view it is relevant which interface is used first.

The presented input and output parameters in Fig. 4 are a summary of the most important parameters. Each interface requires a minimal set of input information. To avoid the accidentally disclose of data, the interface returns as little information as necessary in accordance to the requester's role and the role's scope. This procedure contributes to the assurance of privacy and security. The marketplaces' interfaces are developed to actively contribute to protocol standardization for marketplace in the mobility domain and therefore the protocols parameters are presented. The lessons learned while developing the presented processes and interfaces can be used for the implementation of additional interfaces. This closes the gap of individual and outsourced service clearing in marketplaces which has been appointed as necessary by the interviewed experts.

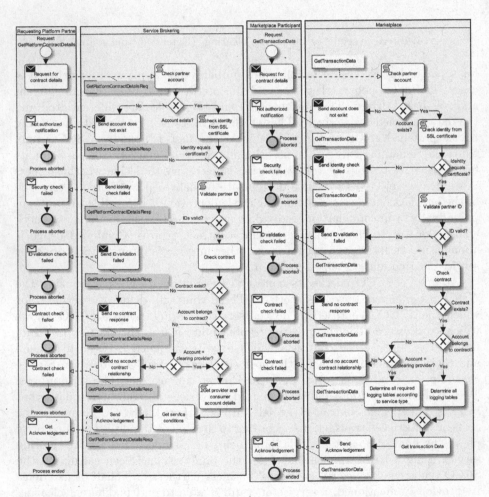

Fig. 5. BPD to retrieve contract data (left) and corresponding transaction data (right). Left figure published in [6].

Figure 5 presents the elaborated business processes to retrieve contract data (left side) and transaction data (right side). The processes are visualized using Business Process Diagrams (BPD). The BPDs show the necessary steps which are identified as necessary to validate, authorize and access data of different tenants from external. The processes have to be implemented by a marketplace operator and provided to its participants. These interfaces are the first of its kind which enable access on internal data from external and solve the current problem issue in respect to service consumption clearing but contributes furthermore to future use cases.

3.4 Interface Implementation

The interfaces have been implemented as web-service interfaces using *inubit*, a Java and XML based BPM application. Its process engine runs on Apache Tomcat and Java Development Kit (JDK) 7. The data model in Fig. 6 shows the input parameters and their data types of the clearing related interfaces (see Table 1). The figure shows that the effort for the interface implementation regarding the message protocol is not challenging due to the manageable number of parameters.

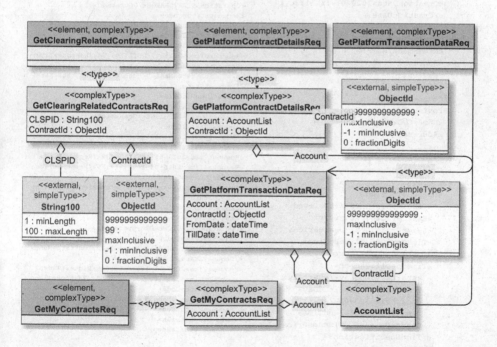

Fig. 6. Clearing interfaces request data model.

The *AccountId* is the identifier used to identify a requester. The *ContractId* is indispensable to determine a specific contract and subsequently retrieve the contract details. The prototypical implementation of the introduced clearing interfaces has proven that these two identifiers are sufficient to determine all account, contract and transaction data. Special roles like SCCO have special identifiers to avoid confusion when using the interfaces. The *GetClearingRelated-Contracts* is an interface that can be used by a SCCO only and thus specifically requires a *SCCOID*. This identifier uniquely identifies (in combination with a certificate) a clearing operator and does not accept an ordinary *AccountId*. All interfaces validate whether an *AccountId* fits to the given *ContractId* or not. This is done to determine a requester's access scope and to prevent contract data from being accidentally exposed. For example, using the *GetTransaction-Data* interface with the same *ContractId* but different *AccountId*s returns either (i) a different data set, (ii) a subset of the former return data or (iii) the very same data. The reason is that the given *AccountId* is (a) set in the contract as business partner or (b) is set as a clearing operator which is allowed to access

```
<GetContractDataResp>
  <ContractDetails>
    <Contract>
      <ContractId>20003</ContractId>
      <ServiceName>ClearingService
      </ServiceName>
      <CreationDate>2015-01-15T11:36.12
      </CreationDate>
    </Contract>
    <ConsumerAddress>
      <Name>EMP</Name>
      <Address>
        <...>
      </Address>
    </ConsumerAddress>
    <ProviderAddress>
      <Name>CPO</Name>
      <Address>
        <...>
      </Address>
    </ProviderAddress>
    <Pricing>
      <UsageBasedPlan>
        <IsDeleted>false</IsDeleted>
        <BasicPrice>0</BasicPrice>
        <UsagePrice>0.2</UsagePrice>
        <TimePeriod></TimePeriod>
        <Currency>EUR</Currency>
        <TimeSpecification>
          <startDay>Monday</startDay>
          <endDay>Tuesday</endDay>
          <startTime>09:00</startTime>
          <endTime>20:00</endTime>
          <usagePrice>0.1</usagePrice>
        </TimeSpecification>
      </UsageBasedPlan>
    </Pricing>
  </ContractDetails>
</GetContractDataResp>
```

```
<GetTransactionDataResp>
  <TransactionDetails>
    <Operation>STARTCHARGE</Operation>
    <RequestStatusCode>
      <sessionStatusName>Success</sessionStatusName>
    </RequestStatusCode>
    <RequestTime>2015-05-15T17:04:47</RequestTime>
    <ResponseTime>2015-07-15T17:04:54</ResponseTime>
    <RequestStatusCode>
      <sessionStatusName>Success</sessionStatusName>
    </RequestStatusCode>
  </TransactionDetails>
  <TransactionDetails>
    <Operation>STARTCHARGE</Operation>
    <RequestStatusCode>
      <sessionStatusName>Success</sessionStatusName>
    </RequestStatusCode>
    <RequestTime>2015-05-18T13:38:32</RequestTime>
    <ResponseTime>2015-05-18T13:38:39</ResponseTime>
    <RequestStatusCode>
      <sessionStatusName>Success</sessionStatusName>
    </RequestStatusCode>
  </TransactionDetails>
  <TransactionDetails>
    <Operation>STARTCHARGE</Operation>
    <RequestStatusCode>
      <sessionStatusName>Success</sessionStatusName>
    </RequestStatusCode>
    <RequestTime>2015-05-19T19:59:55</RequestTime>
    <ResponseTime>2015-05-19T20:00:02</ResponseTime>
    <RequestStatusCode>
      <sessionStatusName>Success</sessionStatusName>
    </RequestStatusCode>
  </TransactionDetails>
</GetTransactionDataResp>
```

Fig. 7. Response of GetContractData (left) and GetTransactionDataListResp (right) (Source: [6]).

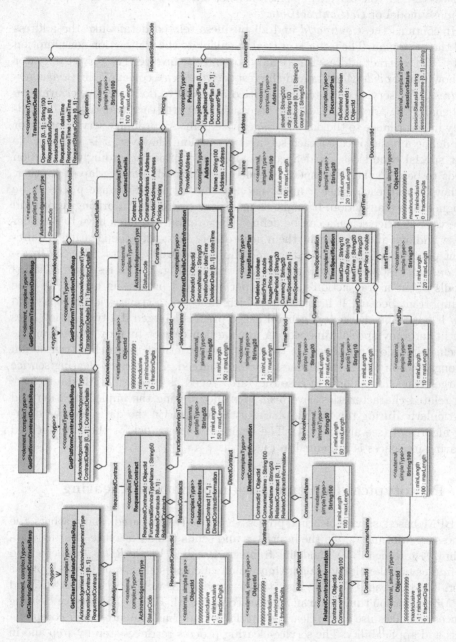

Fig. 8. Clearing interfaces response data model.

this particular data if provided as *SCCOID*. The interfaces need to be designed to detect and differentiate all possible scenarios and respond accordingly as it is necessary in role-driven implementations. The left side of Fig. 7 shows the response model of *GetContractData*.

It contains the *ContractId* and all business related details like the address and the price plan. The latter is critical to calculate the transaction consumption costs. The contract address is necessary to send the invoice. The service operator is shown in *ProviderAddress* and the respective service consumer in *Consumer-Address*. The service operator has entitled a SCCO to do its service-clearing on its behalf. The price conditions are necessary to charge each transaction individually in accordance to the defined usage price plan. An ordinary service operator can define a basic price for each service transaction. However, it is also possible that special time slots are defined to adjust the prices accordingly. Attention needs to be paid that that special defined time slots do not overlap each other. If they overlap each other it might happen that no specific usage price can be assigned to a transaction. It might be necessary to consider a pricing priority in case various price schema can be defined. Future marketplaces have to consider these findings and implement them accordingly to satisfy the clearing requirement pointed out by the literature and the interviewed experts. The *GetTransactionData* response is presented on the right of Fig. 7. The response contains details about a transaction's status, operation and processing time. The time is important because it indicates to which time slot a transaction belongs and how it has to be charged. A general usage price is charged if a transaction does not fit into any special time slot. A transaction cannot fit into more than one specific time slot. All prices are taken from the contract to which a transaction belongs and on which the business partners agreed. The service-clearing service charges transactions only in case they are tagged as successful. All these clearing related characteristics have been identified during the implementation and particularly during the testing phase of the interfaces, the data and the clearing algorithm (see also Sect. 4). The response data model for the implemented clearing interfaces is presented in Fig. 8.

4   Prototypical Implementation of Service Clearing

A BPM based service-clearing process has been elaborated that gathers the necessary information via the interfaces and sends the data to the clearing algorithm. A graphical BRM modeling framework, called *Visual Rules*, has been used to implement the clearing algorithm which calculates the service consumptions cost. The rule based clearing algorithm is provided as web service. The BRM and BPM application run within an Apache Tomcat with JDK 7. The service-clearing processes makes use of the clearing interfaces and thus evaluates their completeness and applicability. The service-clearing process retrieves step-by-step and in an automated fashion all internal marketplace data that is in relation to a specified contract. The service-clearing process complies to the sequence presented in Fig. 4. Once all data is gathered, it is submitted to the service-clearing algorithm which processes the input data accordingly and calculates the final cost. Figure 9 presents the dependance diagram of the service-clearing rule model which depicts the relationships between the developed flow rules and decision tables. Both are graphical representations of java code. The start of the clearing algorithm is the flow rule *Clearing*. This rule calls sub-rules which finally process the input data and return the final price for a set of transactions.

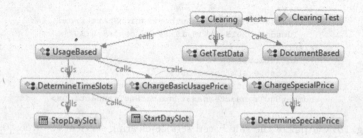

Fig. 9. Dependency diagram of the rule model used for service clearing.

The formula to calculate the final total cost TC is

$$TC = BP \cdot \underset{\{\text{status}(s) \mid s \in S\}}{\chi \ (\text{successful})} + \sum_{Z \in T} \sum_{s \in S_{suc}} \underset{Z}{\chi(\text{ts}(s)) \cdot P(Z)} + \sum_{s \in S_{suc, \, T^c}} UP$$

whereby:

$$S_{suc} \quad = \left\{ s \in S : \text{status}(s) = \text{successful} \right\}$$
$$S_{suc, \, T^c} = \left\{ s \in S_{suc} \mid \forall \, Z \in T : \text{ts}(s) \notin Z \right\}$$

All successful transactions are checked whether they fit into any special time slots (defined in the contract usage price plan) or not. The transactions' usage costs are accumulated. The basic price is charged if at least one transaction is successful. The total cost consists of the sum of all individual transaction costs and the basic price. Figure 10 depicts the output of the service-clearing service after applying the sequence shown in Fig. 4. The data used to calculate the cost is the one presented in Fig. 7.

The value of *numberTransaction* in Fig. 10 indicates how many transactions have been finally processed by the service-clearing BRM based algorithm. The *numberSucTransactions* show how many transaction are actually charged. The *numberTransaction* and *numberSucTransactions* are either equal or the former is higher than the latter. The *transactionList* in Fig. 10 depicts all processed transactions and should be part of any invoice which is sent to a service consumer. This overview contains what price is charged for any specific transaction based on the price plan given in the contract. Two transactions are charged with a special price (0.10 Euro) as they fit into a particularly defined time slot while one matches no specific time lot and thus is charged with the ordinary usage price (0.20 Euro). As no basic price is defined in the contract the overall cost

```
<transactionList>
  <entry>
     Transaction: ChargeAuthorizationStart from:
     Friday, 2015-05-15 17:04:47 costs: 0:20
  </entry>
  <entry>
     Transaction: ChargeAuthorizationStart from:
     Monday, 2015-05-18 13:38:32 costs: 0:10
  </entry>
  <entry>
     Transaction: ChargeAuthorizationStart from:
     Tuesday, 2015-05-19 19:59:55 costs: 0:10
  </entry>
</transactionList>
<totalCost>0.40</totalCost>
<numberTransaction>3</numberTransaction>
<numberSucTransactions>3</numberSucTransactions>
```

Fig. 10. Output of the rule based service-clearing service (Source: [6]).

for the three charged transactions is 0.40 Euro. The *transactionList* also shows
the amount of received transactions and the amount of transactions that have
been used to calculate the costs (three transactions have been processed and
three transactions are considered for the total costs calculation). Various test
sets have been executed to demonstrate the algorithms performance. Test data
was defined and the tests executed without any human interaction. Tests were
executed to check whether the identified interfaces and their design are valid and
fulfill the requirements of individual service clearing or if certain parameters are
still undetected. Furthermore did the tests show what data has to be available
and accessible and what is the data access sequence. The test results are shown
in Table 2.

Table 2. Rule model and algorithm performance measurement (Source: [6]).

Processed data sets	Time slots	Execution time	Read data
500	3	825–845 ms	222 ms
	5	980–1000 ms	
	10	1045–1060 ms	
500	3	1200–1270 ms	360 ms
	5	1300–1400 ms	
	10	1400–1450 ms	
2000	3	1600–1610 ms	572 ms
	5	1710–1735 ms	
	10	1790–1825 ms	
3000	3	1930–2020 ms	703 ms
	5	2120–2190 ms	
	10	2180–2360 ms	

Because the test performance depends on the underlying IT infrastructure and internet connection, a scenario has been chosen that assumes that an electronic marketplace hosts the clearing service of a clearing operator. This was discussed in [22] whereby a marketplace has been appointed as a Software as a Service (SaaS) and extends its capabilities towards the provisioning of a Platform as a Service (PaaS). Transaction data is not queried from a dislocated database but from a local CSV file. This aims to reduce the impact of the underlying infrastructure on the overall performance as the test aims at the performance of the clearing algorithm itself. The transaction data column shows how many test transactions are processed by the rule model. The time slots show how many special time slots have been checked against the transactions. The execution time shows the complete execution time of the rule model, inclusive the time to read the test data, validate the test data and compute the cost calculation algorithm. The *Read Data* column indicates how long it took for the rule model to read and load the test data into the rule before executing the price calculation. Figure 11 visualizes the executed tests and depicts that even though due to the high increase in the number of test cases the execution time increases only moderately.

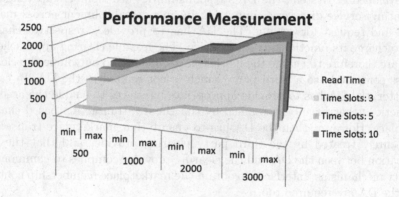

Fig. 11. Graphical performance overview.

One concern during the interface design was the high amount of transaction data and its impact on the performance. It was furthermore expected that the role-based validations have a higher impact on the clearing service performance. However, after the test case with 3000 processed transactions it can be said that the interfaces and their underlying processes are well designed and implemented and the clearing service performance appropriate. The time to retrieve and load the data it into the rule model depends on the type of the data source, how the data source is connected and the connection's properties. In summary, it can be outlined that the combination of BPM and BRM to implement a clearing service is feasible and has a good performance.

5 Service Clearing Within an Ecosystem of Interconnected Marketplaces

A Directory Agent (DA) roams service requests between marketplace participants across their marketplaces' system boundaries. No matter whether a marketplace pays a membership fee or is charged in respect to the amount of roamed requests, the DA does not offer its roaming service and IT infrastructure free of charge. The same is true for the marketplace which charges its participants for its service(s) too. Nevertheless, a marketplace may offers additionally services like Platform as a Service which can be individually contracted and thus is individually charged [5]. It is recommended that each environment (marketplace and DA) charges their participants for their participation and/or for the amount of roamed service requests (or any other contracted service capability) alike the solutions described in Sect. 2. A marketplace may outsources its clearing processes to one of its locally registered SCCOs which then has the responsibility to send the invoices to all participants of a marketplace on behalf of the marketplace. The DA does not have this chance as it does not have to or need to enable service providers to register and to offer service capabilities [5]. In our developed ecosystem, the DA's responsibilities are membership validation, facilitating service discovery, supporting business establishment across marketplaces and request forwarding. The DA has to provide a respective charging plan to charge its functionality and service. The feasible clearing approaches for a DA are therefore to charge the marketplaces itself or to outsource the clearing process completely to a third party which is not related to the DA. In case of the latter, the DA has to provide appropriate interfaces to enable the access on transactions that have been roamed by the DA if a transaction based charging is in place. However, then the DA has to ensure that marketplace transactions are securely treated by the third party clearing provider and that the communication between the DA and the clearing provider complies to commonsense security mechanisms. Interfaces to obtain the marketplace membership contracts from the DA are required too.

The DA does not have to provide the possibility to forward clearing requests between marketplaces. Due to the foreign business relationship establishment between a foreign service provider and consumer (described in detail in [23]), an appointed SCCO can retrieve all the necessary information (contract data and transactions) from its local marketplace which keeps a copy of all business contracts for request roaming purposes. That marketplaces charges its participants and they charge their end customers who actually consumed the service and thus triggered all the service roaming within a marketplace and/or ecosystem. In summary, a DA is not involved in any B2B service consumption clearing between marketplaces entities. The DA is only involved in charging its own participants in accordance to its price plan and membership contracts.

6 Conclusion

During the interviews we identified the urgency for a capable clearing concept within service solutions of the connected mobility domain. Throughout several workshops we identified the parameters which are at least required to perform service consumption clearing and incorporated ideas from the experts. During the design phase of the interfaces we recognized the need for a role-driven concept. Additionally, we figured out that the to be developed interfaces have to be lightweight and encapsulate one specific functionality only. This facilitates their maintenance and re-usage as the functionality might be used in other use-cases. The clearing approach realized within this work satisfies the demands for sophisticated service clearing and access on internal processed and stored data. Our implementation appears to be the first of its kind within the connected mobility domain that supports individual and outsourced service clearing. With that we solved two disadvantages of current mobility solutions as we proposed missing interfaces to access data from outside while at the same time used these interfaces to provide an appropriate service clearing mechanism. It was said by the experts that the connected mobility service solutions are still in their early stages and the budget for new functionality is little. However, the prototypical implementation of the interfaces has shown that the effort for the interface implementation is manageable and the message protocol not too comprehensive. The developed clearing service with its clearing algorithm has proven that the interfaces and chosen parameters for the message protocol are appropriate. This justifies again the previous statement that service clearing can be accomplished with manageable investments. Therefore, the often used argument of the high efforts is not longer valid and service solutions can start to implement an appropriate clearing functionality while using this work as a foundation. Our clearing mechanism can be considered as an unique selling point as well as a critical success factor of future service trading solutions in the mobility domain. Service marketplace architects can use this work as a foundation to discuss, design and implement further interfaces which can be used to realize further use cases like marketplace interconnectivity.

7 Outlook on Future Work

The findings of and experience gained during this work can be used to define and elaborate more interfaces to close various gaps of current service solutions like interconnectivity or protocol standardization. The findings of this work have been used as a foundation to elaborate a concept for interconnectivity between service marketplaces which is already prototypically implemented in [19].

References

1. Thitimajshima, W., Esichaikul, V., Krairit, D.: Developing a conceptual framework to evaluate public B2B E-marketplaces. In: Proceedings of the Pacific Asia Conference on Information Systems 2015 (2015)
2. Strasser, M., Albayrak, S.: The Current Situation and future trends of marketplaces for mobility services : findings from qualitative expert interviews. In: Proceedings of the 5th International Conference on Smart Cities and Green ICT Systems (SMARTGREENS), pp. 21–31 (2016)
3. Gubbi, J., Buyya, R., Marusic, S., Palaniswami, M.: Internet of Things (IoT): a vision, architectural elements, and future directions. Fut. Gener. Comput. Syst. **29**, 1645–1660 (2013)
4. Strasser, M., Albayrak, S.: Conceptual architecture for self-discovering in fragmented service systems. In: 7th International Confrence on New Technologies, Mobility and Security (NTMS), pp. 1–5 (2015)
5. Strasser, M.: Roaming von Echtzeitdaten innerhalb eines Kollektivs von Mobilitätsanbietern und -nutzern: Architektur eines Cloud basierten Marktplatzes für Serviceleistungen im Ökosystem. In: Kolloquium Das neue Auto - elektrisch, automatisiert und vernetzt (2016)
6. Strasser, M., Weiner, N., Albayrak, S.: Individual service clearing as a business service : a capable reference solution for B2B mobility marketplaces. In: Proceedings of the 6th International Conference on Cloud Computing and Services Science (CLOSER), pp. 27–38 (2016)
7. Grieger, M.: Electronic marketplaces: a literature review and a call for supply chain management research. Eur. J. Oper. Res. **144**, 280–294 (2003)
8. Ågerfalk, P., Bannon, L., Fitzgerald, B.: Action in Language. Operational Research Society, Organisations and Information Systems (2006)
9. Legner, C.: Is there a market for web services? In: Nitto, E., Ripeanu, M. (eds.) ICSOC 2007. LNCS, vol. 4907, pp. 29–42. Springer, Heidelberg (2009). doi:10.1007/978-3-540-93851-4_4
10. Heinrich, B., Huber, A., Zimmermann, S.: Make-and-sell or buy of web services. In: 19th European Conference on Information Systems (ECIS) (2011)
11. Buchinger, U., Lindmark, S., Braet, O.: Business model scenarios for an open service platform for multi-modal electric vehicle sharing. In: The second International Conference on Smart Systems, Devices and Technologies, SMART 2013, pp. 7–14 (2013)
12. Turban, E., King, D., Lee, J.K., Liang, T.-P., Turban, D.C.: Business-to-business e-commerce. Electronic Commerce. STBE, pp. 161–207. Springer, Cham (2015). doi:10.1007/978-3-319-10091-3_4
13. Ghenniwa, H., Huhns, M.N., Shen, W.: eMarketplaces for enterprise and cross enterprise integration. Data Knowl. Eng. **52**, 33–59 (2005)
14. Johnson, M.: Critical success factors for B2B e-markets: a strategic fit perspective. Mark. Intel. Plan. **31**, 337–366 (2013)
15. Balocco, R., Perego, A., Perotti, S.: B2b eMarketplaces Aclassification framework to analyse business models and critical success factors. Ind. Manage. Data Syst. **110**, 1117–1137 (2010)
16. Pfeiffer, A., Bach, M.: An E-clearinghouse for energy and infrastructure services in e-mobility. In: Helber, S., Breitner, M., Rösch, D., Schön, C., Graf von der Schulenburg, J.-M., Sibbertsen, P., Steinbach, M., Weber, S., Wolter, A. (eds.) Operations Research Proceedings 2012. ORP, pp. 303–308. Springer, Cham (2014). doi:10.1007/978-3-319-00795-3_44

17. Rust, R.T., Zahorik, A.J.: Customer satisfaction, customer retention, and market share. J. Retail. **69**, 193–215 (1993)
18. Buyya, R., Yeo, C., Venugopal, S.: Market-oriented cloud computing: vision, hype, and reality for delivering it services as computing utilities. In: Proceedings - 10th IEEE International Conference on High Performance Computing and Communications, HPCC 2008, pp. 5–13. IEEE (2008)
19. Strasser, M., Albayrak, S.: Smart city reference model: interconnectivity for on-demand user to service authentication. Int. J. Suppl. Oper. Manage. **3**, 1126–1142 (2016)
20. Luttge, K.: E-Charging API: outsource charging to a payment service provider. In: 2001 Intelligent Network Workshop, pp. 216–222. IEEE (2001)
21. Vidal, N., Iaarakkers, J., Marques, R., Scuro, P., Caleno, F., Matrose, C., Bolczek, M.: Report on Billing and stakeholders architecture and ICTs recommendations. Technical report 241295, G4V (2011)
22. Strasser, M., Albayrak, S.: A pattern based feasibility study of cloud computing for smart mobility solutions. In: Proceedings of the 8th International Workshop on Resilient Networks Design and Modeling (RNDM), pp. 332–338. IEEE (2016)
23. Strasser, M., Mauser, D., Albayrak, S.: Mitigating traffic problems by integrating smart parking solutions into an interconnected ecosystem. In: The twenty first IEEE Symposium on Computers and Communications, pp. 38–43 (2016)

Author Index

Printed in the United States
By Bookmasters